PRINCIPLES OF SOLIDIFICATION AND MATERIALS PROCESSING

PRINCIPLES OF SOLIDIFICATION AND MATERIALS PROCESSING

Volume 1

Editors

R. Trivedi
J.A. Sekhar
J. Mazumdar

Proceedings of Indo-US Workshops organized by
DEFENCE METALLURGICAL RESEARCH LABORATORY
Hyderabad, India

Sponsored by :

THE DEFENCE RESEARCH AND DEVELOPMENT ORGANISATION, India
DEPARTMENT OF SCIENCE & TECHNOLOGY, India
THE OFFICE OF NAVAL RESEARCH, U.S.A.
THE AMERICAN INSTITUTE OF BIOLOGICAL SCIENCES, U.S.A.

TRANS TECH PUBLICATIONS
Switzerland—Germany—UK—USA

Trans Tech Publications
P.O. Box 10
CH-4711 Aedermannsdorf
Switzerland

and

Old Post Road
Brookfield VT 05036
U.S.A.

Crystal Properties and Preparation Vols. 22–25
Pt I (in Volume 1) and Pt. II (in Volume 2)

1990 (ISBN 0-87849-594-0)

Printed in India
by Rekha Printers Pvt. Ltd., New Delhi 110 020

Introduction

It is with great pleasure and a sense of privilege that I write to introduce the two-volume proceedings of the Indo-US workshops on Solidification Principles and Materials Processing held at our Laboratory from January 15–21, 1988.

There is worldwide awareness that improvements in the microstructural condition in a way as to give rise to better properties and performance of our commonly used as well as new generation alloys can only be heralded through a fundamental understanding of the rules that govern microstructure formation. The twin workshops aimed at a step forward in that direction. By encompassing in a single endeavour the understanding of the physics of pattern evolution on the one hand and, on the other, the fundamental engineering principles of process development, the twin workshops have served to emphasize a two-fold approach to designing and optimizing solidified materials and products. The two-volume proceedings document this effort.

I am writing this some months after the event that brought into our midst several specialists from abroad who have made valuable contributions to the field. We in this Laboratory have come to cherish the time that they were here. The publication of the proceedings revives the pleasant memories of our brief association with the visitors and hopefully will also serve to nurture the scientific contacts made.

P. RAMA RAO
Director
Defence Metallurgical Research Laboratory
Hyderabad

Foreword

Directions in Solidification and Materials Processing Research

During the last two decades we have come to realize the importance of advanced material design from a global view of the complex interrelationships between synthesis, structure, property, processing, performance, reliability and durability of materials and systems. While structure-property interplay remains as the dominant concern, synthesis and processing are imperative for research planning in materials science and engineering. An increasing demand on higher systems' performance and cost-effective manufacture has placed an ever-increasing emphasis not only in the United States but in most other industrial nations on design, processing and fabrication of advanced engineered materials. We are seeing the results of this emphasis on basic and engineering research in advanced ceramics, semiconducting and superconducting materials, magnetic and optical materials, polymers, high performance composites, light weight and high temperature metallic materials and bioengineered materials. Concerns of life-cycle cost of systems require developments in better life prediction algorithms, nondestructive evaluation, advanced integrated sensors and computer modeling.

One area of research, related to a better understanding of solidification has made a significant inroad for advanced metallic material design. Nonequilibrium materials are now being processed to near-net shapes with compositions, phases and microstructures well outside the range normally associated with conventional casting and wrought processing methods. Rapid Solidification Processing has provided new insights on the fundamentals of nonequilibrium solidification processes as well as on the design and development of new and improved structural alloy system. For example, the design of light weight high temperature and high strength alloys such as Al-Cr-Zr, Al-Cr-Zr-Mn, Al-Fe-Mo and Al-Fe-Ce, ultra high strength steels such as Fe-Ni-Mo-C-La, high performance permanent magnets such as Nd-Fe-B, amorphous transformer steels such as Fe-B-Si are direct results of our better understanding of the role of solidification on microstructure and properties. There is yet a lot to be done to enhance our knowledge of the effects of thermodynamic and kinetic parameters on crystal nucleation and growth, phase selection and solute redistribution during solidification. Papers presented at this workshop on solidification and novel processing methods represent the forefront of our knowledge in these fields of materials science and engineering and it is hoped that an interaction between research scholars of our two countries will contribute to further our understanding on the behavior of structural and functional materials.

B.B. RATH
O.P. ARORA

Preface

One of the most rewarding experience in any scientific endeavor is to develop and organize a conference that consolidates information in a difficult multidisciplinary area. We are fortunate to have done this for the solidification processing of materials by linking fundamental principles of microstructure evolution with the development of advanced technology.

Presented in this two-volume set is the outcome of the five day Indo-United States meeting held at the Defence Metallurgical Research Laboratory, Hyderabad, India. Scientists from the United States, Switzerland, United Kingdom and France met from January 15 to 21, 1988 and discussed their research results in the form of lecture and poster presentations. The speakers were requested to include a comprehensive review of their topic in the written version, and we are grateful to them for their cooperation in this regard. The poster presentations have been reproduced as **communications** at the end of the proceedings. We hope that this two-volume series will adequately fill the information gap in this multidisciplinary field.

Dr. V.S. Arunachalam, Scientific Advisor to Raksha Mantri (Defence Minister), Government of India, very graciously inaugurated the workshops. Dr. B.B. Rath, Associate Director, Naval Research Laboratory, U.S.A. and Dr. P. Rama Rao, Director, Defence Metallurgical Research Laboratory (DMRL), India, presented overviews of the subject as perceived in the United States and India, respectively.

This workshop could not have been successful without the efforts and care taken by Dr. Mary-Francis Thompson, Director, Special Programs, American Institute of Biological Sciences, and Sri A.K. Chakravorty, Scientist, DMRL. We are deeply grateful to them for their significant contributions to the success of the conference.

The Indian Institute of Metals, Materials Sciences Division, provided us with the forum for this meeting. We would like to thank the department of Science and Technology and the Defence Research and Development Organization, Government of India, for their valuable support from the Indian side. The Institute of Biological Sciences provided support from the U.S. side. Several young scientists of the Defence Metallurgical Research Laboratory contributed their time and energy to ensure a successful meeting. Without their warm help,

x

generously supported by Director, DMRL, the conference could not have
taken place. Finally, we would like to acknowledge the contributions of all
speakers, poster presenters and session chairpersons who made it possible to
exchange ideas which contributed to the success of the conference.

Contents

xii

COMMUNICATIONS

1

Recent Developments in the Theory of Dendritic Solidification

J.S. LANGER

Institute for Theoretical Physics
University of California
Santa Barbara, California 93106, U. S. A.

ABSTRACT

Some recent developments in the theory of dendritic solidification are described. The presentation focusses on attempts to find mechanisms for the selection of dendritic growth rates and tip radii, starting with the marginal-stability hypothesis of about a decade ago, and concluding with the recently discovered solvability principle. Also included in the discussion is a report on comparisons between theory and experiment (inconclusive, at best), and some remarks about the dynamics of sidebranching.

The theory of dendritic solidification has taken some interesting turns in the last year or so. It is possible, but by no means certain, that we now have an accurate idea about what is happening in the simplest kinds of dendritic processes. There are two points about which we can be sure. First, the dendritic mechanism is one that we shall have to understand in a fundamental way if we are to continue to build a science of solidification processing. And, second, that understanding is going to be more subtle and complex than most of us had expected just a few years ago.

My plan in this paper is to review briefly the recent history of the dendrite problem and then to try to communicate at least the general flavor of recent developments. For simplicity, I shall deal almost exclusively with the case of a thermal dendrite, that is, a pure substance solidifying in an undercooled melt. After describing Ivantsov's classic isothermal solution and the puzzle which it presents, I shall make a few remarks about the marginal-stability theory because I think it is important to understand why this intuitively attractive picture of dendrites is, at best, mathematically incomplete. The second half of the paper is devoted to the new, so-called "solvability theory." I shall be concerned mostly with the problem of velocity selection, but shall add some remarks at the end about the origin of sidebranches.

In the conventional thermodynamic model of the solidification of a pure substance from its melt [1] the fundamental rate-controlling mechanism is the diffusion of latent heat away from the interface between the liquid and solid phases. The latent heat that is released in the transformation warms the material in the neighborhood of the solidification front and must be removed before further solidification can take place. This is a morphologically unstable process which characteristically produces dendritic, i.e., tree-like structures. In a typical sequence of events, an initially featureless crystalline seed immersed in an under-cooled melt develops bulges in crystallographically preferred directions. The bulges grow into needle-shaped branches whose tips move outward at constant speed. These primary branches, in turn, are unstable against sidebranching, so that each outward growing tip leaves behind itself a complicated dendritic structure.

The dimensionless thermal diffusion field in this model is conveniently chosen to be

$$u = \frac{T - T_\infty}{(L/c)} \tag{1}$$

where T_∞ is the temperature of the liquid infinitely far from the growing solid, and the ratio of the latent heat L to the specific heat c is an appropriate unit of undercooling. The field u satisfies the diffusion equation

$$\frac{\partial u}{\partial t} = D\nabla^2 u \tag{2}$$

where D is the thermal diffusion constant which we shall take to be the same in both the liquid and solid phases. The remaining ingredients of the model are the boundary conditions imposed at the solidification front. First, there is heat conservation:

$$v_n = -[D\,\hat{n}\cdot\vec{V}u], \tag{3}$$

where $\hat{n}$ is the unit normal directed outward from the solid, v_n is the normal growth velocity, and the square brackets denote the discontinuity of the flux across the boundary. The physically more interesting boundary condition is a statement of local thermodynamic equilibrium which determines the temperature u_s at the two-phase interface:

$$u_s = \Delta - d_0\chi, \tag{4}$$

where

$$\Delta = \frac{T_M - T_\infty}{(L/c)} \tag{5}$$

is the dimensionless undercooling, a measure of the driving force for the processes that we are considering. The second term on the right-hand side of (4) is the Gibbs-Thomson correction for the melting temperature at a curved surface; χ is the sum of the principal curvatures and $d_o = \gamma c T_M / L^2$ is a length, ordinarily of order Ångstroms, which is proportional to the surface tension γ. The latter quantity and, accordingly, d_o may be functions of the angle of orientation of the interface relative to the axes of symmetry of the crystal. In particular, for a cubic crystal in the $(1, 0, 0)$ plane, d_o may be taken to be proportional to $(1 - \alpha \cos 4\theta)$, where θ is the angle just mentioned and α is a measure of the strength of the anisotropy.

One aspect of the experimental situation that has been clear since the earliest quantitative observations of dendritic growth is that the growth rates and shapes of dendritic tips are determined reproducibly by the undercooling Δ. The historic puzzle has been that the capillary length d_o is generally three or more orders of magnitude smaller than the characteristic length scales associated with the dendrites, specifically, the tip radius ϱ or the spacings between sidebranches. If one neglects d_o altogether, however, one arrives at Ivantsov's paradox. Instead of there being a unique growth velocity v and tip radius ϱ at a fixed Δ, as required by experiment, there exists a continuous family of steady-state, shape-preserving solidification fronts—paraboloids of revolution—which satisfy the Ivantsov relation [2].

$$\Delta = p \, e^p \int_p^\infty dy \, \frac{e^{-y}}{y} \tag{6}$$

where $p = \varrho v/2D$ is the thermal Peclet number. The tips of dendrites often do look very paraboloidal, and quantitative experiments [3] generally indicate that the Ivantsov relation (6) is satisfied. But obviously some essential ingredient of the theory is missing.

Over a decade ago, Müller-Krumbhaar and I [4] explored the idea (originally suggested by Oldfield [5]) that the missing element of the theory might have something to do with stability of the growth form. We performed many complicated calculations, but what was left in the end was a relatively simple conjecture that has since been confirmed remarkably well by experiment. In the simplest possible terms, our conjecture was that the tip radius ϱ might scale like the Mullins-Sekerka [1, 6] wavelength $\lambda_s = 2\pi(2Dd_o/v)^{1/2}$. Note that λ_s is the geometric mean of the microscopic capillary length d_o and the macroscopic diffusion length $2D/v$; it is therefore of roughly the right magnitude to characterize dendritic structures. A planar solidification front moving at speed v is linearly unstable against sinusoidal deformations whose wavelengths are larger than λ_s. Therefore, we reasoned, a dendrite with tip radius ϱ appreciably greater than λ_s must be unstable against sharpening or splitting. The dynamical process which leads to the formation of the dendritic tip might naturally come to rest at a state of marginal stability, that is, at a state for which the dimensionless group of parameters

$$\sigma = \frac{2Dd_o}{v\varrho^2} = \left(\frac{\lambda_s}{2\pi\varrho}\right)^2 \tag{7}$$

is a constant, independent of Δ. Moreover, if we take the idea literally and set ϱ equal to λ_s, then the value of this constant should be $\sigma^* \cong (1/2\pi)^2 \cong 0.025$. Remarkably, the assumption $\sigma = \sigma^* = $ constant is consistent with a wide range of experimental observations [3] (when convective effects are eliminated or otherwise taken into account) and the specific value $\sigma^* \cong 0.0195$ for succinonitrile—by far the most carefully studied material—is quite close to the naive prediction.

A development that, at the time, seemed an even more impressive confirmation of the marginal stability theory was the experimental verification of the enhancement of dendritic growth speeds by the addition of small amounts of impurity to the melt [7-9]. Ordinarily, impurities are rejected from the growing solid and therefore, like the latent heat, must diffuse away from the solidification front if the solid is to continue to grow. Because impurity diffusion is much slower than thermal diffusion, it might be thought that the addition of any amount of impurity would diminish the growth speed. The marginal stability theory makes a different prediction, however. The diffusion length $2D/v$ that enters into the definition of λ_s may be visualized as the thickness of the thermal boundary layer that is carried along in the fluid ahead of the solidification front. As this layer becomes thin, λ_s decreases, and the dendrite becomes sharper and faster. Because the diffusion constant for impurities is so small, a layer of rejected impurities is thinner than the corresponding thermal layer, the front is destabilized at shorter length scales, and the growth speed is enhanced. More detailed calculations [7, 9] indicate that the speed goes through a maximum at a fairly small concentration of impurities, beyond which there is a crossover from thermal to impurity control and the growth rate decreases as originally expected. This prediction has been confirmed in experiments by Chopra [8].

What, then, is wrong with the marginal stability theory? It seems that its mathematical foundation has been knocked from under it by the discovery that the Ivantsov family of solutions does not survive in the presence of surface tension [10-15]. A nonvanishing d_o, no matter how small, reduces the continuum of solutions to, at most, a discrete set; and the existence of any solution whatsoever depends on there being some angular dependence of the surface tension, that is, a nonvanishing anisotropy strength α. Thus, the stability calculation that Müller-Krumbhaar and I thought we were performing was unfounded because the family of steady-state solutions whose stability we supposedly were testing did not exist.

All is not lost, however, because the mathematics immediately suggests an alternative selection mechanism, albeit one which has little of the intuitive appeal of marginal stability. A natural guess is that the selected dendrite is the one for which a solution exists. In more formal language, we guess that the condition for solvability of the steady-state equations is equivalent to a condition for the existence of a stable fixed point with a large basin of attraction in this dynamical system. If this conjecture is correct, orderly, steady-state dendritic growth does not occur at all in isotropic materials. In suitably anisotropic systems, there is a denumerably infinite set of solutions, only the fastest (and thus sharpest) of which can be stable. The hypothesis that this unique solution describes the tip of a dynamically selected dendrite has come to be known as the "solvability theory" [12–15].

I do not propose to explain the mathematics of the solvability theory here, but there are several features that do need to be mentioned. In the limit of small Peclet number p, the controlling group of parameters in the theory is the same quantity σ, defined in (7), that appeared in the stability analysis. This happens because one is looking for a small surface-tension-induced correction to the shape of the Ivantsov parabola and, in computing this correction, one encounters an equation quite similar to the one which arises in linear stability theory [4]. (As shown by Pomeau and coworkers [12, 13], linearization is not a necessary ingredient of the argument for solvability.) Technically speaking, σ enters the theory as a singular perturbation; it describes the strength of the curvature effect in (4) and, accordingly, multiplies the highest derivative in the equation for the shape correction once one has reduced this equation to dimensionless form.

There is a very nice way to visualize the effect of this perturbation. In practical numerical calculations [16–18], and also in the analytic approaches that have been applied successfully to this problem [12–15], one can generally assure the existence of some kind of solution by relaxing a boundary condition—most commonly the condition of smoothness at the tip. Suppose one allows the tip to have a cusp of outer angle θ and then, either numerically or analytically, computes what value θ must have in order to achieve a solution at a given value of σ. Because θ must vanish for a physically acceptable solution, a formally exact statement of the solvability condition is

$$\theta(\sigma, p, \alpha) = 0. \qquad (8)$$

We may think of θ as a measure of how close we have come to finding a solution at an arbitrary value of σ. The special values of σ for which (8) is satisfied are denoted $\sigma^*(p, \alpha)$.

If one tries to compute θ by expanding it in powers of σ, one finds that θ vanishes to all orders, a result which would be consistent with the original expectation of a continuous family of solutions. If the calculation is performed more carefully, however, the answer—at small p and zero anisotropy α—has the form

$$\theta(\sigma, 0, 0) \propto \exp\left(-\frac{\text{constant}}{\sqrt{\sigma}}\right), \qquad (9)$$

where the constant is a pure number of order unity. This function has an essential singularity at $\sigma = 0$ and no possible expansion about that point. It is extremely small for small σ, but it does not vanish exactly unless $\sigma = 0$. Thus, an arbitrarily small amount of isotropic surface tension destroys all solutions. For finite anisotropy α, however, the function $\theta(\sigma)$ has the same form as (9) for large σ but oscillates rapidly in the limit $\sigma \to 0$. The largest value of σ at which θ passes through zero occurs at $\sigma = \sigma^* \propto \alpha^{7/4}$, the latter approximation being valid only in the limit of very small α.

The solvability theory has essentially all of the qualitative features that made marginal stability look correct. The group of parameters σ is expected to have a nonzero value in the experimentally relevant limit of small p (i.e. small undercooling Δ). The impurity effect has also been confirmed in the new theory, at least qualitatively [19]. The major difference is the apparent α-dependence, which was missing in the earlier theory. The outstanding question, therefore, is whether solvability is quantitatively correct, and whether the anisotropy effect actually occurs.

The situation is not clear at the time this report is being prepared. Part of the difficulty is that the theoretical evaluation of σ^* for three-dimensional dendrites with finite anisotropies is a numerical challenge that, so far, has been attempted by only one group—Kessler et al. [18, 20]—and even there not with the full cubic anisotropy. The other side of the problem is that the experiments are also difficult, especially if it is necessary to obtain accurate values of α.

At the moment, agreement seems to be poor for succinonitrile. The measured anisotropy strength α is about 0.075 [21], and the corresponding theoretical value of σ^* seems closer to 0.01 [20] than the observed value of 0.02. Glicksman and coworkers currently are trying to make a more accurate measurement of α, which could conceivably be enough larger than originally estimated to bring the theoretical σ^* up to the experimental value. But the argument in favor of the anisotropy effect is further weakened by the observation that σ^* remains about 0.02 for pivalic acid despite the fact that α for this substance is almost ten times that for succinonitrile [21]. On the other hand, no theoretical estimate of σ^* is available for pivalic acid, and the three-dimensional calculation for so anisotropic a material is particularly difficult. D. Meiron [16], who has confirmed the basic principles of the solvability theory by very careful numerical analysis for the relatively simpler two-dimensional situation, reports that the function $\sigma^*(\alpha)$ in that case flattens out and actually decreases at large α. Thus, we cannot really be sure of any inconsistency here.

One very recent and auspicious development is the work of Dougherty and Gollub [22] on NH_4Br dendrites growing in aqueous solution. They have measured tip radii and growth rates, and also have made careful measurements of all other relevant parameters including α, which they estimate to be 0.24 ± 0.06

(about three times that for succinonitrile). (These authors report the anisotropy in terms of the quantity ε_4 which, for four-fold symmetry, is equal to $\alpha/15$.) Solidification in this system is controlled by chemical rather than thermal diffusion, thus diffusion in the solid is much slower than in the liquid. In such an asymmetric situation, Müller-Krumbhaar and I have estimated that σ^* should be about twice as large as in the (almost) symmetric thermal case [4]. Dougherty and Gollub find $\sigma^* \cong 0.08 \pm 0.02$ and report that, in an independent calculation using the above value of α, Kessler and Levine have found $\sigma^* \cong 0.06 \pm 0.02$. This is beginning to look like a quantitative agreement, and one might even say that there is weak evidence here for an effect of anisotropy.

What about the $\alpha^{7/4}$ law? My guess is that the solvability theory is a correct description of a large class of observable dendritic behavior, but that it breaks down at small anisotropy, and that the $\alpha^{7/4}$ law is not physically meaningful. If taken literally at arbitrarily small α, the solvability theory predicts arbitrarily small values of σ^* and, according to (7), tip radii ϱ which are much larger than the stability length λ_s. It is hard for me to see how stable dendritic growth can occur under such circumstances. In this regard, I must report that the most complete stability analyses performed to date [23] indicate that dendritic tips which satisfy the solvability condition (with the largest of the set of allowable σ^*'s) remain linearly stable no matter how small this σ^* may turn out to be. But linear stability analyses for nonlinear, open-ended, dynamical systems of the kind that we are dealing with here can be highly misleading—even the concept of stability is not well defined. For example, we know that dendrites whose tips are dynamically stable still undergo sidebranching instabilities, and that the evolution of the sidebranches is an intrinsically nonlinear phenomenon. Considerations like these lead me to believe that some mechanism other than solvability is operative for dendrites with small anisotropy. Perhaps there exists a nonlinear oscillatory mode. Perhaps old-fashioned marginal stability is right after all. Or, perhaps, regular dendritic growth does not occur in this regime; sufficiently isotropic materials might always form chaotic patterns when they solidify in an undercooled melt.

Finally, I'd like briefly to discuss sidebranching. Until quite recently, I had assumed that the tip of a real three-dimensional dendrite must be weakly—perhaps marginally—unstable against some oscillatory mode of deformation, and that this oscillation must generate the train of sidebranches that seems always to be observed in these systems. However, neither the theorists or the experimentalists have succeeded in finding any evidence for oscillatory tip modes in purely thermal dendrites.

One possibility that I find interesting is that dendritic sidebranches are generated by the selective amplification of noise. To investigate this possibility, we have performed both numerical studies [24] of the "boundary-layer model" [10] (a local version of the solidification problem that was instrumental in the discovery of the solvability theory) and analytic studies [25, 26] of the more realistic model described above. In the latter work, we have linearized the equations of motion about a steady-state solution determined by solvability, and have looked at the response of the solidification front to a localized pulse (say, of heat) applied near the tip. We find that this pulse generates a wavepacket-like deformation whose center moves away from the tip. More precisely, the center of the wavepacket stays at a fixed position along the side of the dendrite as viewed in the laboratory frame of reference, while the tip grows at constant speed away from the perturbation. There are two important features of this wavepacket. First, its amplitude $A(s)$ continues to grow as its center moves away from the tip. More specifically

$$A(s) \approx \exp \left[\frac{\text{const.}}{(\sigma^*)^{1/2}} \left(\frac{s}{\varrho} \right)^{1/4} \right], \tag{11}$$

where s is the distance from the tip to the center of the packet measured along the side of the dendrite. Second, the packet spreads and stretches in such a way that, as it grows, it acquires a sharply defined wavelength which, itself, increases slowly with distance from the tip. This is what is meant by selective amplification. If we look at some fixed distance behind the tip, say, at the point where initially very small deformations have grown out of the linear regime and are big enough to be visible, then we see that only a relatively narrow band of wavelengths has been selected from the original broad-band perturbation. Another way of saying this is that small, noisy perturbations near the tip produce large deformations away from the tip which look very much like sidebranches. I recently have tried to push this analysis far enough to estimate a noise temperature that would be required to generate the sidebranches that are seen experimentally. Purely thermal noise seems too small, but only by one or two orders of magnitude [26]. The important lesson is that the dendrite is an extremely sensitive and selective amplifier of week fluctuations in its environment.

ACKNOWLEDGEMENTS

I thank M. Glicksman, J. Gollub, and D. Meiron for useful discussions and for communicating their results prior to publication.

This research was supported by U.S. Department of Energy Grant No. DE-FG03-84ER45108 and by National Science Foundation Grant No. PHY 82–17853, supplemented by funds from the National Aeronautics and Space Administration.

REFERENCES

1. Langer, J.S. *Rev. Mod. Phys.* **52**, 1 (1980); Lectures in the Theory of Pattern Formation, Les Houches Summer School, 1986, Chance and Matter, J. Souletie, J. Vannimenus and R. Stora, eds. (North Holland, 1987).
2. Ivantsov, G.P. *Dokl. Akad. Nauk SSSR* **58**, 567 (1947).
3. Glicksman, M.E., R.J. Shaefer and J.D. Ayers. *Metall. Trans.* A **7**, 1747 (1976); S.C. Huang and M.E. Glicksman. *Acta Metall.* **29**, 701 and 717 (1981).

4. Langer, J.S. and H. Müller-Krumbhaar. *Acta Metall.* **26,** 1681, 1689, 1697 (1978). H. Müller-Krumbhaar and J.S. Langer. *Acta Metall.* **29,** 145 (1981).
5. Oldfield, W. *Mater. Sci. Eng.* **11,** 211 (1973).
6. Mullins, W.W. and R.F. Sekerka. *J. Appl. Phys.* **34,** 323 (1963); *J. Appl. Phys.* **35,** 444 (1964).
7. Langer, J.S. *Physicochem. Hydrodyn.* **1,** 41 (1980); A. Karma and J.S. Langer. *Phys. Rev.* **A30,** 3147 (1984).
8. Chopra, M. Ph.D. Thesis, Rensselaer Polytechnic Inst., Troy, N.Y. (1983).
9. Glicksman, M.E. *Mater. Sci. Eng.* **65,** 45 (1984); J. Lipton, M.E. Glicksman and W. Kurz. *Mater. Sci. Eng.* **65,** 57 (1984).
10. Ben-Jacob, E., N. Goldenfeld, J.S. Langer and G. Schon. *Phys. Rev.* **A29,** 330 (1984).
11. Langer, J.S. *Phys. Rev.* **A33,** 435 (1986).
12. Pelcé, P. and Y. Pomeau. *Studies in Applied Mathematics,* **74,** 245 (1986).
13. Ben Amar, M. and Y. Pomeau. *Europhys. Lett.* **2,** 307 (1986).
14. Kessler, D., J. Koplik and H. Levine. "Pattern formation for from Equilibrium: The free space Dendritic Crystal" Proceedings of the NATO Advanced Research Workshop on "Patterns, Defects, and Microstructures in Non-Equilibrium Systems", Austin, Texas (March 1986).
15. Barbieri, A., D.C. Hong and J.S. Langer. *Phys. Rev.* **A35** 1802 (1987).
16. Meiron, D. *Phys. Rev.* **A33,** 2704 (1986); also private communication.
17. Ben-Amar, M. and B. Moussallam. *Physica D* (1987).
18. Kessler, D.A., J. Koplik and H. Levine. *Phys. Rev.* **A33,** 3352 (1986); D.A. Kessler and H. Levine. *Phys. Rev.* **B33,** 7867 (1986).
19. Barbieri, A. "Pattern Selection in Realistic Models of Dendritic Solidification, Ph.D. Thesis, Univ. of Calif., Santa Barbara, 1988.
20. Kessler, D.A. and H. Levine. *Phys. Rev.* **A36,** 4123 (1987).
21. Glicksman, M.E. private communication.
22. Dougherty, A. and J. Gollub. *Phys. Rev.* **A38,** 3043 (1988).
23. Kessler, D. and H. Levine. *Europhys. Lett.* (to be published).
24. Pieters, R. and J.S. Langer. *Phys. Rev. Lett.* **56,** 1948 (1986).
25. Barber, M., A. Barbieri and J.S. Langer. *Phys. Rev.* **A36,** 3340 (1987).
26. Langer, J.S. *Phys. Rev.* **A36,** 3350 (1987).

2

Fundamentals of Dendritic Growth

M.E. GLICKSMAN

Materials Engineering Department
Rensselaer Polytechnic Institute
Troy, NY 12180, U.S.A.

ABSTRACT

The subject of dendritic growth pervades many areas of crystal growth and metal solidification. Dendrites are complex, branched crystals that usually form in an uncontrolled manner under conditions of net supersaturation or supercooling. Their highly ramified form permits the occurrence of a three-dimensional redistribution of heat and solute during freezing. This redistribution, called microsegregation, is undesirable in most instances and leads to diminished electrical, mechanical, and chemical properties of the cast material.

The basic concepts of heat and mass transport surrounding a growing dendrite are reviewed for the now classical steady-state model, first proposed by Ivantsov about forty years ago. The presence of a solid-liquid interface is then considered by introducing the physics of capillarity. Older results, such as the maximum velocity operating condition, are briefly discussed for the historical perspective they provide, and for their connection with the more modern concepts of dynamical stability of solid-liquid interfaces, as first introduced into dendritic growth theory by Langer and Müller-Krumbhaar. The development of dendritic scaling laws and the ability to predict with these laws the rudiments of dendritic microstructures is then reviewed in the context of a large, accurate data base on the kinetics and morphological characteristics of several pure "model" systems, such as succinonitrile, ice-water, and pivalic acid—each selected for some special attribute and each rather well characterized with respect to all the relevant thermophysical properties. Finally, the influence of combined solutal and thermal diffusion is discussed, with comparisons made with a few carefully conducted kinetics studies on dilute binary systems.

I. INTRODUCTION

I.1 Importance of dendritic crystal growth

Dendritic growth is perhaps the most common form of solidification especially in metals and other systems that freeze with relatively low entropies of transformation. Dendritic or branched growth in alloys generates microsegregation as well as other internal defects in castings, ingots, and weldments. More subtle effects introduced by the complex dendritic microstructure in solidified materials include crystallographic texturing, hot cracking, suboptimal toughness, and reduced corrosion resistance. Moreover, the dendritic microstructure and its effects may be modified by subsequent heat treatments, but they are seldom fully "erased." As such, the understanding and control of dendritic growth in solidification processing is crucial in order to achieve specific material properties in final products.

Dendritic growth is a coupling of two seemingly independent growth processes: the steady-state propagation of the dendrite main stem and the non-steady-state evolution of dendrite branches. Until recently, the time-dependent features of dendrites were completely ignored and theoretical models of dendrites were limited to the mathematical description of a branchless, paraboloidal needle, which grew at a constant rate in a shape-preserving manner. Furthermore, theoretical studies of dendritic growth have concentrated on the steady-state development of a one-component needle-dendrite growing in a pure melt, wherein the major transport process is heat conduction.

I.2 Steady-state dendritic growth

Despite the fact that dendrite formation seems to involve both steady-state attributes (near the tip) and time-dependent features (side branches), the earliest models employed steady-state descriptions of needle-like, branchless dendrites growing in a shape preserving manner [1–3]. The dendrite was assumed to grow at a constant axial rate, V, into a melt of uniform supercooling, ΔT, such that the surrounding thermal field would appear to be stationary in a coordinate frame traveling with the dendrite tip. The steady-state shape was chosen, *ab initio*, such that the prescribed solid-liquid interface remained at its bulk thermodynamic melting temperature, T_m. Imposition of an isothermal boundary condition retained linearity of the heat flow solution and led to a large class of steady-state "dendrite" solutions which depended on the arbitrarily chosen shape. Ivantsov's solution for the paraboloid of revolution is typical of these linear shape-preserving solutions.

I.3 Ivantsov's transport solution

An exact solution to the non-dimensional temperature distribution was first developed by G.P. Ivantsov in 1947, for the case of a paraboloidal "needle" dendrite, the shape for which was first suggested by Papapetrou in 1935.

Ivantsov [2] modeled dendritic growth in the form of an isothermal paraboloid of revolution with a steady-state tip radius R, growing at a uniform rate of V along the z-axis. This geometry is illustrated in Fig. 1, where (z, r) is a coordinate *moving at the same speed V as the tip*. Ivantsov set the temperature of

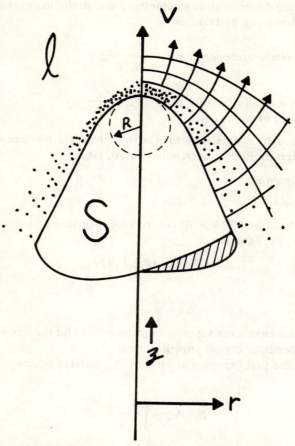

Fig. 1. Sketch of Ivantsov's paraboloid of revolution, having tip radius R and moving at constant velocity V in the z-direction. The arrows represent lines of heat flow from the interface in the moving frame which are orthogonal to the paraboloidal isotherms in the liquid phase.

the paraboloidal s/l interface to T_m, and the temperature far from the dendrite at T_∞, so $\Delta T = T_m - T_\infty$.

Ivantsov correctly suggested that a steady-state solution to the heat flow equations could be obtained in the moving (z, r) frame. He solved in the melt phase (l) $\nabla^2 T + \dfrac{V}{\alpha} \nabla T = \dfrac{1}{\alpha} \dfrac{\partial T}{\partial t} l = 0$, and realized that the heat flow solution for the solid (isothermal paraboloid) was trivial, $T_s (z, r) = T_m$.

If the thermal diffusivity, α, heat capacity, C, and latent heat, L, are chosen as constants, then the temperature equation is conveniently non-dimensionalized as $\nabla^2 \theta + 2 Pe \nabla \theta = 0$, where operators ∇^2 and ∇ are taken with respect to non-dimensional distances z/R and r/R.

The following dimensionless parameters prove useful in developing the heat flow solution first given by Ivantsov

$$\text{(dimensionless temperature scale)} \quad \theta = \frac{T - T_\infty}{L/C} \quad (1)$$

(*Peclet number, α = diffusivity for heat* $\quad Pe = \dfrac{VR}{2\alpha_i}$ $\quad (2)$
in the melt ($i = l$) or solid ($i = s$)

Ivantsov used the *approximations* of an isothermal s/l interface and a uniform far-field temperature in the supercooled liquid, viz.

At the s/l interface : $\qquad\qquad\qquad\qquad \theta = \hat\theta$ $\qquad (3)$

Far from the interface : $\qquad\qquad\qquad\quad \theta = \theta_\infty$ $\qquad (4)$

Ivantsov's solution $\theta(z/R, r/R)$ may be used to relate supercooling, $\Delta\theta$, to the Peclet number, Pe, namely

$$\Delta\theta = Pe \, \exp\,(Pe)\, E_1\,(Pe), \qquad (5)$$

where

$$\Delta\theta = \hat\theta - \theta_\infty, \qquad (6)$$

i.e., $\Delta\theta$ is the dimensionless supercooling responsible for the free energy change "driving" the dendritic crystal growth process.

$E_1\,(Pe)$ is the first exponential integral—a tabulated function, defined here as

$$E_1\,(Pe) = \int\limits_{Pe}^{\infty} \frac{e^{-s}ds}{s} \qquad (7)$$

Figure 2 shows that for a given value of $\Delta\theta$, a unique value of Pe occurs which provides the hyperbolic relationship between V and R. This is also referred to as the "point effect", well known in the theory of diffusion.

The relationship between $\Delta\theta$ and Pe, obtained from Ivantsov's steady-state heat flow solution is illustrated in Fig. 3. Note also that as $\Delta\theta \to 1$, $Pe \to \infty$ and that at small supercooling the behavior is almost linear. This "linearity" is with respect to log-log coordinates. Actually the slope of log $\Delta\theta$ versus log Pe approaches 2 as $Pe \to 0$. It is also important to understand that only the product $V \cdot R$ is specified by Ivantsov's solution, and that a manifold of growth states is possible. Thus, although $V = V(R)$ is known, *the unique, or operating state of the dendrite is not specified by transport theory.*

This is perhaps understandable based on the fact that the only length scale appearing in the transport problem, *per se*, is the diffusion length, α/V. A second independent length scale is required to go further and determine the unique growth state $V(\Delta\theta)$, $R\,(\Delta\theta)$ for a dendritic crystal growth system.

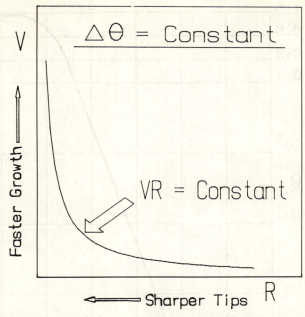

Fig. 2. Velocity vs. tip radius for a fixed value of $\Delta\theta$, obtained from Ivantsov's heat-flow solution for a paraboloid of revolution.

II. INFLUENCE OF THE SOLID-LIQUID INTERFACE

II.1 Need for capillarity effects

Equation (5) relates $\Delta\theta$ to the Peclet number, $Pe = VR/2\alpha$, where R is the radius of curvature of the dendrite tip, and α is the thermal diffusivity of the melt. The inverse of eqn. (5), although not expressible in terms of known functions, does establish that $V \cdot R = f (\Delta\theta)$, which provides an infinite range of hyperbolic solutions, i.e., unbounded values for V and R for a given value of $\Delta\theta$. Clearly, the transport solutions for the steady-state dendrite lack uniqueness when the only physical length scale introduced into the problem is the characteristic, but unknown, diffusion distance, α/V, or, alternatively, the equally unknown dendrite tip radius of curvature, R. This limitation was recognized over twenty years ago by Bolling and Tiller [4], who then introduced a non-linear boundary condition into the problem which effectively places an upper bound on V and a lower bound on R. Bolling and Tiller suggested that local thermodynamic equilibrium along the dendrite surface requires that the melting temperature, T_e, is a function of the mean interfacial curvature, k, namely, $T_e = T_m - \Gamma k$, which is the well known Gibbs Thomson equation. Here $\Gamma = \gamma\Omega/\Delta S$; γ is the solid-liquid surface energy; Ω is the molar volume of the solid; and ΔS is the molar entropy of fusion. The Gibbs-Thomson equation requires, therefore, that the dendrite grows with a *non-isothermal* interface.

16

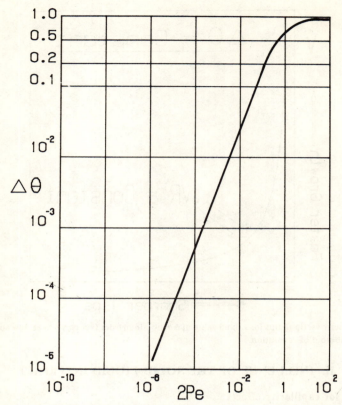

Fig. 3. Log-log plot of $\Delta\theta$ vs. $2Pe$ from Ivantsov's solution. The slope of this plot approaches 2 for small values of Pe and $\Delta\theta$, while $Pe \to \infty$ as $\Delta\theta \to 1$.

II.2 Non-isothermal dendritic growth

Introduction of the non-isothermal temperature boundary condition raises a severe difficulty, inasmuch as the steady-state shapes which were treated as a class of shape preserving solutions by Horvay and Cahn [2] no longer simultaneously satisfy both energy conservation and the non-isothermal equilibrium temperature boundary condition. Several approximate theories were developed in which the interface shape was chosen to satisfy either energy conservation [5–7] or the non-isothermal condition [8, 9], and a decade ago, a self-consistent theory was finally developed [10] which determined the dendrite shape as part of the solution and satisfied both physical requirements. All of these non-isothermal theories shared the common result that the values of V and R, which constitute possible operating states, lie along curves with a maximum in the value of V. Figure 4 shows two typical V versus R relationships at a fixed value of the supercooling.

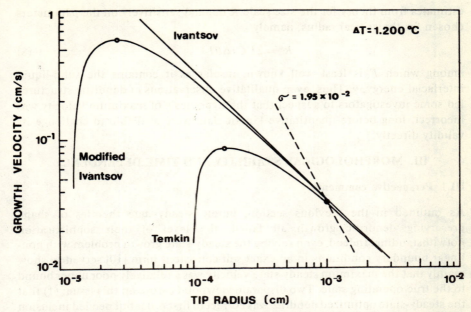

Fig. 4. Velocity (logarithmic scale) vs. dendrite tip radius (logarithmic scale) at a fixed supercooling T of $1.200°C$. The isothermal (Ivantsov) solution is characterized by $VR = $ constant, whereas the nonisothermal (Temkin and modified Ivantsov) solutions show progressive departures from the line $VR = $ constant as the tip radius decreases. $VR^2 = $ constant (---) represents the condition of marginal morphological stability with a separation constant σ^* of 1.95×10^{-2}, close to those discussed in the text.

II.3 Maximum growth rate hypothesis

The value of V at the maximum was selected as the probable operating state over the manifold of the possible operating states. Note that there were never compelling reasons used to incorporate within these theories the hypothesis that the maximum velocity is the unique velocity characterizing axial dendritic growth. As shown in Fig. 4 the states of maximum velocity for non-isothermal models have tip radii such that the Peclet number at $V = V_{max}$ is one-half that of the isothermal Ivantsov model. The tip radius for the Modified Ivantsov model is just twice the critical radius R*, where R* is the radius which depresses the dendrite tip temperature by an amount equal to the supercooling. Without a temperature difference between the interface and the supercooled melt, latent heat cannot be transferred and $V = 0$. R* is easily estimated from the Gibbs-Thomson relationship by finding the value of $k^* = 2/R^*$ which reduces the tip temperature, T_m, to T_∞. Figure 4 corresponds to a dimensionless supercooling $\Delta\theta = 0.05$, which is relatively small, yet as seen in this figure $R^* \simeq 10^{-5}$ cm. Even Temkin's analysis predicts a tip radius of only 10^{-4} cm, which is still much smaller than the observed scale of dendrites at such small supercoolings. Of course, estimates from theory for the size scale R depend sensitively on the parameters chosen for the critical radius, namely

$$R^* = 2\Gamma C/\Delta\theta L, \qquad (8)$$

among which Γ is least well known, insofar as it contains the solid-liquid interfacial energy, γ. Thus, even qualitative observations of dendritic structures led some investigators to suspect that the hypothesis of maximum velocity was incorrect, long before quantitative kinetic data were available to challenge its validity directly.

III. MORPHOLOGICAL STABILITY AND TIME DEPENDENCE

III.1 Perspective commentary

As outlined in the previous section, purely steady-state theories of shape preserving dendritic growth all failed, the level of their sophistication notwithstanding. Indeed, even solving the steady-state growth problem with non-linear boundary conditions in an exact self-consistent form [10] served to show mainly that the maximum steady-state velocity was a relatively poor upper bound to the true operating state. Two disparate viewpoints arose on this issue: (1) that the steady-state optimized dendrite was correct to first order but needed inclusion of nonsteady-state features such as side branching [17, 18], and (2) that dendritic growth was intrinsically time-dependent and unstable [19, 20–21]. Analyses based on the first point of view showed tha the steady-state needle dendrites were unstable when tested for morphological stability using linear perturbation methods fashioned after those first used by Mullins and Sekerka [21] and by Voronkov [23]. An unfortunate aspects of these approaches was that the steady-state dendrite shapes themselves were only approximations, and therefore were intrinsically unstable *without* perturbation. Consequently, viewpoint (1) only served to emphasize the deficiency of steady-state approaches, and did not lead to new insights into the problem. The second viewpoint was originally proposed a decade ago by Oldfield [19] who was the first investigator to stress that the size of a dendrite tip might be selected through a balance of destabilizing forces arising from diffusion by stabilizing forces arising from capillarity. He found by simple logic and some numerical analysis that

$$V R^2 \simeq 100 a \Gamma C / L, \qquad (9)$$

which, as we shall show, is remarkably close to the results obtained later by linear perturbation analysis. Oldfield also demonstrated through computer generated cinematography that such a dendrite was actually a fully time-dependent object, with branches emanating as waves from a *nearly* steady-state tip. The numerical character of Oldfield's work, unfortunately, prevented its wide acceptance at that time.

III.2 Dynamical theory of dendrites

The proper estimation of size scales for morphologically unstable systems begins with Mullins and Sekerka's ideas that any Fourier component of a perturbed planar interface represented as $\delta = \delta_0 \exp{(i\omega x)}$ is subject initially to a time dependence described by $\delta(t) = \delta \exp{(\sigma t)}$, where σ, in general, is a complex eigenvalue of the linearized dynamical equations of the interface motion [21]. The quantity ω is the Fourier component's wave number, t is time, and δ_0 denotes the initial (small) amplitude of this component at $t = 0$. If the real part of σ is negative, then the perturbation decays to zero amplitude, whereas if the real part of σ is positive then δ grows exponentially. If ω is purely imaginary, then on average the amplitude of δ remains equal to δ_0 and the interface is deemed to be marginally stable. The condition of marginal stability for a pure material growing from its supercooled melt may be shown to be [21]

$$Re\,(\sigma) = 0 = -\Gamma \omega^{*2} - \bar{G}, \tag{10}$$

where $\bar{G}$ is the average thermal gradient (weighted by the thermal conductivities of each phase) and ω^* is the wave number of the marginal perturbation. We now adopt the remarkable suggestion originally made by Langer and Müller-Krumbhaar [22] that the wavelength, $\lambda^* = 2\pi\omega^{*-1}$, of the marginal perturbation sets the scale of the dendrite tip radius, R. Moreover, the average temperature gradient acting on the tip may be found from the transport solution of the *unperturbed*, steady-state, shape preserving dendrite. For example, the paraboloid of revolution, as described by Ivantsov's solution [2], is an isothermal shape with a zero gradient within the solid phase and with a dimensionless temperature gradient in the melt phase ahead of the tip $G_l = -2Pe$. Again, Pe, the Péclet number, can be related to the supercooling through eqn. (5), and thus the average dimensionless gradient $G = -Pe$. The gradient G can be dimensionalized to $\bar{G}$ by rescaling by the characteristic temperature L/C divided by the dendrite tip radius, R. Thus, the average temperature gradient at the dendrite tip is

$$\bar{G} = \frac{-Pe}{R}\frac{L}{C} = \frac{-VL}{2\alpha C} \tag{11}$$

If eqn. (11) is substituted into eqn. (10) we find that the marginally stable state $Re(\sigma) = 0$ occurs when $R = \lambda^*$, which after some rearrangement yields the condition for growth

$$VR^2 = 4\pi^2 \alpha C \Gamma / L, \tag{12}$$

which, except for a slight difference in the numerical coefficient, agrees with Oldfield's expression, eqn. (9).

Equation (12) is the central result obtained from dynamical analysis of dendrite tip motion. If we recall the definition $Pe = VR/2\alpha$, then eqn. (12) may be recast in an especially convenient form, namely

$$Pe = \frac{2\pi^2 \, C\Gamma}{RL} \qquad (13)$$

Now, Ivantsov's transport solution, eqn. (5), may be written in operator form as $\Delta\theta = Iv[Pe]$, where $Iv[\ ]$ represents the series of operations carried out on the right-hand side of eqn. (5). We can formally invert eqn. (5) to stress that Pe is some function of $\Delta\theta$, viz.,

$$Pe = Iv^{-1}[\Delta\theta], \qquad (14)$$

although the inversion operator $Iv^{-1}[\ ]$ cannot be represented exactly by any known series of algebraic or transcendental operations. Nonetheless the inversion operator Iv^{-1} exists, if only as a graph or an asymptotic expansion. Equations (13) and (14) can now be combined, eliminating explicit dependences on Pe, to yield the operating state of the dendrite under marginally stable dynamical conditions. Specifically, we find that

$$R = \frac{2\pi^2}{L} \, \frac{C\Gamma}{Iv^{-1}[\Delta\theta]}, \qquad (15)$$

and

$$V = \frac{1}{4\pi^2} \left(\frac{2\alpha \Delta S L}{\gamma \Omega C} \right) \{ Iv^{-1}[\Delta\theta] \}^2 \qquad (16)$$

IV. SCALING LAWS AND OBSERVATIONS

IV.1 Analysis

The coefficient $4\pi^2$ appearing in eqns. (15) and (16) is often defined as $(\sigma^*)^{-1}$. Also the parameter grouping

$$\frac{C\Gamma}{L} = \frac{C\gamma\Omega}{L\Delta S} \equiv d_o, \qquad (17)$$

where d_o is called the capillary length scale. d_o is about 1 nm in size for many common materials, and represents the second physical length scale introduced.

Thus, we can rewrite eqns. (15) and (16) in a scaled form:

$$\frac{d_o}{R} = \sigma^* \, Iv^{-1}[\Delta\theta], \qquad (18)$$

and

$$\frac{V}{V_0} = \sigma^* \, \{ Iv^{-1}[\Delta\theta] \}^2, \qquad (19)$$

where V_0 is the "characteristic" velocity of the dendrite forming system given by

$V_0 = \dfrac{2a\Delta S L}{\gamma \Omega C}$. Note that the *scaled* radius only depends on the dimensionless supercooling, $\Delta\theta$, and σ^*, as does the *scaled* velocity. All the materials parameters appear either in d_o or V_0.

The predicted operating point (V, R) of dendrites can be expressed in a form especially convenient for checking against experiments, namely

$$V \cong 0.018 \frac{\Delta S \alpha L}{\Omega \gamma C} \Delta\theta^{2.5}, \tag{20}$$

$$R \cong 55 \frac{C\gamma\Omega}{L\Delta S} \Delta\theta^{-1.25}. \tag{21}$$

These expressions are accurate for *moderate* values of $\Delta\theta$. In typical experiments, $0.01 < \Delta\theta < 0.3$. At extremely *small* values of $\Delta\theta$, $V_0 \sim \Delta\theta^2$, and $R_0 \sim \Delta\theta^{-1}$. At *large* values of $\Delta\theta$, the local equilibrium assumption breaks down and the theory, as presented here, fails.

Several careful experiments (Fig. 5) conducted by Glicksman, Schaefer, and Ayers [11], Huang, and Glicksman [24], and by Fujioka [15], confirm this result to within experimental error and the current knowledge of the thermophysical parameters. To date, three systems have been checked:

  a) Succinonitrile
  b) ice/water
  c) pivalic acid

Figure 6 shows that in a transparent model system such as succinonitrile (SCN) the images of the dendrite tip do indeed scale with the supercooling. Also, Fig. 8 shows that the dendrite tip radius, R, scaled to the critical wavelength, λ^*, is of unit order, where

$$\lambda^* = \frac{d_0}{\sigma^* Iv^{-1}[\Delta\theta]} \tag{22}$$

That is, $R \cong \lambda^*$, which is the fundamental scaling arising from the dynamical theory.

Another interesting scaling relationship for dendrites can be obtained from the stability analysis by inserting eqn. (8) into eqn. (15) and then solving for the ratio of the tip radius to the nucleation or critical radius, R/R^*. We find

$$\frac{R}{R^*} = \frac{\pi^2 \, \Delta\theta}{Iv^{-1}[\Delta\theta]}, \tag{23}$$

which by virtue of eqns. (5) and (14) and the definition of $\sigma^* = 1/4\pi^2$ may be rewritten in the forms

$$\frac{R}{R^*} = \frac{Iv[Pe]}{4\sigma^* Pe} = \frac{\exp(Pe) E_1[Pe]}{4\sigma^*}. \tag{24}$$

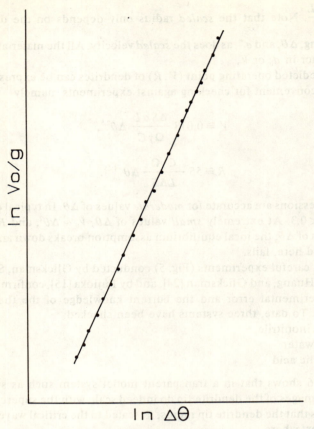

In Vo/g

In Δθ

Fig. 5. Experimental values of growth velocity, V_o (normalized by the characteristic velocity of the material, $g = 0.018 \dfrac{\Delta S a L}{\Omega \gamma C}$) vs. dimensionless undercooling, $\Delta \theta$. The slope of this log-log plot is about 2.5, which agrees with the result obtained from a marginal stability criterion at the tip (eq. IV. 4 in the text).

In the range of small supercoolings ($\Delta \theta << 1$), specifically where the value of Pe is sufficiently small that $E_1(Pe) \to -\text{In } Pe$, eqn. (24) becomes

$$\frac{R}{R^*} (\Delta \theta << 1) = \frac{-\ln Pe}{4\sigma^*}. \tag{25}$$

The value of R/R^* predicted from eqns. (23) to (25) over the typical range of experimentally useful "small supercoolings" ($10^{-3} < \Delta \theta < 0.1$) is of the order of 100, clearly indicating that marginally stable dendrites ought to grow with their tip radii much larger than R^*, which is a morphological scaling law at variance with the steady-state theories that predict small multiples of R^*. Figure 7 shows

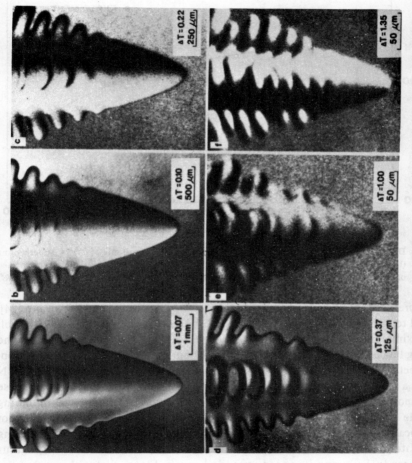

Fig. 6. Tip morphologies of succinonitrile dendrites growing at various supercoolings. Photographic magnifications have been adjusted to produce constant apparent tip radii. It should be noted that at increasing supercoolings the side branches amplify more rapidly and encroach on the steady state region near the tip.

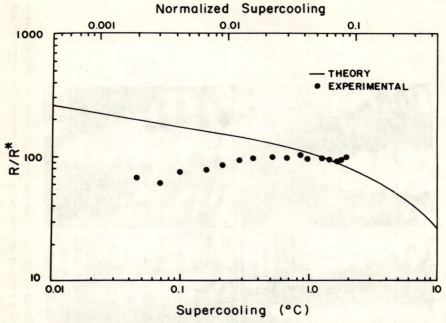

Fig. 7. Dendrite tip radius R scaled to the critical radius R^* (logarithmic scale) vs. supercooling (logarithmic scale):———, theory; •, experimental. Data, based on measurements performed on succinonitrile, show that the operating states of dendrites occur at large multiples of R^*.

measurements of SCN dendrite tip radii, obtained by Huang and the author [24], scaled to the critical radii, R^*. These data shows that the value of R/R^* is well approximated by eqns. (23) to (25). The precise fit of the data in Fig. 7 to the scaling law is limited at very small supercoolings by the presence of convection. Convection effects, relative to transport by diffusion, become negligible in SCN above a supercooling of about 1 K [29], and, as shown in Fig. 7, a rather close correspondence occurs near and above $\Delta T = 1 \, K$. It is unfortunate that obtaining reliable morphological data above a supercooling of 2 K is, at present, not technically feasible with SCN. This is due simply to the difficulty in obtaining adequate photographic resolution of the dendrite tip structure as the tip radius decreases to 1 μm and its speed exceeds 1 mm/s. Figure 6 also shows that the series of SCN dendrite images growing at increasing levels of supercooling decline in optical quality as the supercooling increases. Lappe [30] has confirmed many of the morphological and kinetic measurements originally reported by Huang [31], and has further demonstrated that at a supercooling of 2 K, or beyond, a growing SCN dendrite cannot be resolved optically to permit an accurate measurement of the tip radius. Figure 6 also shows another effect which might be explained by the scaling law shown in Fig. 7 and described in eqns. (20) and (21), namely, that the dendrites are not self-similar, except for the tip itself. Inspection

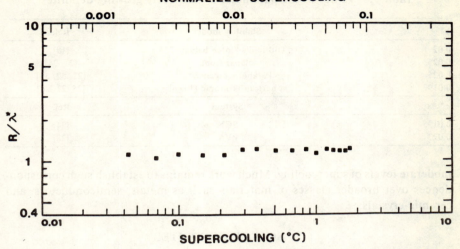

NORMALIZED SUPERCOOLING

SUPERCOOLING (°C)

Fig. 8. Dendrite tip radius R scaled to the wavelength of the marginal eigenmode λ^* (Longarithmic scale) vs. supercooling (logarithmic scale). The scaling law demonstrated here remains valid over at least two decades in supercooling, i.e. $R/\lambda^* = 1.2$.

of the micrographs in Fig. 6 show that the side branches intrude on the tip as the supercooling increases, due primarily to the faster amplification rate of the marginal eigenmode, ω^*. It would be of extreme interest to follow this trend into the regime of large supercoolings and rapid solidification where a great deal of current research interest is focused. For the present, we can accept that the ratio R/R^* should decrease at large supercoolings, with the dendrite becoming commensurately less stable. Eventually, the interfacial molecular attachment rate will become rate limiting, causing the interface to depart radically from local equilibrium and from the scaling laws based on local equilibrium. This remains as an interesting topic of research.

Finally, the fundamental assertion of Langer and Müller-Krumbhaar is supported by the data in Fig. 8 where the scaling law $\lambda^* = R$ is shown to hold for SCN over two decades of supercooling. Specifically, $R = 1.2 \, \lambda^*$ in Fig. 8 which is equivalent to an error of 20% in the value of $\sigma^* = 1/4\pi^2$. As shown in Table 1, the analysis of tip stability based on spherical harmonics almost provides the σ^* value needed, viz., $\sigma^* = 0.021$. Although the differences among the σ^* values appearing in Table 1 seem modest, the detailed physical assumptions employed in each stability analysis are markedly different. For example, the planar front model bears little geometrical similarity to a dendrite tip and cannot include factors such as crystal symmetry or anisotropy. For the spherical harmonic model, by comparison, only the BCC system (SCN) and an fcc high anisotropy system (PVA) have been shown to display kinetics and tip morphologies which are in quantitative agreement with the dynamical theory of dendritic growth at low to

Table 1. Values of the parameter σ^* for freely growing dendrites

σ^*	Stability model	Ref.
0.02	Oldfield's "force balance"	[19]
0.0253	planar front	[21, 24]
0.025	Parabolic eigenstate	[22, 32, 33]
0.0192	spherical harmonic $(1 = 6)$	[25-27, 20]
σ^* exp.	System	Ref.
0.0195	SCN	[24]
0.022	PVA	[28]

moderate levels of supercooling. Much work remains to establish such correspondences over broader classes of materials such as metals, semiconductors, and ceramic crystals.

V. SOLUTE DENDRITES

V.1 Importance of alloy solidification

Most materials are solidified as alloys, so that solute diffusion and constitutional supercooling are encountered, in addition to thermal supercooling as already introduced for pure substances. Virtually all castings are formed from alloy dendrites, and, indeed, a typical cast microstructure is delineated by strong microsegregation of alloying additions surrounding the dendritic crystals.

A brief outline is presented covering the solidification fundamentals of binary alloy dendritic growth. The theory, as formally presented, covers the general situation of coupled heat and mass transfer at the solid-liquid interface. Both thermal and constitutional effects are included, so the case described is that of an equiaxed alloy dendrite, of the sort often encountered in the central zones of slowly frozen castings.

V.2 Combined transport model

The basic transport model of Ivantsov eqn. (5) can be used to describe solute transport as well as heat transport. We merely define a dimensionless supersaturation,

$$U = \frac{C_l^* - C_0}{C_l^*(1 - k_0)}, \tag{26}$$

where C_l^* is the concentration in the liquid phase just ahead of the dendrite tip; C_0 is the nominal alloy concentration established far from the dendrite; k_0 is the equilibrium distribution coefficient. Thus we may re-write eqn. (5) for alloys as

$$U = Pc \ \exp \ (Pc) \ E_1 \ (Pc), \tag{27}$$

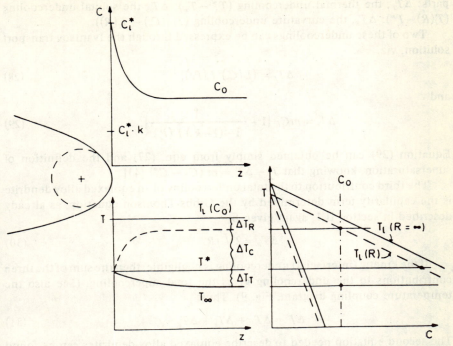

Fig. 9. Solute concentration profile and thermal fields ahead of a growing alloy dendrite. The portion of the phase diagram at the right shows the capillary shift in equilibrium concentrations due to the interfacial radius of curvature, R. The temperature–distance plot illustrates the capillary, solutal and thermal contributions to the effective undercooling in the liquid phase.

where Pc is the chemical Péclet number [c.f. Pe, which is the thermal Peclet number]. By analogy with $Pe \equiv VR/2\alpha$, we define $Pc = VR/2D$, and D is the chemical diffusivity of the solute in the molten phase.

Figure 9 shows the various relationships concerning the distribution of solute, $C(Z)$, and temperature, $T(Z)$, ahead of a dendrite. The key assumptions required in arriving at these distributions are:

• The dendrite grows steadily into an undercooled melt of constant undercooling.
• The shape of the dendrite tip is described by a paraboloid of revolution (isothermal and isoconcentrate tip).
• The heat and mass transport in the solid and liquid is controlled by diffusion only (i.e., convection plays a negligible role).

The melt undercooling, ΔT, determines the driving force for the dendrite growth. In pure materials ΔT can be set approximately equal to the thermal undercooling. In the solute case the situation becomes more complicated as shown in Fig. 9. The total undercooling ($\Delta T = T_l - T_\infty$) is subdivided into three parts: ΔT_T, the thermal undercooling $(T^* - T_\infty)$; ΔT_c, the solutal undercooling $(T_l(R) - T^*)$; ΔT_R, the curvature undercooling $(T_l(C_o) - T_l(R))$.

Two of these undercoolings can be expressed through the Ivantsov transport solution, viz.

$$\Delta T_T = (L/C_p)\, I_v(Pe), \tag{28}$$

and

$$\Delta T_c = mC_0 \left\{1 - \frac{1}{1-(1-k_0)\,I_v(P_c)}\right\}. \tag{29}$$

Equation (29) can be obtained simply from eqn. (27) and the definition of supersaturation, knowing that $T_l - \Delta T_c = m\,(C_0 - C_l^*)$ [43].

The third contribution to the total undercooling of an equiaxed alloy dendrite is the capillarity term determined by the Gibbs-Thomson equation, as already described in section II.1, again given by

$$\Delta T_R = 2\, \Gamma/R. \tag{30}$$

If the kinetic (interfacial) undercooling is negligible, then the sum of the three contributions to the undercooling gives the total supercooling. (See also the temperature coupling diagram, Fig. 9). Thus,

$$\Delta T = \Delta T_T + \Delta T_c + \Delta T_R. \tag{31}$$

The second equation needed to describe equiaxed alloy dendrites can be found from the condition of marginally stable tip growth [22, 32]. This criterion (the same one as used for pure materials) postulates that a dendrite tip grows at the margin of stability, i.e. the radius at the tip is equal to the wavelength of the fastest growing instabilities to be formed under the local growth conditions. To simplify the problem we use the lower limit of the perturbation wavelength of a plane front [44]. This condition has been shown to correspond closely to the growth conditions encountered in dendrites of pure SCN (pure thermal dendrites) [24].

At low Peclet numbers (corresponding to low undercoolings, as encountered in castings) the instability wavelength λ^* is given by [44].

$$\lambda^* = R = \left(\frac{\Gamma}{\sigma^*(mG_c - \bar{G})}\right)^{1/2} \tag{32}$$

Here σ^* is a stability constant on the order of $1/(4\pi^2) \simeq 0.025$; m is the liquidus slope, assumed to be constant; G_c and $\bar{G}$ are the concentration gradient and conductivity-weighted temperature gradient at the interface, respectively.

The concentration gradient at the tip of the dendrite is determined from the transport solutions (eqn. 27) to be

$$G_c = -\frac{2\,P_c\,C_l^*(1 - k_0)}{R}, \tag{33}$$

and

$$G_l = -\frac{2\,Pe\,\Delta H}{C_p\,R}.\tag{34}$$

The conductivity-weighted average temperature gradient is $\bar{G} = (k_s G_s + k_l G_l)/(k_s + k_l)$ where k_s and k_l are the thermal conductivities of solid and liquid, and G_l and G_s are the thermal gradients in liquid and solid. With equal thermal conductivities in liquid and solid, and with the isothermal Ivantsov dendrite ($G_s = 0$) one obtains for the mean temperature gradient ahead of the tip

$$\bar{G} = -\frac{Pe\,\Delta H}{C_p\,R}.\tag{35}$$

Substituting eqns. (33) and (35) in (32) yields an expression for the tip radius

$$R = \frac{\Gamma}{\sigma^*}\left[\frac{-\Delta H\,Pe}{C_p} - \frac{2\,P_c\,m C_o\,(1 - k_o)}{1 - (1 - k_o)\,I_v(P_c)}\right]^{-1},\tag{36}$$

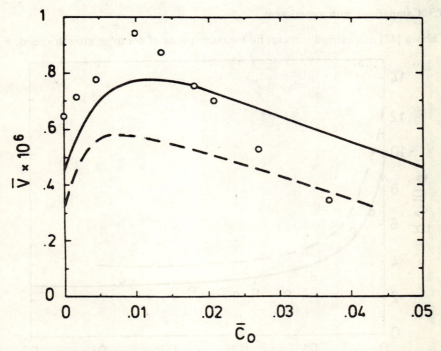

Fig. 10. Variation of dimensionless tip velocity with dimensionless concentration for succinonitrile–acetone mixtures with an undercooling of 0.5K. Points: Experimental values; solid line: Lipton-Glicksman-Kurz (LGK) model [36]; Dashed line: Karma-Langer (KL) model [37].

and one for the capillarity undercooling at the tip,

$$\Delta T_R = 2\sigma^* \left[\frac{-\Delta H \, Pe}{C_p} - \frac{2P_c \, mC_o \, (1-k_o)}{1-(1-k_o) \, I_v(P_c)} \right]. \tag{37}$$

The solutal Péclet number is simply related to the thermal Péclet number by $P_c = Pe \, (\alpha/D)$ (with α the thermal and D the solutal diffusivity), so the only unknown appearing in eqn. (31) is the product $(R \cdot V)$, which appears in the Peclet numbers.

Finally, the growth rate can be calculated from the definition of the Peclet number

$$V = 2\alpha Pe / R. \tag{38}$$

Note that eqn. (31) with eqns. (28), (29), (37), and (36) define completely the growth problem at low Péclet numbers, i.e., for a given composition, C_o, and undercooling, ΔT, the radius, R, and the growth rate, V, can be calculated *uniquely* by simultaneously satisfying the transport equations for heat and solute and the condition of marginal stability. These equations constitute a coupled non-linear set which must be solved by iterative methods on a computer.

V.3 Comparison with experiment

Chopra [42] has carried out careful measurements of dendrite growth speed, V,

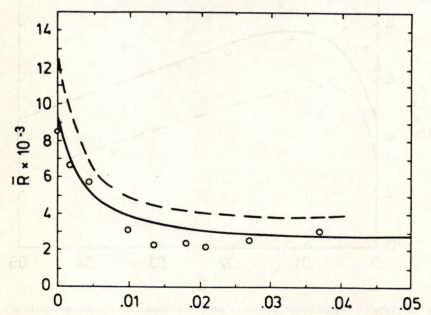

Fig. 11. Relationship between dendrite tip radius and dimensionless concentration for the system described in Fig. 10.

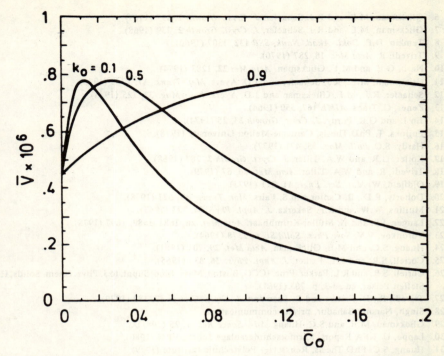

Fig. 12. Velocity versus solute concentration for succinonitrile alloys with solutes having different distribution coefficients k_o, as predicted by the LGK model [36]. Mean undercooling is 0.5K.

and dendritic tip radius, R, as functions of solute concentration C_o. Figures 10 and 11 compare Chopra's observations on acetone-succinonitrile alloys with the predictions of eqns. (36) and (37).

The general trends appear correct, with a maximum appearing in V at small concentrations and a sudden drop in R occurring at small solute concentrations. Figure 12 shows what the model predicts for acetone-succinonitrile alloys assuming different values for k_o. Other details have been discussed regarding this model of alloy dendrite growth and the interested reader is referred to reference [45].

REFERENCES

1. Fisher, J.C. As referenced by Bruce Chalmers, *Principles of Solidification*, John Wiley and Sons, New York, p. 105 (1964).
2. Ivantsov, G.P. *Dokl, Akad. Nauk. SSR* **58**, 567 (1947).
3. Horvay, G. and J.W. Cahn. *Acta Met.* **9**, 695 (1961).
4. Bolling, G.F. and W.A. Tiller. *J. Appl. Phys.* **32**, 2587 (1961).
5. Sekerka, R.F., R.G. Seidensticker, D.R. Hamilton and J.D. Harrison. *Investigation of Desalination by Freezing* (Westinghouse Res. Lab. Rep.), Ch. 3 (1967).

32

6. Glicksman, M.E. and R.J. Schaefer. *J. Cryst. Growth* **1**, 297 (1967).
7. Glicksman, M.E. and R.J. Schaefer. *J. Cryst. Growth* **2**, 239 (1968).
8. Temkin, D.E. *Dokl. Akad. Nauk, SSR* **132**, 1307 (1960).
9. Trivedi, R. *Acta Met.* **18**, 287 (1970).
10. Nash, G.E. and M.E. Glicksman. *Acta Met* **22**, 1283 (1974).
11. Glicksman, M.E., R.J. Schaefer and J.D. Ayers. *Met. Trans.* **A7**, 1747 (1976).
12. Schaefer, R.J., M.E. Glicksman and J.D. Ayers. *Phil. Mag.* **32**, 725 (1975).
13. Zener, C. Trans. *AIME* **167**, 550 (1964).
14. Jin I. and G.R. Purdy. *J. Cryst. Growth* **23**, 25 (1974).
15. Fujioka, T. PhD Thesis, Carnegie-Mellon University (1978).
16. Hardy, S.C. *Phil. Mag.* **35**, 471 (1977).
17. Kotler, G.R. and W.A. Tiller. *J. Cryst. Growth* **2**, 287 (1968).
18. Trivedi, R. and W.A. Tiller. *Acta Met.* **26**, 67 (1979).
19. Oldfield, W. *Mat. Sci. Engr.* **11**, 211 (1973).
20. Doherty, R.D., B. Cantor and S. Fairs. *Met. Trans.* **A9**, 621 (1978).
21. Mullins, W.W. and R.F. Sekerka. *J. Appl. Phys.* **34**, 323 (1963).
22. Langer, J.S. and H. Müller-Krumbhaar. *Acta Met.* **26**, 1681-1689, 1697 (1978).
23. Voronkov, V.V. *Sov. Phys. Solid St.* **6**, 2378 (1964).
24. Huang, S.C. and M.E. Glicksman, *Acta Met.* **29**, 701 (1981).
25. Coriell, S.R. and R.L. Parker. *J. Appl. Phys.* **36**, 632 (1965).
26. Coriell, S.R. and R.L. Parker. Proc. ICCG, Boston, Mass. 1966, Suppl. to J. Phys. Chem. Solids, H. Steffen Peiser, ed., J-3, p. 703 (1967).
27. Trivedi, R., H. Franke and R. Lacmann. *J. Cryst. Growth* **47**, 389 (1979).
28. Singh, Narsing Bahadur. private communication.
29. Glicksman, M.E. and S.C. Huang. *Adv. Space Res.* **1**, 25 (1981).
30. Lappe, U. KFA Report, Kernforschungsanlage Julich, FRG (1980).
31. Huang, S.C. PhD Thesis, Rensselaer Polytechnic Institute (1979).
32. Langer, J.S. and H. Müller-Krumbhaar. *J. Cryst. Growth* **42**, 11 (1977).
33. Langer, J.S. *Rve. Mod. Phys,* **52**, (1), 1 (1980).
34. Burden, M.H. and J.D. Hunt. *J. Cryst. Growth* **22**, 99 (1974).
35. Witzke, S., J.P. Riquet and F. Durand. *Acta Met.* **29**, 365 (1981).
36. Fredricksson, Hasse. In: *Materials Processing in the Reduced Gravity Environment of Space*, Guy E. Rindone ed., Elsevier, Amsterdam, p. 619 (1982).
37. Trivedi, R. and W.A. Tiller. *Acta Met.* **26**, 679 (1978).
38. Langer, J.S. *Phys. Chem. Hydrodyn.* **1**, 41 (1980).
39. Lindenmeyer, C. PhD Thesis, Harvard University (1959).
40. Glicksman, M.E. Narsingh Bahadur Singh and M. Chopra. In: *Materials Processing in the Reduced Gravity Environment of Space,* Guy E. Rindone, ed., Elsevier, Amsterdam, p. 461 (1982).
41. Kurz, W., J. Lipton and M.E. Glicksman. unpublished work (1983).
42. Chopra, M. PhD Thesis, Rensselaer Polytechnic Institute (1983).
43. Lipton, J., M.E. Glicksman and W. Kurz. *Mater. Sci. Eng.* **65**, 57-63 (1984).
44. Kurz, W., and D.J. Fisher, *Acta Metall.* **29**, 11-20 (1981).
45. Lipton, J., M.E. Glicksman and W. Kurz. *Met. Trans.* **18A**, 341-345 (1987).

3

Directional Solidification of Alloys

R. TRIVEDI

Ames Laboratory, U.S. Department of Energy and
the Department of Materials Science and Engineering,
Iowa State University, Ames, IA 50011, U.S.A.

ABSTRACT

Directional solidification (DS) technique is extensively used for producing single crystals of alloys. The DS process, which leads to the columnar crystal formation in casting and welding has also been recognized for a long time. However, the high degree of control of the solidification process that is possible with the DS technique has now led to its use in producing structural and nonstructural components of the advanced engineering devices. In this paper, the basic ideas on DS are presented with an emphasis on the science and technology of various microstructural manipulations that are possible by changing experimentally controlled parameters. Specifically, the formation of planar, cellular and dendritic fronts is discussed and the processing parameters and the alloy design required for the selection of a specific microstructure are presented.

INTRODUCTION

Solidification is one of the key techniques for materials processing. The properties of the solidified materials depend strongly on the microstructure which is governed by the conditions under which the material is solidified and by the phase diagram. Consequently, to obtain a product which meets specific requirements, it is important that we identify the processing parameters for a given alloy that give the desired microstructure.

The development of microstructure depends on the manner in which the latent heat of fusion is carried away from the advancing solid-liquid interface. Two distinctly different heat flow conditions may be present which give rise to either the free growth or the constrained growth behavior [1–3]. In free growth, where the solidification occurs in an undercooled melt, the latent heat is dissipated through the liquid. In constrained growth the latent heat is carried away through the solid and the temperature gradient in the liquid ahead of the interface is positive. This constrained growth condition gives rise to directional solidification (DS).

The microstructural development under rapid solidification conditions is discussed in detail by Kurz and Trivedi (in this volume). We shall thus restrict ourselves only to normal solidification conditions. Under normal solidification conditions, two major types of microstructures, viz., dendritic and eutectic microstructures, may form when the solidification occurs in an undercooled melt. In directional solidification, in addition to dendritic and eutectic interfaces, a planar, cellular or cellular dendritic interface morphologies may also be present. Which of these microstructures actually form in a given alloy depends upon the growth rate and temperature gradients imposed upon the system. We shall first discuss the DS technology and then establish the relationship between the processing parameters and the microstructure.

DIRECTIONAL SOLIDIFICATION PROCESS

The basic ideas of the DS technology are illustrated schematically in Fig. 1. Two major aspects are: (1) the heater, insulater and a cold chamber to establish a required temperature gradient along the sample, and (2) a mechanism to solidify the alloy. The part of the sample in the heater is liquid, whereas the part in cold chamber is solid. The solid-liquid interface is present between the hot and the cold zones. The insulater is used to provide a unidirectional heat flow profile such that the heat flow lines near the interface are downwards and parallel to the axis of the sample tube. Once the desired temperature gradient is established, the solidification of the alloy is achieved by one of the following three techniques:

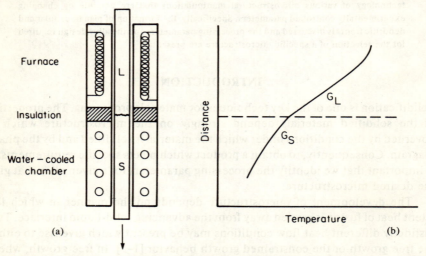

Fig. 1. Schematic illustration of the basic directional solidification process. The furnace, insulator and water-cooled chamber produce a required temperature gradient, shown in (b).

(1) The tube which contains the alloy is moved downwards into the cold chamber at a constant rate. After some initial transient, the solid-liquid interface will also move at the same rate as the tube. (2) The alloy is held stationary, and the furnace insulator-cold chamber assembly is moved upwards at a constant rate. (3) Both, the alloy tube and the temperature gradient assembly are held stationary, but the temperatures of the furnace and the cold chamber are reduced synchronously at a constant rate such that the temperature gradient is not altered. Since the temperature at a given point changes linearly with time, the solid-liquid interface moves at a constant rate to maintain a specific interface temperature.

In the DS process the temperature gradient, G, at the interface, the velocity of the interface, V, and the alloy composition can be varied independently so that the effect of each of these variables on the microstructural evolution can be precisely characterized. Because of this ability to isolate each experimental variable, the DS technique is a very powerful tool to study the fundamental aspects of the correlation between the microstructure and the processing variables. For example, many microstructural features have been correlated with cooling rate, which is equal to GV. Thus, DS studies will be able to sort out those characteristics which have similar dependence on G and V so that they could be described by one variable which is the cooling rate. As seen in the later sections, not all microstructural characteristics can be described by the cooling rate only.

Since both the temperature gradient and the growth rate can be precisely controlled, a high degree of microstructural control is achieved in the DS technique. Thus, the DS process is ideally suited for the production of single crystal devices. A single crystal is obtained by either starting with a seed crystal of required orientation, as shown in Fig. 2a, or by using an angular or a helical grain selector which allows only one grain to emerge as the tortuous path excludes other grains (Figs. 2b and 2c). The DS technique, under large G and low

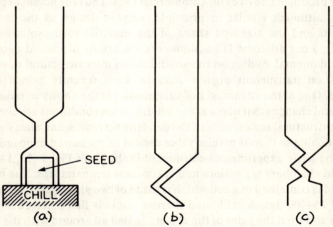

SEED

CHILL

(a)   (b)   (c)

Fig. 2. Selected techniques for growing single crystals. (a) seeding technique, (b) angular grain selector, and (c) helical grain selector.

V, may give rise to a planar interface growth which allows the crystal to form with a uniform composition. Alternately, cellular or dendritic interface, produced from the same grain also forms a single crystal except that such a crystal exhibits a microsegregation profile which can be removed by the subsequent annealing of the crystal.

The DS technique is also useful when an aligned grain structure is desired. A directionally solidified alloy will exhibit a columnar grain structure in which all the grain boundaries are parallel to the growth direction. The absence of transverse boundaries significantly improves the high temperature creep properties so that the DS technique is used to produce turbine blades with a columnar grain structure [4–9].

The importance of directional solidification in the design of molds is evident in the use of localized chills to avoid centerline shrinkage problem in castings. Also, the higher temperature gradient near the chills reduces the length of the mushy zone so that the feeding problems are reduced significantly. Directional solidification conditions are also present in the welding processes.

Another important application of the DS process is *in situ* processing of composite materials. Both, lamellar and rod eutectics can be directionally solidified to obtain aligned structures. The eutectic spacing is inversely proportional to the square root of velocity, so that a finer eutectic with improved strength can be obtained by directionally solidifying the eutectic alloy at higher rates [10–11]. Furthermore, off-eutectic alloys can often be directionally solidified to obtain a eutectic microstructure so that the volume fractions of the two phases can also be varied to obtain optimum properties [2, 12].

DIRECTIONAL SOLIDIFICATION TECHNIQUES

The DS technique is used for fundamental studies on microstructural evolution and also for producing devices for commercial uses. Thus the actual experimental techniques, although similar in principle, vary in design as the temperature requirements and the size and shape of the material to be solidified change significantly. Four different DS equipments are briefly discussed here.

For fundamental studies on the solidification microstructural development, DS studies on transparent organic material have recently gained increasing importance. One of the advantages of such studies is the ability to observe *in situ* the dynamical changes that occur as the solidification conditions are altered. Also, some microstructural scales, such as the dendrite tip radius, can alter significantly during quenching so that its precise value cannot be measured in opaque metallic systems. The basic experimental equipment is shown in Fig. 3 [13, 14]. The hot and the cold chambers are maintained at constant temperatures. The material to be solidified is contained in a cell which is made of two glass plates, separated by a distance of 25–400 μm. A calibrated thermocouple is first inserted between the glass plates and then the sides of the cell are sealed all around with the exception of two small openings for filling the cell. After the cell is filled, these openings are

Fig. 3. Experimental apparatus to study directional solidification of thin samples.

sealed. The cell, which is commonly referred to as the Hele-Shaw cell, is then placed on a translating stage. The material is directionally solidified by moving the stage at a constant rate by the use of a precision screw which is driven by a compumotor. The solid-liquid interface can be directly observed *in situ* through a microscope positioned between the hot and the cold chamber.

For a given temperature gradient and composition, the changes in interface shapes with velocity can be studied [15]. Figure 4 shows such observations in the pivalic acid-ethanol system. Four distinct interface morphologies are observed as the interface velocity is increased [15, 16]. At low velocities, a planar interface is present. As the velocity is increased above some critical velocity, the planar interface becomes unstable and forms a cellular structure. These cells have small curvature and the amplitude of the cells is of the same order as the cell spacing. As the velocity is increased further, a change from cellular to cellular dendritic structure occurs. Cellular dendrites are elongated cells in which the amplitude of the cells is significantly larger than the cell spacing. Also, the tip of cellular dendrites is much sharper and the tip region approaches a parabolic shape. With a further increase in velocity, dendritic structures are formed which are characterized by the presence of sidebranches.

The DS apparatus to study metallic systems is shown in Fig. 5 for the case in which the alloy sample is held stationary and the temperature gradient assembly is

38

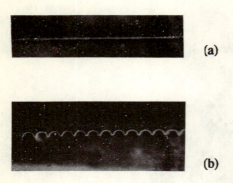

(a)

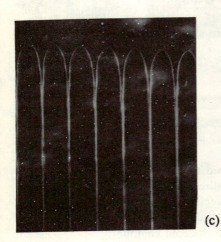

(b)

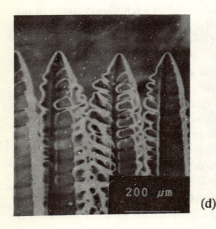

(c)

(d)

Fig. 4. The effect of increasing growth rate on interface morphology in pivalic acid 0.076 wt% ethanol system which is directionally solidified at $G = 2.98$ K/mm. (a) a planar interface at $V = 0.2$ μm/s, (b) a cellular interface at $V = 1.0$ μm/s, (c) a cellular-dendritic interface at $V = 3.0$ μm/s, and (d) a dendritic interface at $V = 7.0$ μm/s.

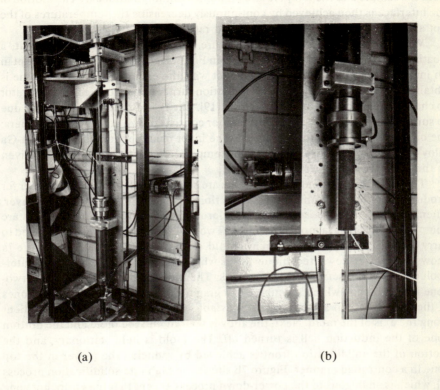

(a)   (b)

Fig. 5. Directional solidification apparatus in which the alloy is held stationary and the temperature gradient assembly is moved at a constant rate. (a) Complete system (left); (b) Close-up showing heat sink, lava insulator and cooling assembly (right).

moved upwards at a constant rate [17]. In order to ensure a uniform temperature in the hot zone, a long cylinder of a high conductivity material such as silver is placed around the sample. A split cylindrical furnace, placed around the high conductivity cylinder of silver, is not shown in Fig. 5 for the clarity of the figure. When low temperature gradients are desired, an oil bath is added between the cold chamber and the insulator. The entire temperature gradient assembly is attached to a drive plate which is moved upwards by a high precision screw driven by a computer-operated motor. A precise alignment of the sample and the temperature gradient assembly is achieved by attaching these components to a shaft. Also, in order to obtain a uniform motion, the friction effects at the moving contacts are minimized by the use of ball bearings. A more detailed description of this equipment is given by Mason [17]. A variation in the above system is to keep both the sample and the temperature gradient assembly stationary. The motion of the interface is then achieved by continuously decreasing the temperatures of the hot and the cold chambers at some fixed rate [18].

When DS studies of high temperature materials are to be carried out, a floating zone technique is used, as shown in Fig. 6. Here the solid sample is kept in an inert environment without any contact with the enclosure. A molten zone is obtained by using a high frequency induction furnace and a conical concentrator to channel the energy into a small region [19]. The liquid zone remains stable due to surface tension forces. The cold chamber consists of two concentric chambers: the outer chamber is water-cooled, and the inner chamber contains liquid In-Ga alloy for the rapid removal of heat. This temperature gradient assembly is driven in the same way as that shown in Fig. 5.

The above experimental equipments are suitable for laboratory studies or for producing materials of uniform cross-sections, such as single crystals. However, when the DS technique is to be used for producing engineering devices that are complicated in shape, a mold is required. Two different processes are developed to carry out directional solidification in a mold, as shown in Fig. 7a and 7b. Figure 7a shows the power-down technique [7] in which the mold is placed on a water-cooled chill which serves as the heat sink. The mold is then placed inside a two-zone induction coil which is coupled to a graphite suscepter. Initially, both zones of the induction coil are energized to establish a desired temperature gradient along the axis of the mold. Next, the alloy is poured into the mold and the bottom zone of the induction cell is turned off. The mold is held stationary, and the motion of the solidification front is achieved by reducing the power in the top zone in a controlled manner. Figure 7b shows the high rate solidification process [8], which is analogous to the power-down process except that the entire hot zone is kept at the desired temperature and the solidification is achieved by the gradual withdrawal of the mold from the furnace.

PLANAR INTERFACE GROWTH

Two important aspects of solidification microstructures are the formation of different phases and the solute segregation pattern that result from the non-equilibrium solidification process. Both of these aspects are determined by the temperature and the composition conditions at the interface. The temperature of the interface is always below the equilibrium liquids temperature, and the compositions in the solid and liquid at the interface are governed by the phase diagram and by the Gibbs-Thompson equation for a nonplanar interface. Note that, although solidification occurs far from equilibrium conditions, a local equilibrium between the solid and the liquid at the interface is approached for most metallic systems that are solidified at low velocities. The ratio of equilibrium compositions in the solid and liquid at a planar interface, as given by the phase diagram, is known as the solute distribution coefficient, k. For the phase diagram

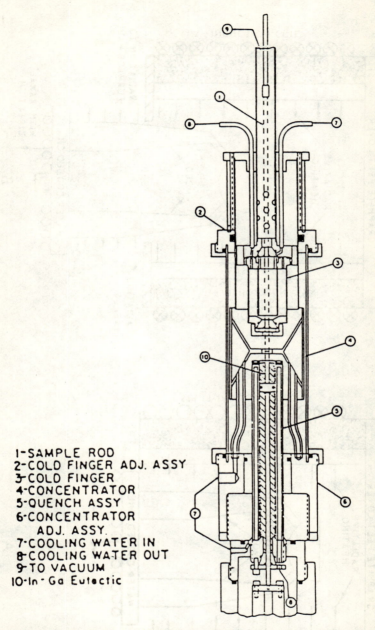

1-SAMPLE ROD
2-COLD FINGER ADJ. ASSY
3-COLD FINGER
4-CONCENTRATOR
5-QUENCH ASSY
6-CONCENTRATOR
   ADJ. ASSY.
7-COOLING WATER IN
8-COOLING WATER OUT
9-TO VACUUM
10-In-Ga Eutectic

Fig. 6. The floating zone technique to directionally solidify high temperature or reactive alloys.

42

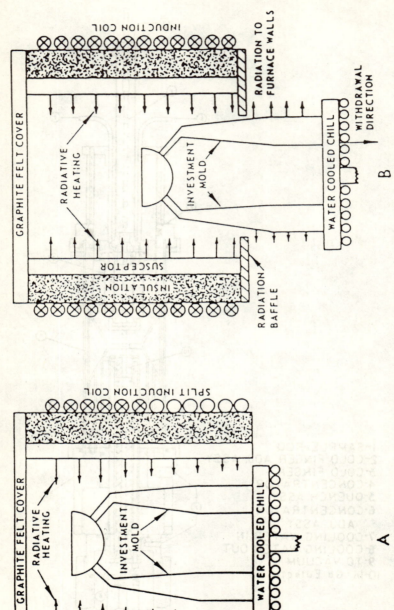

Fig. 7. A schematic illustration of directional solidification of an alloy in the mold. (A) powder-down process, and (B) high rate solidification process [5].

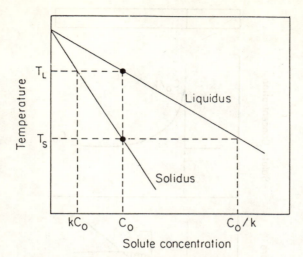

Fig. 8. A schematic phase diagram with constant k.

shown in Fig. 8, k is constant and independent of temperature. If the solidus and liquidus are linear in the composition range of interest, but they do not intersect at the zero solute value, then it is advantageous to use an alternate definition of the solute partioning effect. In this case, the ratio of the liquidus and the solidus slopes is used to define the solute partitioning [20].

Under steady-state growth conditions, when the interface velocity is equal to the externally imposed velocity, the temperature and the compositions at the interface are fixed if no convection effects are present in the liquid. The precise values of the temperature and compositions at the interface are dictated by: (1) processing conditions, such as V and G, (2) the phase diagram, and (3) the morphology of the solid-liquid interface. In the following section we shall examine the steady-state growth of a planar interface.

Consider an alloy of composition, C_o, which is directionally solidified at velocity, V, under the temperature gradients G_L and G_S in the liquid and the solid, respectively, at the interface. We shall consider the case in which the diffusion of solute in the solid is negligible and there are no convection effects in the liquid. From Fig. 8, it is seen that the first solid that forms has a composition equal to kC_o. The solute that is rejected by the solid piles up in liquid ahead of the interface. When the steady-state is reached, the solid and the liquid compositions at the interface are given by C_o and C_o/k, respectively, and the interface temperature corresponds to the solidus temperature if interface kinetics effects are negligibie. The steady-state solute build-up in liquid is shown in Fig. 9. The composition in liquid, $C_L(Z)$, is given by [1]

$$C_L(Z) = C_o + C_o \, [1 - k)/k] \, \exp{(-VZ/D)}, \qquad (1)$$

where D is the diffusion coefficient of solute in liquid and $Z = 0$ defines the

44

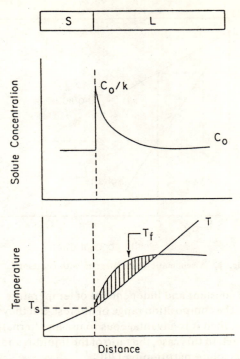

Fig. 9. The steady-state composition profile in the liquid when no convection effects are present in the liquid. The variation in actual temperature, T, and the liquidus temperature corresponding to the local concentration in liquid, T_f, are shown. The shaded region is the liquid which is supercooled, i.e. $T_f > T$.

position of the interface. A small transient effect is present during the initial solidification, but once the steady-state is reached, the solid composition is equal to the average alloy composition, C_o. A solute conservation requires that the composition of the solid, near the end of the freezing process, be higher than C_o.

A planar interface growth, under no convection effects in the liquid, would give rise to a uniform composition of solid except in small initial and final transient regimes. Such planar front solidification is very useful since it gives rise to a solid with homogenous composition over most of the sample. It is observed that the planar front solidification can occur only under low velocity and high temperature gradient conditions. We shall now examine the criteria which allow the planar front growth.

Tiller et al. [21] postulated that a planar interface will be unstable if the liquid region near the interface is supercooled and that this supercooling increases with distances from the interface. Thus, if a small perturbation forms on a planar interface it will see a larger supercooling, and thus it will grow faster. Under a positive temperature gradient condition in the liquid, the supercooling can be present due to the solute effect. This solute-driven supercooling is referred to as the constitutional supercooling. We now define supercooling as:

$$S(Z) = T_f(Z) - T(Z), \tag{2}$$

where $T_f(Z)$ is the equilibrium liquidus temperature corresponding to the composition $C_L(Z)$, and $T(Z)$ is the imposed temperature in the liquid. For a linear liquidus with slope m_L,

$$T_f(Z) = T_M + m_L C_L(Z), \tag{3}$$

where T_M is the melting point of the pure metal. Substituting the value of $C_L(Z)$ from eqn. (1), we obtain:

$$T_f(Z) = T_M + m_L C_o - \Delta T_o \exp(-VZ/D), \tag{4}$$

where $\Delta T_o = m_L C_o (k-1)/k$ is the equilibrium freezing range of the alloy of composition C_o. The imposed temperature gradient in liquid is:

$$T(Z) = T_M + m C_o/k + G_L Z \tag{5}$$

The variation in $T_f(Z)$ and $T(Z)$ are shown in Fig. 9. A constitutionally supercooled zone exists if $T_f > T$ at any location. This is shown by the hatched region in Fig. 9. Substituting the values of $T_f(Z)$ and $T(Z)$ from eqns. (4) and (5) into eqn. (2), we obtain:

$$S(Z) = \Delta T_o [1 - \exp(-VZ/D)] - G_L Z. \tag{6}$$

According to the constitutional supercooling criterion, a planar interface will be stable if

$$[\partial S(Z)/\partial Z]_{Z=0} < 0. \tag{7}$$

From eqn. (6), this condition for the planar interface stability can be written as:

$$G_L/V > \Delta T_o/D \tag{8}$$

Thus, a critical value of the ratio G_L/V is required for growth with a planar interface.

The effects of G_L and V on the presence of a constitutionally supercooled zone are illustrated in Figs. 10a and 10b. For a given velocity the variation in $T_f(Z)$ is fixed. Thus, the value of G_L determines if a constitutionally supercooled zone is present. Line 1 in Fig. 10a shows the existence of a supercooled region, whereas line 2 shows that no supercooling is present in liquid. The critical value of the gradient, shown as $(G_L)_{cr}$, is given by $[\partial T_f(Z)/\partial Z]_{Z=0}$ which is equal to $m_L G_C$, where G_C is the concentration gradient in the liquid at the interface.

For a given temperature gradient, the effect of velocity on the presence of a supercooled zone occurs through the change in solute profile which becomes sharper as the velocity is increased. The variation in T_f with Z is thus altered. This

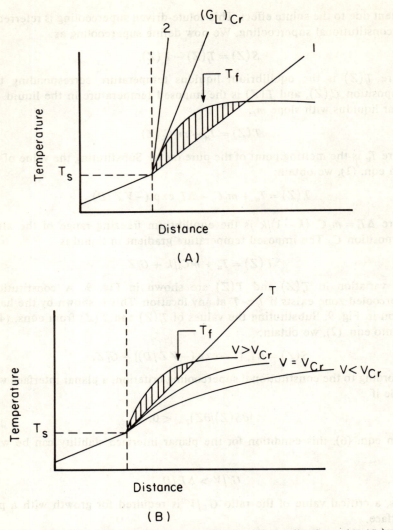

Fig. 10. The effect of G and V on constitutional supercooling. (A) A gradient larger than $(G_L)_{cr}$ gives a stable planar interface (line 2), whereas gradients smaller than $(G_L)_{cr}$ give a constitutionally supercooled region (line 1). (B) The effect of velocity changes the diffusion distance in liquid and thus changes the variation in T_f with distance.

is illustrated in Fig. 10b. At the critical velocity, V_{cr}, the value of $m_L G_c$ becomes equal to G_L. When $V > V_{cr}$, a constitutionally supercooled region exists in the liquid and the planar interface becomes unstable.

The constitutional supercooling criterion gives a good physical insight into the conditions which give rise to a stable planar interface. However, it ignores the thermal gradient in the solid, the latent heat generated during the solidification, and the additional solid-liquid interface area which must be created if the interface becomes non-planar. Furthermore, it gives only the critical condition for instability, and it does not predict the wavelength of the unstable interface when $V > V_{cr}$. These aspects were subsequently examined by Mullins and Sekerka [22], who carried out a linear stability analysis of a planar interface.

The basic ideas behind the linear stability analysis are illustrated in Fig. 11. A planar interface is perturbed into a sinusoidal profile with wavelength, λ, and amplitude, δ. The amplitude is considered to be infinitesimal so that the effects of higher orders of δ are negligible. Both the thermal and the solute diffusion fields around the perturbed interface are determined by taking into account the effect of interfacial energy on the equilibrium temperature and concentrations at the interface. The velocities of different parts of the interface are evaluated by using

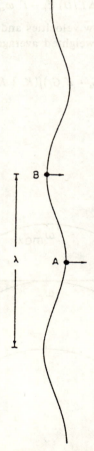

Fig. 11. A sinusoidally perturbed interface with wavelength, λ, and amplitude, δ. The stability analysis determines the conditions when the velocity at point A is less than that at point B.

48

the thermal and the solute flux balance at the interface. If the velocities at points A and B in Fig. 11 are V_A and V_B, respectively, then the planar interface will be stable if $V_A > V_B$. In terms of the sinusoidal profile, the conditions $V_A < V_B$ and $V_A > V_B$ are equivalent to $\dot{\delta}/\delta < 0$ and $\dot{\delta}/\delta > 0$, respectively, where $\dot{\delta}$ is the rate of change of amplitude with time.

A schematic variation in the amplification rate, $\dot{\delta}/\delta$, with the wavenumber, $\omega = 2\pi/\lambda$, as predicted by the linear stability analysis, is shown in Fig. 12 for different velocities. The values of temperature gradient and composition are kept constant. A planar interface is stable only if $\dot{\delta}/\delta < 0$ for all values of ω. The critical velocity and the critical wavenumber to which the planar interface becomes unstable can be obtained from Fig. 12. The values of V_{cr} and ω_{cr} are obtained when $\dot{\delta}/\delta = 0$ and $\partial(\dot{\delta}/\delta)/\partial\omega = 0$. The neutral stability condition, when $\dot{\delta}/\delta = 0$, gives

$$-G + (V \Delta T_o/D) \, \xi_c - \Gamma \, \omega_{cr}^2 = 0, \tag{9}$$

where ξ_c is close to unity for low velocities and it goes to zero at a very high velocity. G is the conductivity-weighted average temperature gradient at the interface, which is given by:

$$G = (K_s G_s + K_s G_L)/(K_s + K_L). \tag{10}$$

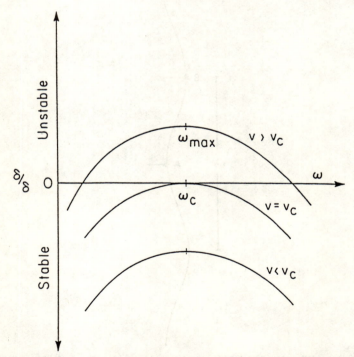

Fig. 12. The variation in the amplification rate of a sinusoidal perturbation as a function of wavenumber, ω.

K_L and K_s are the thermal conductivities of liquid and solid, respectively. The value of ω_{cr}, for low velocities, is given by [23]:

$$\omega_{cr} = \left[\frac{k\Delta T_o}{2\Gamma D^2}\right]^{1/3} V_{cr}^{2/3} \qquad (11)$$

substitution of equation (11) into equation (9) gives the value of V_{cr} which, for small velocities, can be simplified to give:

$$V_{cr} = (GD/\Delta T_o) + (GD/\Delta T_o) \ [\Gamma k^2/2D\Delta T_o]^{1/3} \ V_{cr}^{1/3} \qquad (12)$$

The neutral stability condition, given by equation (9), shows that the temperature gradient and the surface energy effects stabilize the planar interface, whereas the solute effect favors instability. The effect of increased interfacial area for a perturbed interface increases V_{cr} as seen from the second term on the right hand side of eqn. (12). At low velocities, this second term is generally less than 10% of the first term so that one may approximate eqn. (12) as

$$V_{cr} \simeq GD/\Delta T_o. \qquad (13)$$

Equation (13) is similar to the stability condition derived from the constitutional supercooling condition. The major difference is that the temperature gradient which controls the stability is G and not G_L. Consequently, eqn. (13) is often referred to as the modified super-cooling criterion. This difference in the temperature gradient term can be quite significant in some cases. For example, under most casting conditions, G_L is very small compared to G_S, so that the temperature gradient in the solid, and not the constitutional super-cooling in liquid, that would be important for the stability of the planar interface. It is indeed possible that, for $G_L = 0$ and large G_s, a stable planar interface can exist even if a supercooled liquid region is present ahead of the interface.

A number of experimental studies has been reported in the literature to examine the conditions for the planar interface stability [24–30]. Walton et al. [24] and Cole and Winegard [25] carried out directional solidification studies of several Sn-Pb alloys under different G/V conditions. They decanted the liquid and examined the solid-liquid interface morphology. They obtained a linear relationship between the composition and the critical G/V ratio at which a planar-to-cellular transition was observed. The shape of the line gave a reasonable value of the diffusion coefficient. Jamgotchian et al. [29] determined V_{cr} in the Bi-Sb system, and they found V_{cr} to be within 60–86% of the theoretical value. A more detailed study of planar interface stability was carried out by Eshelman and Trivedi [30] in the transparent succinonitrile-acetone system. Since it is not possible to obtain a steady-state planar interface at V_{cr} which becomes unstable with time, Eshelman and Trivedi [30] studied the instability of the planar interface as the velocity was changed from just below to just above V_{cr}. They found that some dynamical effects are present when the interface velocity increases with time. When these effects were into account, they found an excellent agreement between the experimental results and the linear stability analysis result for V_{cr}.

The linear stability analysis is based on the infinitesimal change in the shape of the interface. If the perturbation in the shape of the interface is large, it is possible that a planar interface may become unstable below V_{cr}. Such a planar to cellular bifurcation is called subcritical [31, 32], and its existence has been predicted for alloy systems with $k < 0.45$. Experimental studies in several transparent organic systems [16, 30, 33] have comfirmed the existence of subcritical bifurcation.

CELLULAR AND CELLULAR-DENDRITIC GROWTH

When an alloy is directionally solidified beyond the threshold of planar interface stability, a cellular or a cellular-dendritic interface is formed. The cellular structure which forms just above the threshold conditions has two important characteristics. First, the length of the cell is very small, and it is of the same order of magnitude as the cell spacing, as seen in Fig. 4b. Second, the tip region of the cell is broader and the cell has a larger tip radius. Thus, a significant solute build-up occurs ahead of the interface with some solute diffusing in the lateral directions. When the growth conditions are further from the planar interface stability condition, a cellular-dendritic structure forms (Fig. 4c). In this case, the cell tip becomes sharper and assumes a nearly parabolic shape which is somewhat similar to the dendrite tip shape. Thus, the term cellular-dendrite is used to characterize this interface morphology. As the cell tip becomes sharper, more solute transport occurs laterally so that the intercellular region becomes richer in solute. This lateral solute transport results in a larger two-phase region so that cellular-dendritic structures are characterized by a large amplitude to spacing ratio. This increase in lateral solute transport also causes a significant increase in primary spacing at the cellular to cellular-dendritic transition [16].

The lateral redistribution of solute that occurs with nonplanar interfaces gives rise to microsegregation patterns in the solidified material. Since the temperature at the cell tip is higher than that at the cell base, the solute content of solid at cell tip will be smaller than that at the cell base for phase diagrams with $k < 1$. Thus, if we examine the transverse section, a periodic solute profile will be observed that has a periodicity of the intercellular spacing. This microsegregation profile is approximately characterized by the Reighley or Scheil equation [1–3] for cellular-dendritic structures, which is:

$$C_s = kC_o/f_L^{1-k}, \qquad (14)$$

where C_s is the composition in solid and f_L is the volume fraction of liquid present. The value of f_L is one at the cell tip and zero at the cell base. The above equation is obtained under the assumption that the diffusion in solid is negligible, so that it predicts C_s to be infinity at the base of the cell. Some solid state diffusion is generally present, and if it is taken into account, equation (14) is modified as [34]:

$$C_s = kC_o \left[\frac{1 + \alpha k}{f_L + \alpha k} \right]^{1-k}, \tag{15}$$

where $\alpha = 4D_s t_f / \lambda^2$, where D_s is the diffusion coefficient of solute in the solid, t_f the local solidification time and λ the intercellular spacing. The local solidification time is the time from beginning to end of solidification so that it is equal to the length of the cell divided by the velocity.

If the diffusion in solid is small, a significant solute enrichment occurs in the intercellular liquid region. Thus, depending on the phase diagram, it is possible to nucleate a new phase of higher composition in the intercellular region. For example, in a system that exhibits a eutectic phase, the intercellular region may form a eutectic when the liquid composition reaches the eutectic composition, C_E, or the solid composition of the single phase reaches kC_E. The volume fraction of eutectic phase, f_E, can be estimated from eqn. (14) as:

$$f_E = (C_E / C_o)^{\frac{1}{k-1}} \tag{16}$$

Figure 13 shows a cellular microstructure with intercellular eutectic in the Pb-Sn system.

Equation (16) shows that the fraction of eutectic that is formed depends on composition of the alloy and the phase diagram, i.e. C_E and k. It is independent of the processing parameters. If diffusion in the solid phase is considered, the processing parameters still do not appreciably alter the value of f_E.

For single phase solidification, the microsegregation effect is generally characterized by the segregation ratio (SR) which is defined as the ratio of the maximum solid composition (at the cell base) to the minimum solid composition (at the cell tip). Thus, in a transverse section, SR is the ratio of solute

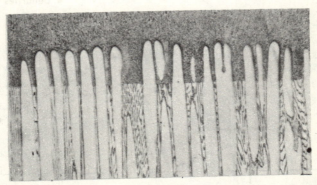

Fig. 13. The formation of a cellular structure with intercellular eutectic in a directionally solidified Pb-20 wt. pct. tin alloy at $V = 1.2$ μm/s and $G = 31$ K/mm.

52

concentrations at the cell boundary and at the cell center. The value of SR is given by:

$$SR = 1 + (lG_m/\Delta T_t), \qquad (17)$$

where l is the length of the cell, G_m the average temperature gradient in the two-phase region and ΔT_t the cell tip undercooling. For a given G_m, SR will be smaller for cellular structures where l is small and ΔT_t large. However, SR will increase as the cellular-dendrites form since l becomes large and ΔT_t decreases upon cellular to cellular-dendritic transition.

The microsegregation that is observed in cellular solidification can be removed by homogenization treatment at high temperatures. The time required to homogenize the solid is given by:

$$t = 0.47 \; \lambda^2/D_s. \qquad (18)$$

The time to homogenize can be reduced if D_s is large, i.e. homogenization temperature is higher, and the intercellular spacing is small. The variation in λ with velocity is shown in Fig. 14. A minimum in λ is observed just before the cellular to cellular-dendritic transition.

A satisfactory theoretical model to predict cellular spacing is not yet available. However, cellular-dendrite spacing model has been proposed by Hunt [36], which utilizes the mass balance concept along with the use of the Scheil equation for solute redistribution. Hunt obtained the following relationship between λ, cell tip radius, R, and processing parameters:

$$G\lambda^2/Rk\Delta T_o = 5.66 \; [(C_l/C_o) - (GD/Vk\Delta T_o)], \qquad (19)$$

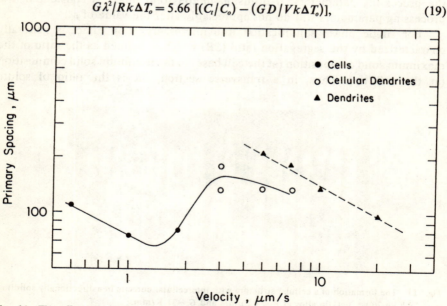

Fig. 14. The effect of velocity on primary spacings of cellular, cellular-dendritic and dendritic structures in pivalic acid-ethanol system which is directionally solidified at $G = 3$ K/mm.

where C_t is the solute concentration in liquid at the cell tip. The value of R was obtained by Trivedi [37] by using the marginal stability or the solvability condition. This Hunt-Trivedi model was found to agree with the experimental results on cellular-dendritic spacing in the pivalic acid-ethanol system [16].

DENDRITIC GROWTH

A transition from cellular-dendritic to dendritic microstructure occurs when the ratio (G/V) becomes smaller. The word "dendrite" is of Greek origin, and it refers to a branched structure. Thus we restrict the terminology of dendrite to those structures which exhibit side-branches. Figure 4d shows a dendritic structure with secondary branches. However, tertiary and quarternary branches are often present in the dendritic structure.

Dendrites have been found to grow in specific crystallographic directions, so that a preferred orientation of grains is obtained in the directionally solidified alloys. For example, in cubic metals, the preferred growth directions are <001>, so that not only the primary dendrite but also all sidebranches tend to form in these directions. Thus, a cubic material will exhibit a four-fold symmetry in the side-branch formation with respect to the axis of the previous branch. The primary dendrites will tend to grow along the <001> direction which makes the smallest angle with the heat flow direction. This formation of preferred grain orientation by the DS process is used to produce magnets in iron alloys since the magnetic induction is higher in the <001> directions. The preferred grain orientation is obtained by directionally solidifying an ingot from bottom to top by placing a severe chill at the bottom of the ingot.

The preferred growth direction for dendrite exists due to the presence of interface kinetic anisotropy. In contrast, cellular structures generally grow in the direction of heat flow. Cellular-dendritic structure often tend to grow in the direction somewhere between the heat flow and the <001> direction. The exact orientation of cellular-dendrites depends upon the magnitude of the kinetic anisotropy in the system. Figure 15 shows a directional solidification microstructure in which a cellular interface was first stabilized. The velocity was then increased to form a dendritic structure. A sharp change in the orientation is observed as the transition from cellular to dendritic structure occurs.

Precise conditions under which dendritic structures form are not yet well-established. However, dendritic structures often form when $V \gg V_{cr}$. V_{cr} decreases with increasing solute content of the alloy so that most commercial alloys generally solidify with velocities significantly larger than V_{cr}. Thus dendritic structures are commonly observed in commercial materials processed by the solidification techniques. Thus, it is important that all aspects of dendritic solidification be well understood so that one can design an alloy and determine processing parameters which produce an appropriate dendritic microstructure that gives desired properties to the material.

As in the case of dendritic cells, the nonplanar dendritic interface gives rise to

Fig. 15. The change in the growth direction when a steady-state cellular interface is suddenly transformed into a dendritic interface by increasing the velocity.

microsegregation patterns and may lead to the formation of different phases in the interdendritic regions where the liquid composition may increase significantly for $k < 1$. In addition, the complex branched structure of dendrites is prone to the formation of porosity and shrinkage voids because of the difficulty of feeding through the intricate interdendritic channels.

We shall examine dendritic microstructural characteristics in terms of important length parameters. These are: (1) the length of the dendrite, L, (2) primary dendrite spacing, λ, (3) secondary dendrite arm spacing, λ_2, (4) the diameter of the dendrite core, d, (5) the spacing of the first secondary sidebranch from the dendrite tip, λ_p, and (6) the dendrite tip radius. These microstructural parameters are shown in Fig. 16. We shall how examine these characteristic length scales and correlate their magnitudes with the processing parameters, G and V, and the composition of the alloy.

1) Dendrite length

The length of the dendrite is important because it characterizes the extent of the two-phase mushy zone. Since L/V is the time it takes from the beginning to the end of solidification, large dendrite lengths increase the total solidification time which allows the microstructure to coarsen. Also, the sidebranches of dendrites can get detached if convection effects are present in the interdendritic region. These detached sidebranches act as a nuclei for the formation of equiaxed crystals. In castings, the detached sidearms are carried into the central region of the casting by the convection effects in the mold, and they give rise to the equiaxed zone. When these detached sidebranches remain between the dendrites, a fine grain structure between columnar dendrites is formed, which is

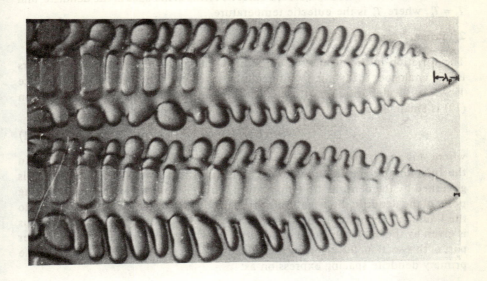

Fig. 16. Various microstructural scales of the dendritic micro-structure: The length of the dendrite, L; the primary spacing, λ; the secondary arm spacing, λ; the diameter of the core, d; the distance between the dendrite tip and the first observable secondary branch, λ_p; and the dendrite tip radius, R.

known as freckles, and these freckles are undesirable if an oriented grain structure is desired.

In castings, the formation of long dendrites make it difficult to feed since transport of liquid from the tip to the base of dendrites is required as the base region solidifies. The length of the dendrite is given by:

$$L = (T_t - T_b)/G_m, \tag{20}$$

where T_t and T_b are the temperature at the tip and at the base of the dendrite, respectively. Under low velocity conditions, the dendrite tip temperature is close to the liquidus temperature. The dendrite base temperature, however, depends upon the magnitude of the diffusion in the solid phase, and it is generally below the solidus temperature. For dilute solutions, with appreciable diffusion in the solid, $T_b \simeq T_s$. This condition is generally valid for solutes which occupy interstitial sites in the solid. Thus, the length of the dendrite is directly proportional to the equilibrium freezing range of the alloy, and inversely proportional to the temperature gradient in the mushy zone. Thus, for a given alloy, higher temperature gradients will give smaller dendrite lengths.

If the alloy system exhibits eutectic phase formation, then a eutectic may form in the interdendritic region when the composition of the liquid reaches the eutectic composition. In this case, a eutectic forms at the base of the dendrite, and $T_b = T_E$, where T_E is the eutectic temperature.

2) Primary dendrite spacing

The variation in the primary dendrite arm spacing with processing parameters is very analogous to that for the intercellular spacing, given by equation (19). Since $V >> V_{C_r}$ for dendritic growth, the last term on the right hand side of equation (19) is small. Thus, for dendritic growth, we obtain:

$$G\lambda^2/Rk\Delta T_o = A(C_i/C_o),$$
(21)

where A is a constant whose value is somewhat larger than the constant factor in the intercellular spacing equation. The composition at the dendrite tip, C_i, is governed by the dendrite tip temperature, which is determined by the operating point of the dendrite. For small velocities, the dendrite tip temperature is close to the liquidus temperature, so that $C_i/C_o \simeq 1$. The value of the dendrite tip radius is given by the operating point of dendrite growth, and it is discussed in the later part of this section. Substituting the value of R, one obtains for low velocities the primary dendrite spacing expression as:

$$\lambda = A'G^{-1/2}V^{-1/4}\Delta T_s^{1/4},$$
(22)

where $\Delta T_s = k\Delta T_o$ is the freezing range of an alloy with composition equal to that at the dendrite tip in the solid. The constant A' is equal to $A[D\Gamma/\sigma^*]^{1/2}$, in which σ^* is the operating parameter which has a constant value for a given system. Equation (22) shows that the primary dendrite spacing decreases as G and V are increased and ΔT_s is decreased. Note that the dependence of G and V is not the same so that, according to eqn. (22), λ cannot be characterized by the cooling rate. Furthermore, λ is often represented to vary at $C_o^{1/4}$. However, as pointed out by Trivedi and Kurz [20], the variation in λ is proportional to $(\Delta T_s)^{1/4}$ and this variation depends upon the phase diagram. Thus, it is possible for λ to increase, decrease or remain constant as C_o increases if the phase diagram shows corresponding variations in the values of ΔT_s with C_o.

Several experimental studies have been reported in the literature on the variation in λ with G, V and C_o [35-48]. The dependence of λ on G, as predicted by eqn. (22) has been confirmed by all experimental studies in which convection effects are not present. The variations in λ with V and C_o, however, are not always found to follow eqn. (22). The main reason for this discrepancy is the assumption that a dendritic interface can be represented by a smooth interface. A proper model must include the effect of sidebranches on primary dendrite spacing [49].

3) Secondary dendrite arm spacing

Secondary dendrite arm spacing, λ_2, is the most important parameter of dendritic structures from the technological viewpoint. The microsegregation pattern that develops in the directionally solidified material has the periodicity of λ_2. Also, the increased composition (for $k < 1$) in the liquid between the secondary arms may form a second phase which can significantly influence the properties of the solidified alloy.

Secondary spacings in the solidified alloy are nearly constant. However, a significant coarsening of the secondary arm spacing occurs in the mushy zone. The coarsening of the higher order branches than the secondary branch is quite rapid so that the solidified material is largely characterized by λ_2. Figure 17 shows the formation of secondary branches near the tip of the dendrite. The initial sidebranches have a uniform spacing, $\lambda_2(o)$, and this spacing increases with distance away from the tip, or with time, $\lambda_2(t)$. Directional solidification experiments have shown that $\lambda_2(o)$ scales with the dendrite tip radius, so that:

$$\lambda_2(o)/R = \text{constant.} \tag{23}$$

The value of the constant is found to depend on the properties of the system. For succinonitrile-acetone system [50–51] the value of the constant is found to be 2, whereas that for the pivalic acid-ethanol system [52] is 3.8.

The coarsening of $\lambda_2(t)$ occurs due to two reasons. First, the initial, finer wavelengths is not stable during growth, and a larger wavelength evolves with time. This wavelength selection mechanism is very analogous to that observed for the instability of a planar interface leading to the wavelength section of cellular patterns [53]. This change in wavelength is governed by the diffusional instability and surface energy may not be reduced during coarsening. The second reason for the coarsening of λ_2 is governed by the reduction in the interfacial area.

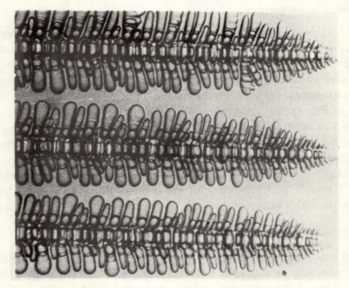

Fig. 17. A directionally solidified succionitrile −4 wt % acetone system showing the coarsening of secondary branches with the distance behind the dendrite tip.

Both the diffusive instability and the surface energy reduction process require diffusion of solute. Thus, coarsening of secondary spacings occurs in the two-phase region where the diffusivity in the liquid is significantly larger than that in the solid. The extent of the change in spacing depends upon the time that a secondary branch spends in the two-phase region. Thus, the final spacing, $\lambda_2(t_f)$, depends upon the local solidification time. Strictly speaking, this local solidification time for secondary branches is the difference between the time of formation of the first secondary branch and the time when it reaches the base of the dendrite. If the distance between the dendrite tip and the first secondary branch is small compared to the length of the dendrite, we may write:

$$t_f = (T_t - T_b)/G_m V. \tag{24}$$

Thus, t_f is in inverse proportion to the cooling rate through the two-phase region. Since T_f is related to the cooling rate, the secondary branch spacing can be correlated with the cooling rate. This relationship is obtained as [2]:

$$\lambda_2(t_f) = a\,(\text{cooling rate})^{-1/3} \tag{25}$$

Experimental studies have been carried out to study the variation in $\lambda_2(t_f)$ with the cooling rate [2], and the value of the experiment has been found to vary between 1/2 and 1/3. Thus, larger cooling rates would give finer secondary arm spacing. Since the periodicity of microsegregation is characterized by λ_2, a finer λ_2 is generally desired to minimize the annealing time required to homogenize the alloy [see equation (18)].

4) Dendrite core diameter

A careful examination of Fig. 17 shows that the base of the secondary branches is nearly parallel to the dendrite axis after a small distance behind the dendrite tip. We shall denote the diameter of this core as d. Although there is some variation in composition along the core diameter, the microsegregation is not as pronounced as that between the secondary branches. Thus, the value of d is important in influencing the properties of the material.

Detailed experimental studies on the effect of G, V and C_o on d have been carried out by Esaka and Kurz [51] in the succinonitrile-acetone system. They observed that d decreases with the increase in velocity and with the increase in composition. The value of d was independent of temperature gradient at low velocities. However d decreased with the increase in temperature gradient at higher velocities. The values of d were found to be roughly between 6 R and 10 R.

5) The first secondary branch location

The distance between the dendrite tip and the location of the first observable secondary branch, λ_p, is important in understanding the physics of the sidebranch formation. Also, if this distance is large, the total solidification time for the secondary branches will be less than t_f.

Experimental studies on λ_p have been carried out by Somsoomuk and Trivedi [54] and Esaka and Kurz [51] in the succinonitrile-acetone system. λ_p is found to decrease with the increase in V, G and C_o. The effect of G is not very pronounced except for high G values. The decrease in λ_p with V or C_o occurs since R also decreases with V and C_o. For high compositions, the sidebranches are sufficiently close to the dendrite tip to cause a significant deviation in the parabolic shape of the dendrite front. The precise variations in λ_p with processing conditions also depend upon the properties of the system. A significantly larger λ_p is observed in the pivalic-acid-ethanol system compared to that in the succinonitrile-acetone system under the same values of V, G and C_o.

6) Dendrite tip radius

Once the alloy has solidified, the dendrite tip radius is unimportant. However, the dendrite tip radius is the most important parameter which governs various microstructural scales of the dendritic structure. All other characteristic lengths of the dendritic structure are not only related to the dendrite tip radius, but their evolution is controlled by the dendrite tip radius.

A detailed theoretical model of dendrite tip radius is quite complex and this subject is discussed in detail by Langer in the first paper. We shall thus give only highlights of the theoretical models. These theoretical models are developed to understand the following three experimental observations: (1) the dendrite tip radius remains constant during growth under steady-state conditions, (2) the dendrite tip radius has a unique value under given G, V and C_o in a given system [55], and (3) the dendrite tip region assumes a steady-state shape which is close to a parabola [15].

Ivantsov [56] first showed that a parabolic shape, under isothermal and isoconcentrate interface conditions, is invariant with time during growth, and for this shape he developed the relationship between the dendrite tip undercooling and radius of curvature for a given velocity. For this case, one may write the total interface undercooling, ΔT, which is the difference between the liquidus temperature and the dendrite tip temperature, as

$$\Delta T = \Delta T_t + \Delta T_s \tag{26}$$

where ΔT_t and ΔT_s are the thermal and solutal undercoolings at the dendrite tip, respectively. For directional solidification, the ΔT_t term is generally quite small compared to the ΔT_s term under dendritic growth conditions. When ΔT_t is significant, cellular growth is observed. Thus, for dendritic growth at low velocities, $\Delta T \simeq \Delta T_s$, or

$$\Delta T = k\Delta T_o \, Iv(p)/[1-(1-k) \, Iv(p)], \tag{27}$$

where $I(p) = p \exp (p) \, E_1 (p)$ is the Ivantsov function, and $p = VR/2D$ is the solute peclet number. The above equation predicts that, for a given V, the interface undercooling is a function of the dendrite tip radius. The interface undercooling goes from 0 to ΔT_o as the radius varies from 0 to ∞.

The above model does not predict a unique dendrite tip radius for a given velocity. Subsequent theoretical developments [57-63] extended the Ivantsov model to include the interfacial energy and the attachment kinetics effects. Thus, equation (26) was modified as:

$$\Delta T = \Delta T_t + \Delta T_s + \Delta T_c + \Delta T_k, \tag{28}$$

where ΔT_c and ΔT_k are the interface undercoolings due to capillarity and kinetics, respectively, which are given by:

$$\Delta T_c = 2\Gamma/R, \text{ and} \tag{29}$$

$$\Delta T_k = V/\mu \tag{30}$$

where $\Gamma = (\gamma + \gamma")/\Delta S$, in which $\gamma"$ is the second derivative of surface energy with respect to orientation, and μ is the interface kinetic coefficient which is also a function of orientation.

The inclusion of the surface energy and the interfacial kinetics effects cause the interface conditions to deviate from the isoconcentrate value so that a parabolic tip will not give a steady-state solution. It was first assumed that the deviation from the parabolic shape will be sufficiently small if the capillarity and the kinetic effects are small compared to the diffusion effect. Under this assumption, one obtains the following dendrite tip undercooling result for isotropic interface properties:

$$\Delta T = \frac{k\Delta T_o \, Iv(p)}{1 - (1-k) \, Iv(p)} + \frac{2\Gamma}{R} + \frac{V}{\mu} \tag{31}$$

This result also predicts infinite solutions between ΔT and R for a given undercooling, except that the undercooling increases at very low R values due to capillarity effects. Thus, ΔT versus R relationship gives a minimum value, as shown in Fig. 18.

Initially, it was assumed that the dendritic growth operates at the radius value which gives the minimum in the undercooling. This assumption thus gives a rationale for the observation of the unique dendrite tip radius for a given velocity.

Langer amd Müller-Krumbhaar [64] proposed a details stability analysis of the dendrite tip and they investigated the conditions under which the dendrite tip will be stable with respect to infinitesimal fluctuations in its shape. They found that only a small band of radius values will be stable. All radii values below the minimum undercooling radius value were found to be unstable. Also, radii value above some critical value, which we shall call marginally stable radius, will be unstable since the dendrite tip at these higher radii values will result in tip-splitting. Experimental results agreed precisely with the prediction of the marginal stability hypothesis [55]. Under this condition, the relationship between the dendrite tip radius and the velocity, for small velocities, is given by:

$$VR^2 = \Gamma D/\sigma^* k\Delta T_o \tag{32}$$

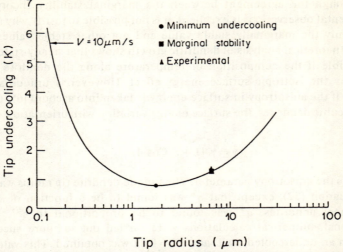

Fig. 18. Theoretical relationship between the dendrite tip radius and the dendrite tip undercooling for given values V, G and C_o.

where σ^* is a constant and equal to about 0.02. α^* is called the operating parameter for the dendritic growth. The above relationship explains quite accurately the experimental results in the succinonitrile-acetone system [51, 55], in which precise value of dendrite tip radius can be measured *in-situ*, Fig. 19.

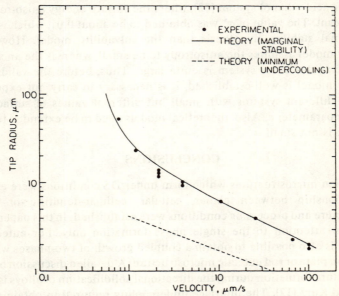

Fig. 19. A comparison of experimentally measured dendrite tip radii with the theoretical models **based** on the marginal stability criterion and the minimum undercooling hypothesis.

62

Although the agreement between the marginal stability theory and the experimental observations is excellent, it is not possible to justify why the system selects only the marginally stable value and not other stable values. A more detailed theoretical analysis of dendritic growth shows that steady-state growth is not possible if the composition or temperature along the dendrite growth is altered by the isotropic surface energy effect. However, a unique solution is obtained if the anisotropy in surface energy is taken into account [65–68]. In this case, for cubic dendrites, the surface energy variation with orientation, θ, is given by:

$$\gamma = \gamma_0 [1 + \varepsilon \cos 4\theta], \tag{33}$$

where ε is the anisotropy parameter. The unique dendrite tip radius was found to obey equation (32), except that σ^* was found to be a function of ε. For two dimensional dendrites, σ^* was found to be proportional to $\varepsilon^{7/4}$. For three dimensional, numerical calculations were carried out for pure succinonitrile growth in an undercooled melt, and $\sigma^* = 0.02$ was obtained. This value of σ^* is remarkably close to the value obtained from the marginal stability theory. The term "solvability condition" is used to characterize the model which gives a unique dendrite tip radius when anisotropic surface energy is included in the model.

The solvability condition shows that σ^* is constant for a given system but that it has different constant values for different systems. In order to check this conclusion, detailed directional solidification experiments were carried out in the privalic acid-ethanol system [52], where the surface energy anisotropy effects are significant. The value of σ^* was obtained to be about 0.01, which is closer to the marginal stability hypothesis than the solvability model. However, the solvability model assumes the anisotropy to be small, whereas the anisotropy in the pivalic acid-ethanol system is quite large. Thus, before the validity of the solvability model is well-established, it is necessary to carry out experimental studies in different systems with small but different values of surface energy anisotropy parameter ε. Also, theoretical models need to be extended to the case in which ε is not small.

CONCLUSIONS

Solidification microstructures which form under DS conditions were examined. The relationship between planar, cellular, cellular-dendritic or dendritic microstructure and processing conditions were established. In this paper, we have limited our attention to the single phase formation only. For eutectic alloy systems, it is also possible to obtain a coupled growth of two-phases which may give rise to a plate or rod eutectic microstructure. A detailed discussion of eutectic microstructure formation during the directional solidification of alloys is given by Trivedi and Kurz [12]. The interface undercooling required to obtain a specific microstructure is also discussed by Kurz and Trivedi in this volume.

In this paper we have emphasized the fundamental principles which govern directional solidification microstructures. In actual applications of the DS technique to the industrial production of engineering devices, other complicated aspects must also be considered:

1) We have considered the case of unidirectional heat flow only. Such simplified heat flow pattern may not be possible for objects with complex shapes. In this case, a detailed numerical analysis of heat flow problem is required to establish the location of the interface and the local values of G and V at any location of the interface.

2) We have considered a material of uniform cross-section only. In many practical situations, the cross-section of the object to be produced by the DS process may change with distance. In such a case, although the velocity at which the mold is removed from the furnace is constant, the interface velocity may change significantly as the cross-sectional area changes. This may give rise to formation of solute segregation bands when the cross-section changes suddenly. In such a case, the mold withdrawal rate needs to be programmed such that the interface velocity remains constant.

3) We have not considered the effect of fluid flow during solidification. If convection effects are present in the liquid, they will significantly influence the microstructural scales. The effect of fluid flow on interface morphology is discussed by Coriell and McFadden in this volume.

ACKNOWLEDGEMENTS

The author would like to acknowledge many valuable discussions with W. Kurz, D. Pearson and J. Sekhar. This work was carried out at Ames Laboratory. Ames Laboratory is operated for the U.S. Department of Energy by Iowa State University under contract no. W-7405-ENG-82. This work was supported by Office of Basic Energy Sciences, Division of Materials Sciences.

REFERENCES

1. Chalmers, B. *Principles of Solidification*, Wiley, New York (1964).
2. Flemings, M.C. *Solidification Processing,* McGraw-Hill, New York (1974).
3. Kurz, W. and D.J. Fisher. *Fundamental of Solidification*, Trans. Tech. Publ, Aedermannsdorf, Switzerland (1986).
4. Giamei, A.F., E.M. Kraft and F.D. Lemkey. In: *New Trends in Materials Processing,* American Society for Metals, Metals Park, OH p. 48–97 (1976).
5. Versnyder, F.L. and M.E. Shank. *Mat. Sci. Eng.* **6,** 213 (1970).
6. Erickson, J.S., C.P. Sullivan and F.L. Versnyder. In: *High Temperature Materials in GAs Turbines,* Elsevier, Amsterdam, p. 315 (1974).
7. Versnyder, F.L., R.B. Barrow, B.J. Piearcey and L.W. Sink. *Trans. Intern. Vacuum Metal Conf.,* American Vacuum Society, NY, p. 391 (1967).
8. Erickson, J.H. W.A. Owczarski and P.M. Curran. *Metal Prog.* **99,** 58 (1971).
9. Giamei, A.F. and F.L. Versnyder. *Proc. of 4th Bolton Landing Conf.*, Claitor's Publ. Div., Baton Rouge, Los Angeles, p. 3 (1974).
10. Jackson, K.A. and J.D. Hunt, *Trans. AIME* **236,** 1129 (1966).

64

11. Trivedi, R., P. Magnin and W. Kurz. *Acta Metall.* **35**, 971 (1987).
12. Trivedi, R. and W. Kurz. In: *Materials Processing of Eutectic Alloys,* AIME, Warrendale, PA (1988).
13. Hunt, J.D., K.A. Jackson and H. Brown. *Rev. Sci. Instrum.* **37**, 805 (1966).
14. Mason, J.T. and M.A. Eshelman. IS-4906, Ames Laboratory, Ames, IA (1986).
15. Somboonsuk, K. J.T. Mason and R. Trivedi. *Metall. Trans.* **15A**, 967 (1984).
16. Eshelman, M.A., V. Seetharaman and R. Trivedi. *Acta Metall.* **36**, 1165 (1988).
17. Mason, J.T. IS-4817, Ames Laboratory, Ames, IA (1982).
18. Burden, M.H. and J.D. Hunt. *J. Cryst. Growth* **22**, 99 (1974).
19. Carder, K. MS Thesis, Iowa State University, Ames, IA (1986).
20. Trivedi, R. and W. Kurz. *Metall. Trans.* (to be published).
21. Tiller, W.A., K.A. Jackson, J.W. Rutter and B. Chalmers. *Acta Metall.* **1**, 428 (1953).
22. Mullins, W.W. and R.F. Sekerka. *J. Appl. Phys.* **35**, 444 (1964).
23. Langer, J.S. *Rev. Mod. Phys.* **52**, 1 (1980).
24. Walton, D., W.A. Tiller, J.W. Rutter and W.C. Winegard. *Trans. AIME* **203**, 1023 (1955).
25. Cole, G.S. and W.C. Winegard. *J. Inst. Met.* **92**, 322 (1963).
26. Sato, T., K. Shibata and G. Ohira. *J. Cryst. Growth* **40**, 69 (1977).
27. Sato, T. and G. Ohira. *J. Cryst. Growth* **40**, 78 (1977).
28. Shibata, K., T. Sato and G. Ohira. *J. Cryst. Growth* **44**, 419 (1978).
29. Jamgotchian, H., B. Billia and L. Capella. *J. Cryst. Growth* **62**, 539 (1983).
30. Eshelman, M.A. and R. Trivedi. *Acta Met.* **35**, 2443 (1987).
31. Wollkind, D.J. and L.A. Segel. *Phil. Trans. Royal Soc. London* **268A**, 351 (1970).
32. Caroli, B., C. Caroli and B.J. Roulet. *J. Physique* **43**, 1767 (1982).
33. DeCheveigne, S., C. Guthmann and M.M. Lebrun. *J. Cryst. Growth* **73**, 242 (1985).
34. Brody, H.D. and M.C. Flemings. *Trans. AIME* **236**, 615 (1966).
35. Mason, J.T., J.D. Verhoeven and R. Trivedi. *Metall. Trans.* **15A**, 1665 (1984).
36. Hunt, J.D. In: *Solidification and Casting of Metals,* The Metals Society, Book 192, London, p. 3 (1979).
37. Trivedi, R. *Metall. Trans.* **15A**, 977 (1984).
40. Jakobi, H. and K. Schwerdtfeger. *Metall. Trans.* **7A**, 811 (1976).
41. Okamoto, T. and K. Kishitake. *J. Cryst. Growth* **29**, 131 (1975).
42. Young, K.P. and D.H. Kirkwood. *Metall. Trans.* **64**, 197 (1975).
43. Spittle, J.M. and D.M. Lloyd. In: *Solidification and Casting of Metal,* The Metal Society, Book 192, London p. 15 (1979).
44. Bell, J.E.A. and W.C. Winegard. *J. Inst. Met.* **92**, 357 (1963).
45. Mason, J.T., J.D. Verhoeven and R. Trivedi. *J. Cryst. Growth* **59**, 516 (1982).
46. Mason, J.T., J.D. Verhoeven and R. Trivedi. *Metall. Trans.* **15A**, 1665 (1984).
47. Miyata, Y., T. Suzuki and J.I. Uno. *Metall. Trans.* **16A**. 1799 (1985).
48. Jamgotchian, H., B. Billia and L. Capella. *J. Cryst. Growth* **64**, 338 (1983).
49. Esaka, H., W. Kurz and R. Trivedi. In: *Materials Processing,* The Inst. of Metals, London (1988).
50. Trivedi, R. and K. Somboonsuk. *Mat. Sci. Eng.* **65**, 65 (1984).
51. Esaka, H. Sc.D. Thesis, Ecole Polytechnique, Lausanne (1986).
52. Trivedi, R. and J.T. Mason. unpublished work, Ames Laboratory, Ames, IA (1987).
53. Eshelman, M.A. and R. Trivedi. *Scripta Metall.* **22**, (1988).
54. Somboonsuk, K. and R. Trivedi. *Scripta Met.* **18**, 1283 (1984).
55. Somboonsuk, K. and R. Trivedi. *Acta Metall.* **33**, 1051 (1985).
56. Ivantsov, G.P. *Dokl. Akad. Nauk SSSR* **58**, 567 (1947).
57. Temkin, D.E. *Sov. Phys. Dokl.* **5**, 609 (1960).
58. Bolling, G.F. and W.A. Tiller. *J. Appl. Phys.* **32**, 2547 (1961).
59. Koller, G.R. and L.A. Tarshis. *J. Cryst. Growth* **2**, 222 (1968).
60. Trivedi, R. *Acta Metall.* **18**, 287 (1970).
61. Trivedi, R. *J. Cryst. Growth* **49**, 219 (1980).
62. Kurz, W. and D.J. Fisher. *Acta Metall.* **29**, 11 (1981).

63. Esaka, H. and W. Kurz. *J. Cryst. Growth* **72**, 578 (1985).
64. Langer, J.S. and H. Müller-Krumbhaar. *J. Cryst. Growth* **42**, 11 (1977).
65. Meiron, D. *Phys. Rev.* **33A**, 2704 (1986).
66. Langer, J.S. and D.C. Hong. *Phys. Rev.* **34A**, 1462 (1986).
67. Kessler, D.A. and H. Levine. *Phys. Rev. Letters* **57**, 3069 (1986).
68. Saito, Y., G Goldbeck-Wood and H. Müller-krumbhaar. *Physica Scripta* (in press).

4

Fundamentals of Rapid Solidification Microstructures

W. KURZ AND R. TRIVEDI[†]

Department of Materials Engineering,
Swiss Federal Institute of Technology,
Lausanne, Switzerland

[†] *Ames Laboratory, USDOE and the Department of*
Materials Science and Engineering
Iowa State University
Ames, Iowa 50011, U.S.A.

ABSTRACT

The different effects and microstructures which may be observed under rapid solidification conditions are presented. Specific attention is given to the differences in process characteristics for laser surface treatment and melt atomization. The two mechanisms which control complete supersaturation of the crystallization product, i.e., nonequilibrium solidification and planar front growth beyond absolute stability are discussed. It is shown that in dendritic as well as in eutectic alloys plane front single phase solidification may be obtained when the interface temperature drops below solidus. Finally comparison of the growth temperature—growth rate relationships for different phases and different growth forms allow the construction of microstructure (and phase) selection diagrams the latter being useful for a rational alloy development for rapid solidification processing.

INTRODUCTION

Rapid solidification processing (RSP) has become a field of great scientific and technological interest because of the potential of producing new materials with hitherto unknown properties. Metallic alloys can be obtained containing highly metastable phases, different compositions and extremely fine microstructures.

For many years the formation of these structures has been treated in a rather empirical manner, for example by measuring spacings of dendrites or eutectics and relating them to cooling rates. The state of research on rapidly solidified microstructures up to 1983 has been reviewed by Jones [1]. Around that time a more mechanistic approach to rapid solidification microstructures was developed. The influence of nucleation in highly undercooled liquids was treated in detail by Perepezko [2] and Flemings and Shiohara [3]. As far as growth is concerned, Boettinger, Coriell et al. have brought attention to a series of important phenomena in RSP [4–7].

Nearly all the growth models developed earlier have used the assumption of small Péclet numbers, i.e. the characteristic microstructural scales are always smaller than the diffusion distance. However, under conditions of RSP this simplification is not valid any more. In order to make the microstructure modelling more quantitative, the present authors have relaxed the low Péclet number approximations, and have developed a general framework for treating growth of dendrites and eutectics under low and high imposed growth rates and under low and high undercoolings of the melt [8–14].

In this paper only those aspects concerned with growth of microstructures are treated. Special attention is given to similarities and differences between columnar and equiaxed growth; the first type of growth characterizing rapid solidification processes like surface treatment with a high power laser and the second type of growth determining microstructure formation in highly undercooled melts as observed for example in melt atomization processes.

After a short presentation of the important characteristics of these processes, the principles of dendritic and eutectic growth models will be developed and finally applied to the construction of microstructure selection diagrams.

PROCESS CHARACTERISTICS

The solidification microstructures comprise two major classes of growth morphologies: dendrites and eutectics. Depending on the process these structures form either under an imposed temperature field which constrains the displacement of the isotherms (columnar growth, $G > 0$) or in an undercooled melt leading to a type of equiaxed growth ($G < 0$). These different microstructures are shown schematically in Fig. 1.

Even if there are many routes to a rapidly solidified product, there are only two RS-processes which differ fundamentally from each other according to the above classification, i.e. production of pure columnar or pure equiaxed structures. These are (Fig. 2):

- laser-beam treatment producing columnar growth
- melt atomization process producing equiaxed structures

There are however many other processes where both types of growth behavior might be observed as in melt spinning and splat quenching.

Figure 2 shows the essential difference between these two processes. One can see that in surface treatment the solid–liquid interface is subject to a large range of

columnar   equiaxed

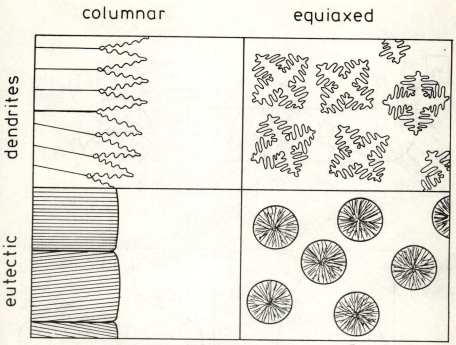

Fig. 1. Solidification microstructures of alloys.

growth rates $(0 < V < V_{max})$ where the maximum in growth rate, V_{max}, under rapid scanning rates of the beam V_b, is always smaller than V_b. This is the case when the corresponding Péclet number, $Pe \ (= V_b r / 2a)$ is larger than one (where r is the beam radius and a the thermal diffusivity of liquid). If Pe is smaller than one, $V_{max} = V_b$. The minimum in growth rate $(V = 0)$ at the bottom of the melt pool (permitting epitaxial growth) corresponds to a maximum of the temperature gradient in liquid at the interface. As the solidification progresses, the temperature gradient at the interface decreases with time due to a rapid decrease in the superheat of the melt.

The other case of solidification of an undercooled drop shows quite different behavior as can be seen in Fig. 2. The growth rate (depending on the supercooling, ΔT) is maximum at the beginning of the transformation (nucleation) and the temperature gradient in the liquid at the interface is negative. After some time, due to recalescence of the untransformed liquid, growth will slow down and the final rate will depend on the external heat flux.

These differences in solidification conditions will have a profound influence on the microstructures observed as is shown later in this paper.

It must be stated however that all the models presented above assume pure diffusion and steady state growth morphologies. Very often a quasi-steady state may be assumed but caution is necessary when the steady state theory is applied as

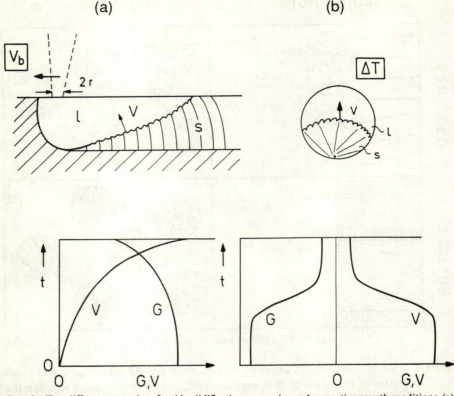

Fig. 2. Two different examples of rapid solidification processing and respective growth conditions. (a) surface treatment with high energy beams (columnar growth); (b) melt atomization (equiaxed or undercooled solidification).

the actual growth conditions may vary drastically with time. Furthermore strong convection found in the melt pool due to large surface temperature gradients may alter some of the predictions.

SUPERSATURATION

One of the most prominent phenomenon observed in RSP is complete supersaturation of the crystal. This result may be obtained in two different ways: through plane front growth beyond the limit of absolute stability and/or solute trapping due to loss of local equilibrium.

i) Plane front growth

Morphological stability of the solid liquid interface has been analyzed in detail by Mullins and Sekerka [15]. These authors showed that steady state plane front growth might be obtained (1) at low rates under an imposed positive temperature gradient below the limit of constitutional undercooling, V_c

$$V_c = \bar{G}D/\Delta T_o \qquad (1)$$

where $\bar{G}$ is the conductivity weighted average temperature gradient at the solid/liquid interface, D is the solute diffusion coefficient in liquid and ΔT_o is the equilibrium melting range of the alloy; (2) at high rates, above the limit of absolute stability, V_a

$$V_a = \frac{D\Delta T_o}{\Gamma k}, \qquad (2)$$

where Γ is the ratio of surface energy to volumetric melting entropy and k the solute partition coefficient. Figure 3a shows these two limits and the corresponding microstructures formed between these limits; cells close to the transition velocities and dendrites in between. Each time a steady state plane front growth is observed the crystal will grow with the composition of the melt, C_o, due to the establishment of a steady-state boundary layer in the liquid. This solid composition is independent of the value of the growth rate dependent partition coefficient.

As has been shown by the authors [9] absolute stability does also exist in undercooled melts where the undercooling limit for absolute stability $\Delta T_a'$ is given by the corresponding growth rate, V_a'

$$V_a' = V_a + \frac{a\,\theta_t}{\Gamma} \qquad (3)$$

where V_a' is now the sum of the solutal and thermal limit of absolute stability and a is the thermal diffusivity, $\theta_t = \Delta H_f/C_p$, where ΔH_f and C_p are the latent heat

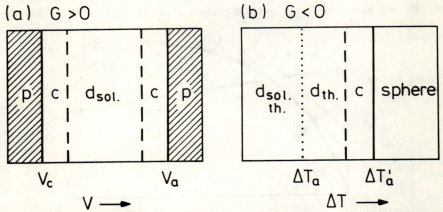

Fig. 3. Conditions for obtaining planar front growth and cells or dendrites. (a) columnar, (b) equiaxed growth.

and specific heat of liquid, respectively. As shown in Fig. 3b (see also section 4), at undercoolings up to ΔT_a (corresponding to V_a of eqn. 2), solutal and thermal dendrites will be obtained. Between ΔT_a and $\Delta T'_a$ only thermal dendrites will grow and beyond $\Delta T'_a$ stabled spherical growth is predicted.

ii) Solute trapping

The other mechanism producing a supersaturated crystal is the deviation from local equilibrium at high interface rates. In fact, when the diffusion distance D/V becomes smaller than the atomic jump distance, a_o equilibrium partition of solute at the interface cannot be expected any more. Therefore the partition coefficient will approach unity at large values of the growth rate as described by [16]

$$k_v = \frac{k + Pi}{1 + Pi} \qquad (4)$$

where k and k_v are the equilibrium and the growth rate dependent partition coefficients, respectively, and $Pi\ (= a_o V/D_i)$ represents an interface Péclet number, where D_i is the interface diffusion coefficient.

Complete supersaturation via plane front growth becomes more difficult with increasing alloy composition, C_o (eqn. 2; note that $\Delta T_o\ \alpha\ C_o$ for a phase diagram with constant k). On the other hand, solute trapping does not seem to depend strongly on alloy composition (even if a_o will be a function of composition). Therefore replacing $\Delta T_o = C_o m(k-1)/k$ in eqn. 2 and setting $V_a = D_i/a_o$ (i.e. $P_i = 1$) one obtains [17]

$$C_{crit} \cong k^2 \Gamma / [m(k-1)a_o] \qquad (5)$$

where m is the liquidus slope. C_{crit} is the composition below which supersaturation is obtained by absolute stability when $V_a < D_i/a_o$ and above which loss of local equilibrium will control the supersaturation (Fig. 4). In the latter case the

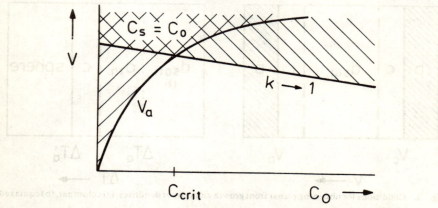

Fig. 4. Conditions for obtaining completely supersaturated solid (columnar growth).

condition $k = 1$ can only be obtained when the interface temperature, T^*, falls below T_o, the temperature of equal Gibbs free energy of both phases [2, 4].

DENDRITE GROWTH

As can be seen in Fig. 3 cellular/dendritic growth is observed between V_c and V_a in the case of columnar growth and up to $\Delta T'_a$ for equiaxed growth. The essential difference in these two cases is shown in Fig. 5. For columnar dendrites the heat source imposes a positive temperature gradient at the dendrite tip and one can assume it to be constant across the interface when the thermal diffusivities of both phases are similar and the latent heat term is small. In equiaxed growth latent heat is rejected into the liquid building up a negative temperature gradient. In the solid the temperature gradient is close to zero in this case.

The problem of modelling dendrite tip growth is twofold. First, one has to solve the transport problem and second the operating point must be determined.

The corrected transport equation can be written in the usual form of the temperature continuity equation (see for ex. [18]), i.e.

$$\Delta T - \Delta T_r - \Delta T_k = \Delta T_t + \Delta T_c \tag{6}$$

where ΔT is the total undercooling and the other four contributions are due to capillarity (r), to attachment kinetic (k) to heat flow (t) and to solute diffusion (c). The LHS of eqn. 6 characterizes the total driving force for diffusion and the RHS the splitting of the irreversible diffusion terms. Both diffusional undercoolings, ΔT_t and ΔT_c, can be obtained by assuming the dendrite tip to be of the form of a paraboloid of revolution. The solution to this problem has been given by Ivantsov [19] and can be written in the general form

$$\Omega_j = I_v(P_j); \ (j = t, c) \tag{7}$$

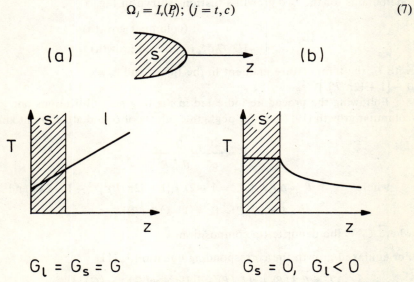

Fig. 5. Temperature fields at the tips of (a) columnar, and (b) equiaxed dendrites.

where Ω is the dimensionless undercooling as defined below, P is the Péclet number of the tip (ratio of the tip radius, R, to the diffusional distance $2D/V$ or $2a/V$), and $I_v(P)$ is the Ivantsov function as defined for example by Kurz and Fisher [20]. The latter function has a simple approximate form for its inverse such that eqn. 7 may also be written as [21]

$$P_j \cong \frac{-\Omega_j}{\log\Omega_j} ; (j = t, c) \qquad (8)$$

The solutal Péclet number, P_c, is related to the thermal one, P_t, by the diffusivity ratio (a/D)

$$P_c = P_t \left(\frac{a}{D}\right) \qquad (9)$$

For metals, a/D is typically between $10^2 - 10^4$ so that $P_c >> P_t$.

For the determination of the operating point several analyses show that the tip radius, R, follows a relationship which can be obtained by stability equation, [22–24]:

$$R = \left\{ \frac{1}{\sigma^*} \left(\frac{\Gamma}{m(G_c)_{\text{eff}} - (G)_{\text{eff}}} \right) \right\}^{\frac{1}{2}} \qquad (10)$$

Here σ^* is a stability constant (which is of the order of $1/4\pi^2$ for several systems studied), $(G_c)_{\text{eff}}$ and $(G)_{\text{eff}}$ are effective concentration gradient and temperature gradient as defined in [9].

Assuming equal thermal diffusivities and conductivities in liquid and solid one obtains for the two growth situations shown in Fig. 5

$$(G)_{\text{eff}} = \begin{cases} G_L & ; \text{(columnar growth)} \\ 1/2 \ G_L\xi_L; & \text{(equiaxed growth)} \end{cases} \qquad (11)$$

with G_L the temperature gradient in the liquid and $\xi_L = 1 - [1 + (2\pi/P_t)^2]^{-1/2}$.

Following the procedure indicated in the original publications one gets for columnar growth [10] (here for negligible effects of G and attachment kinetics).

$$V = \frac{1}{\sigma^*} \frac{D \ \Gamma}{\theta_c \xi_c R^2} \qquad (12)$$

where $\qquad \theta_c = \Delta T_o k A, \ \xi_c = 1 - 2k/\{[1 + (2\pi/P_c)^2]^{1/2} - 1 + 2k\}$ and

$$A = C_L^*/C_o = [1 - (1-k) I_v (P_c)]^{-1}, \qquad (13)$$

where C_L^* is the dendrite tip composition.

For equiaxed growth the corresponding equation is [11]

$$\Delta T = \theta_t I_v(P_t) + \theta_c I_v (P_c) + 2[\Gamma \sigma^* \{m(G_c)_{\text{eff}} - (G)_{\text{eff}}\}]^{1/2} \qquad (14)$$

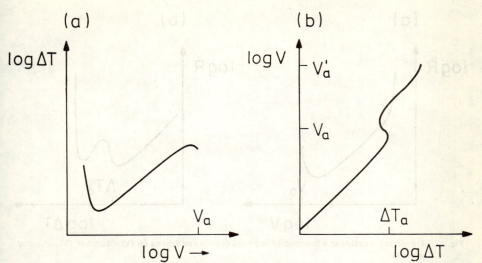

Fig. 6. Growth rate-undercooling relationship for (a) columnar, and (b) equiaxed growth.

where $(G_c)_{\text{eff}} = -2P_c\theta_c\xi_c/R$ and $(G)_{\text{eff}} = -P_t\theta_t\xi_L/R$. Equations 12 and 13 permit the calculations of tip radius, tip composition and undercooling for columnar growth when V is given, and eqns. 13–14 allow calculations of tip radius, tip composition and growth rate of equiaxed growth for a given ΔT. The corresponding behavior of the complex equations is shown schematically in Figs. 6 to 8.

Figure 6a shows the columnar dendrite tip undercooling as a function of growth rate. At low rates the high undercooling which decreases with V corresponds to cellular growth. Beyond the minimum, dendritic growth is observed and close to V_a again cells (Fig. 3a).

The growth rate, V, as a function of undercooling, ΔT, for equiaxed alloy dendrites is shown in Fig. 6b. At low undercooling the growth rate increases monotonically until the value given by the first term in eqn. 3 (corresponding to eqn. 2) has been reached. There the dendrite would reach absolute stability if the negative temperature gradient would not destabilize the interface. Above V_a, but below V_a', the dendrites accelerate drastically as they become purely heat flux controlled and $a \gg D$.

A similar behavior can be observed with respect to R as a function of V or ΔT (Fig. 7). The large flat part of the curve in Fig. 7a corresponds to dendritic growth while the other sharply rising branches indicate cellular growth. The drastic change in the R-ΔT relationship at ΔT_a in Fig. 7b is again related to a growth rate larger than V_a and the final steep increase in R corresponds to V_a'. Figure 8 shows finally the composition in the solid of the dendrite tip which corresponds to the trunk composition for the case of negligible back diffusion.

Experiments on Ag-Cu seem to agree well with the predicted composition-

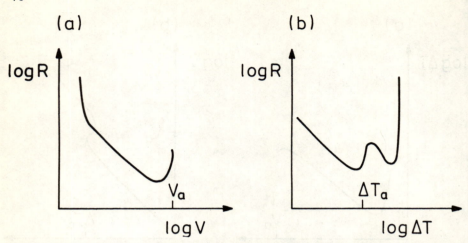

Fig. 7. Dendrite tip radius as a function of imposed growth conditions for (a) columnar, (b) equiaxed growth.

growth rate relationship [14]. Figure 9 gives an overview of the liquid/solid tip-composition and tip-temperature as a function of the growth rate for columnar growth. In this case planar front growth is obtained well before k approaches 1 and the final interface temperature is therefore predicted to be above T_o.

EUTECTIC GROWTH

As eutectic alloys contain generally substantial amounts of solute their growth is controlled by solute diffusion. Due to the high thermal diffusivity of metals one

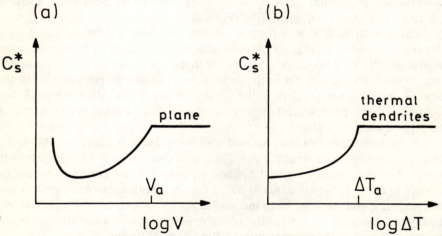

Fig. 8. Dendrite tip (trunk) composition as a function of growth conditions for (a) columnar, and (b) equiaxed growth.

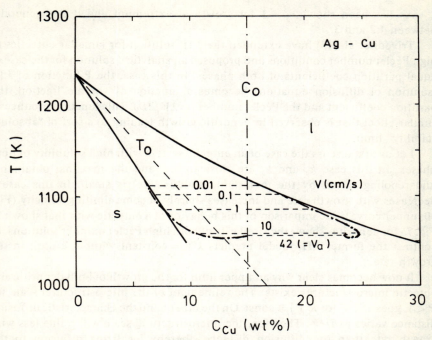

Fig. 9. Solidification path in a single laser trace showing the interface temperature and liquid/solid composition of the dendrite tip as a function of growth rate for columnar growth [14].

does not need to consider thermal boundary layers. Therefore a distinction between columnar and equiaxed growth in the eutectic growth models is unnecessary.

Eutectic growth has been treated in detail by Jackson and Hunt [25] for low Péclet number conditions. Their solution can be given in analogous form to eqn. 6. Neglecting attachment kinetics and thermal undercooling,

$$\Delta T - \Delta T_r = \Delta T_c \tag{15}$$

where $\Delta T_c = K_1 \lambda V$ and $\Delta T_r = K_2/\lambda$, with λ the interphase spacing and K_1 and K_2 constants [25, 26]. The operating point for regular eutectics is close to the extremum (i.e. the minimum undercooling value), while irregular eutectics like Fe-C or Al-Si grow, at least at small growth rates, far from extremum [20]. These different operating points can be described by different constants ϕ in the equations [26]

$$<\lambda>^2 V = \phi^{\,2} K_2/K_1 \tag{16}$$

$$<\Delta T>^2/V = [\phi + (1/\phi)^2 K_1 K_2 \tag{17}$$

where $<\lambda>$ and $<\Delta T>$ are mean values of the corresponding quantities.

As has been shown, $\phi = 1$ for growth at extremum and it lies generally between 1.2 and 3.

Trivedi et al. [13] have extended the J.H. solution for lamellar eutectics to high Péclet number conditions and proposed an analytical solution for the case of equal partition coefficients of both phases. In this case, the P-function of J.H. (solution of diffusion equation) becomes a function of volume fraction, the partition coefficient and the Péclet number ($= \lambda V / 2D$). This function produces a similar effect that is observed in dendritic growth leading to a sort of "absolute stability" limit.

Let us first discuss the case of an eutectic system with high solubility of both phases. In this case k_a and k_b are close to one and the maximal obtainable undercooling (given by the metastable solidus line) is small. In this case λ decreases with growth rate and increases sharply at some limiting velocity (Fig. 10, upper curve). A comparison of this behavior of a eutectic with that shown in Fig. 7a for dendrites shows the similarity of both high Péclet number solutions. In contrast the former J.H. model predicts $\lambda^2 V = $ constant without a limit in the growth rate.

It now becomes clear why an upper limit for the growth of both dendritic and eutectic microstructures exists: The refinement of the microstructural scale for $P < 1$, goes as $\lambda^2 V$ (or $R^2 V$) $\cong$ const. On the other hand the characteristic diffusion distance varies as D/V. Therefore the microstructural scale will refine less with growth rate than the diffusion distance, thereby localizing diffusion to the interface when $P > 1$. One could also say that diffusion becomes more and more unidimensional.

When the solid solubility of an eutectic alloy is small ($k \to 0$) than a large undercooling eventually builds up at the interface. In this case the diffusion coefficient will drop and produce a maximum in the $V(\lambda)$ relationship (Fig. 10,

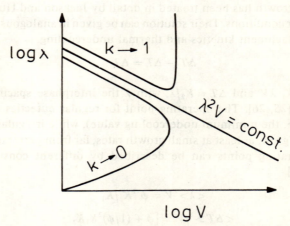

Fig. 10. Eutectic spacing versus growth rate according to the classical model ($\lambda^2 V = $ const. [25]) and to a recent model from the authors [13] for $k = k_a = k_b$ at the limits of $k = 1$ and $k = 0$.

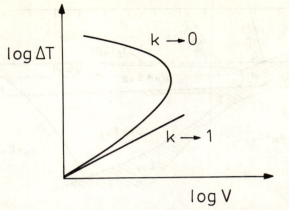

Fig. 11. Eutectic undercooling as a function of growth rate for different k-values.

lower curve). There is therefore a limiting velocity for eutectic growth above which such a growth form cannot be obtained any more.

Figure 11 represents the eutectic undercooling-growth rate relationship indicating the typical nose of the inverse TTT diagram when $k \to 0$. For large k-values the diffusion coefficient has no effect on the relationship.

MICROSTRUCTURE SELECTION DIAGRAMS

Dendritic or eutectic microstructures can make up the whole microstructure but they can also grow together in one alloy producing a mixture of these morphologies. Also, sometimes several stable and metastable phases can be present. In order to predict the phases and the corresponding microstructure which will form for a given condition a comparison of the different growth temperatures or rates will be useful (at least as long as a complete stability analysis of this problem cannot be done). As has been proposed already by Tammann and Botschwar [27] the phase and morphology with the highest rate or temperature will be the observed one. Using the models developed by the authors one can now make predictions of complex microstructures that form under slow and rapid solidification conditions [26]. Figure 12 shows first results of such calculations for the Al-Fe system [28]. The interest for such types of diagrams is evident.

CONCLUSION

In conclusion it can be said that our present modelling allows us to make reasonable predictions of scale and composition of microstructures under all solidification conditions. We can start now more systematic work on the effect of alloy composition and processing conditions on phases and microstructures.

Finally let us summarize which microstructures would be expected in the two processes mentioned in the beginning, i.e., in laser surface remelting and in powder atomization (Fig. 13).

80

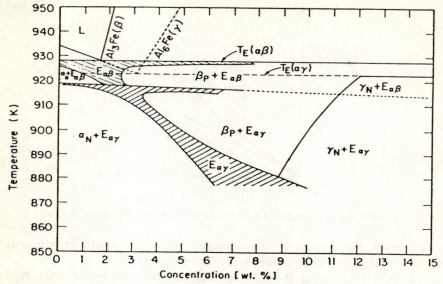

Fig. 12. Theoretical microstructure map for the Al-Fe system considering the stable Al₃Fe and the metastable Al₆Fe phase [28].

Assuming that $V_{max} > V_a$ the following microstructural sequence is expected for a single phase alloy:

Laser treatment. After a thin layer of epitaxial planar front growth (very large G/V) the interface will break down into cells which quickly become very fine. There might be an intermediate range of dendritic growth before the interface becomes cellular again and reaches, close to the surface, the absolute stability condition. Therefore in laser surface treatment the surface has either the highest supersaturation or (for $V < V_{max}$) the finest structure.

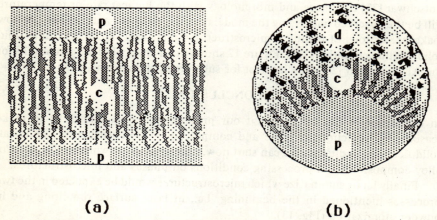

(a)   **(b)**

Fig. 13. Characteristic microstructures for the two process types of figure 2.

Melt atomization. Here the solidification sequence is very different. After nucleation has produced a crystal, it will proceed either with a stable spherical interface (if $\Delta T > \Delta T_a'$) or with a completely homogeneous dendritic morphology (if $\Delta T_a' > \Delta T > \Delta T_a$). In both cases the first solid will be structureless after complete solidification. After the initial rapid growth the drastic slow down shown in Fig. 2 will produce a coarse cellular or dendritic pattern in the remaining volume. In the case of $\Delta T < \Delta T_a$ (which will often be the case in systems with small k-values like in Al-Fe) one will observe an initial very fine cellular (also called microcellular) zone and then a coarser cellular/dendritic structure.

REFERENCES

1. Jones, H. *Mater. Sci. Eng.* **65**, 145 (1984).
2. Perepezko, J.H. *Mater. Sci. Eng.* **65**, 125 (1984).
3. Flemings, M.C. and Y. Shiohara. *Mater. Sci. Eng.* **65**, 157 (1987).
4. Boettinger, W.J. In: *Rapidly Solidified Amorphous and Crystalline Alloys*, B.H. Kear and B.C. Giessen (eds.), Elsevier, New York (1982).
5. Boettinger, W.J., S.R. Coriell and R.F. Sekerka. *Mater. Sci. Eng.* **65**, 27 (1984).
6. Boettinger, W.J., D. Shechtman, R.J. Schaefer and F.S. Biancaniello. *Metall. Trans.* **15A**, 55 (1984).
7. Boettinger, W.J. and S.R. Coriell. In: *Science and Technology of the Undercooled Melt*, P.R. Sahm, H. Jones and C.M. Adam (eds.), Martinus Niehoff, Utrecht (1986).
8. Trivedi, R. *J. Cryst. Growth*, **73**, 289 (1985).
9. Trivedi, R. and W. Kurz. *Acta Metall.* **34**, 1663 (1986).
10. Kurz, W., B. Giovanola and R. Trivedi. *Acta Metall.* **34**, 823 (1986).
11. Lipton, J., W. Kurz and R. Trivedi. *Acta Metall.* **35**, 957 (1987).
12. Trivedi, R., J. Lipton and W. Kurz. *Acta Metall.* **35**, 965 (1987).
13. Trivedi, R., P. Magnin and W. Kurz. *Acta Metall.* **35**, 971 (1987).
14. Kurz, W., B. Giovanola and R. Trivedi. *J. Cryst. Growth* (in press).
15. Mullins, W.W. and R.F. Sekerka. *J. Appl. Phys.* **35**, 444 (1964).
16. Aziz, M.J. *J. Appl. Phys.* **53**, 1158 (1982).
17. Jones, H. to be published.
18. Lipton, J., M.E. Glickman and W. Kurz. *Metall. Trans.* **18A**, 341 (1987).
19. Ivantsov, G.P. *Dokl. Akad. Nauk SSSR* **58**, 567 (1947).
20. Kurz, W. and D.J. Fisher. *Fundamentals of Solidification*, Trans Tech Publ, 4711 Aedermannsdorf, Switzerland (1986).
21. Kurz, W. and R. Trivedi. In: *Solidification Processing*, H. jones (ed.), The Institute of Metals, London (1988).
22. Langer, J.S. and H. Müller-Krumbhaar. *Acta Metall.* **26**, 1681, 1689, 1697, (1978).
23. Ben-Amar, M. and Y. Pomeau. *Europhysics Letters* **2**, 307 (1986).
24. Saito, Y., G. Goldbeck-Wood and H. Müller-Krumbhaar. *Physica Scripta* (in press).
25. Jackson, K.A. and J.D. Hunt. *Trans. AIME* **236**, 1129 (1966).
26. Trivedi, R. and W. Kurz. In: *Solidification Processing of Eutectic Alloys,* D. Stefanescu and G.J. Abbaschian (eds.), AIME, Warrendale, PA (1988).
27. Tammann, G. and A.A. Botschwar. *Z. Anorg. Chemie* **157**, 26 (1926).
28. Gremaud, M., W. Kurz and R. Trivedi. to be published.

5

High Pressure Induced Slow and Rapid Solidification

G.S. REDDY AND J.A. SEKHAR

Defence Metallurgical Research Laboratory
Po: Kanchanbagh, Hyderabad 500 258, India

ABSTRACT

This paper reviews our recent results on solidification under pressure. We first discuss the various categories and techniques of pressure application during solidification. These include techniques where pressures vary from megapascals to gigapascals and where application times vary from as low as 100 ms to the order of hours. Next we discuss the various forms of processing that are available with each technique. Concomitantly with pressure application; (i) there is a rise in the melting point for all metals which contract on solidification, (ii) adiabatic heat is generated, and (iii) the heat transfer coefficient of at the metal die interface approaches a value of 10^5 W/m^{2}K. All these determine the phase and morphology of the solidified material and are systematically discussed. The equivalance of pressure induced solidification and cooling induced solidification is discussed with an illustrative example. The rapid pressure application (RPA) technique to cause rapid solidification in bulk volume is discussed next and all the classic signatures of rapid solidification, namely, metastable phases and morphological refinement are demonstrated with the RPA technique. Moderate pressure ($\simeq$ 200 MPa) application during solidification is also a technique to induce different cooling rates during solidification. This effect is harnessed to study transformation temperatures in aluminium and aluminium-manganese icosahedral alloys to determine the validity of available theories on dynamic nucleation. Results from the study are discussed to highlight the potential of the technique.

1. INTRODUCTION

With the advent of energy concious, net shaped, directly solidified technologies, an awareness of the limitations of common and new solidification processes is increasingly being felt. Consequently many efforts are underway to extend the range of variables which affect solidification. One such variable which has received little or no attention is pressure. For the past three years we have been paying some attention to this variable [1–5] and this article is an effort to condense our understanding in this area. We have discovered that the effect of pressure on solidification is a scientifically fascinating and a technically challenging problem, with the possibility of greatly extending the rich variety of already available solidification microstructures.

In this paper we first present a brief theoretical treatment. Next we discuss the experimental techniques that have been employed. After this we discuss the various effects on solidification at low pressures (~ 200 MPa) and then discuss very high pressure effects, including the rapid pressure application technique to induce rapid solidification in large volumes. We then conclude the article with a novel concept of utilizing moderate pressures to demonstrate a new technique to study *in situ* nucleation in large volumes.

2. BRIEF THEORY

Solidification affected by pressure application may be divided into three categories: (i) where low pressure (less than 0.2 GPa) is applied during casting [6-9] (here the pressure mainly influences the heat transfer coefficient at the mold-metal interface and the porosity level in the casting [7]; (ii) where high pressure is applied to adiabatically drive solidification, and (iii) where the high pressure is applied instantaneously to cause undercooling and subsequent rapid solidification [1]. The last technique has come to be called RPA (rapid pressure application) and is distinct from category (ii) by the time scale in which pressure is applied (typically less than 200 ms).

Concomitantly with pressure application on a liquid contained inside a die three important processes occur: (i) the melting point (or liquidus) rises for all metals which contract on solidification; notably the rise in temperature with pressure has been shown to be nearly linear for most metals [10] and examples are given in Table 1, (ii) there is an adiabatic heat release which is mildly nonlinear; the magnitude of heat release for 3 GPa pressure application on a liquid metal initially at T_{mo}, is again given in Table 1, and (iii) the heat transfer coefficient at the mould-metal interface approaches a value of 10^5 W/m^{2}K with increasing pressure [7, 9].

Solidification under category (i) is not directly driven by pressure (pressure only allows for efficient heat removal) and is not considered further. Categories (ii) and (iii) are now compared.

2.1 Adiabatic heat

There is an adiabatic heat release during pressure application on a liquid. Simultaneously there is cooling by the die walls. Detailed analysis of this aspect in large volumes will be presented elsewhere but some trends emerge by considering that the liquid volume displays Newtonian behaviour; i.e., there are no temperature gradients in the liquid during pressurization. This necessarily limits

Table 1. Pressure related constants for various metals all material constants are from smithells [12]

Metal	ΔV m³/kg 10⁶	Amount of melting point increase with pressure (K/GPa) calculated by the Clausius-Clapeyron on equation	$\beta/P_i \cdot C_{pl}$ (GPa)⁻¹	Adiabatic temp. rise in the liquid for 3 GPa pressure (K)	Pt (GPa)	Adiabatic temp. rise in the liquid for a Pt pressure application, (K)
Al	26	62.4	.0387*	114	5.7	231
Pb	3	74.7	.0600*	118	2.1	81
Sn	3	25.4	.0561*	93	9.1	336
Fe	6	31.8	.0237*	133	11.2	549
Zn	7	43.7	.0314*	68	5.2	124
Ag	4	47.0	.0357*	139	7.8	396
Mg	25	64.7	–	–	4.0	–

The value of β for liquids is listed in reference (11). However, the pressure dependence of β is not available. For these calculations β has taken to be 10^{-4} K⁻¹.

the Biot number to less than 0.01, requiring at least one dimension of the liquid volume to be small but helps in simplifying the mathematics. The heat balance equation for a cooling sphere with pressure application can be written as

$$V_1 \varrho_1 C_{pl} \, dT/dt = - Ah \, (T - T_f) + V\beta T \, dP/dt \tag{1}$$

here the last term in the RHS arises on account of the adiabatic heat release. Equation (1) can be recast in dimensionless form (similar to manipulations made in reference [13] as

$$dT^*/dF_o + (3Bi - Ai) \, T^* + Bi = 0 \tag{2}$$

The adiabatic pressure number Ai is process dependent and will determine whether the adiabatic heat release influences the solidification behaviour. We have already fixed the Bi as equal to 0.01 and if $Ai = 0.001$ (i.e. $Ai \ll 3Bi$) then the rate of pressure application and the adiabatic heat rise is too slow to influence solidification. For most metals this implies that even a rapid pressure rise of 10^2 GPa/s will be too slow to affect heat flow. However $Ai \, \alpha R^2$, whereas $Bi \, \alpha R$ and for large volumes there will be gradients in the liquid which the pressure application will tend to increase. It will be shown elsewhere that reasonably large volumes can aslo be rapidly pressurized without significant adiabatic heat rise. The above calculations pertain to pressurization of a liquid. For solids, $\beta = 10^{-5}$ and the amount of adiabatic temperature rise is small.

2.2 Interface velocity

It the time scale of the pressurization process is such that all the adiabatic heat release is removed much faster than the rate of pressurization and if solidification is occurring simultaneously with the pressure application (no nucleation barrier) then the complete liquid will transform to a solid (quasi-adiabatically) after a pressure application which raises the melting point by H/C_{pl}. We call this pressure P_t, given by $H^2/T_{mo} \Delta V . C_R$. This quantity is also tabulated in Table 1. It is worth noting here that practical difficulties involving containment may limit the achievement of pressure above 5 GPa in large liquid volumes.

The rate of category (ii) solidification can be determined for the Newtonian case and can be related to the rate of pressurization. For a sphere of radius R solidifying from T_{mo}, we can write (assume $C_{pl} = C_{Ps} = C_P$ and that solidification is concentric from centre to surface; i.e.; $r = 0$ to $r = R$)

$$4\pi r^2 \, H \, dr/dt = (4\pi R^3 . \, C_P T_{mo} \Delta V /3H) \, dP/dt \tag{3}$$

equation (3) may be rearranged to

$$r^2 dr/dt = (R^3/3) \, (C_p/H) \, (T_{mo} \Delta V /H) \, dP/dt \tag{4}$$

For a 100 μm radius aluminium sphere and $dP/dt = 10^2$ GPa/s the solid-liquid interface velocity dr/dt is plotted in Fig. 1. Two important conclusions may be drawn from eqn. 4: (1) the rate of solidification is directly proportional to a dimension (radius) of the material for a fixed dP/dt, and (ii) the time of complete solidification is inversely proportional to dP/dt.

For category (iii) type solidification, the transformation is not simultaneous with pressure application. Pressure is employed as a means of obtaining high undercooling. If the rate of pressurization is much higher than the nucleation rate [1] then effectively the liquid volume will build up some pressure before the beginning of solidification. We consider pressure application on a liquid initially at T_{mo} and category (iii) solidification. If at any given instant no adiabatic heat is removed by the die and no solidification has occurred, the net undercooling ΔT_u (temperature) at any given pressure P will be given by the difference between the melting point at the high pressure and the actual temperature of the liquid which has risen on account of the adiabatic heat release. This can be written as

$$\Delta T_u = (\Delta V /H - \beta/\varrho_1 C_{pl} - \beta^2 P/2 \, \varrho_l^2 \, C_{pl}^2 \, P T_{mo} \tag{5}$$

Adiabatic heat removal, such as given in eqn. (2), will increase the undercooling. After nucleation the solidification is driven by this undercooling ΔT_u. This is category (iii) solidification. Again invoking Newtonian conditions, a constant die temperature and linear growth of the interface [13, 14] we can write for the dimensionless interface velocity as

$$(1/3) \, d^2 r^*/dF_o + (Bi + \varepsilon r^{*2}) \, dr^*/dF_o - \varepsilon Bi \, St = 0 \tag{6}$$

The initial value of interface controlled velocity as in reference [13] is given by

$$dr^*/dF_o = \varepsilon St \, n \tag{7}$$

Several forms of eqn. (6) have been solved previously [13, 14]. The

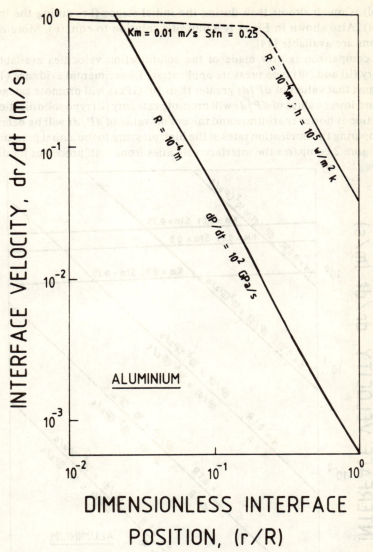

Fig. 1. Interface velocity for a sphere solidifying concentrically from $r = 0$ to $r = R$ for both category (II) and category (III) type pressure solidification. The interface velocity is plotted against dimensionless radius r/R. The dashed part of the curve for category (III) a close approximation.

growth of the interface occurs in two regimes. Initially the growth is interface controlled and the growth rate is high. After a dimensionless fraction of solid f equal to the dimensionless undercooling Stn has formed, the growth is controlled by heat flow removal. The growth in the second regime (i.e. during the heat flow

88

control) is much slower than during the initial stages (i.e. during the interface control). Also shown in Fig. 1 is a plot of a solution to eqn. (6). More rigorous solutions are available [14].

A comparison is now made of the solidification velocities available from category (ii) and (iii) type pressure application. Experimental evidence (1) tends to suggest that values of dP/dt greater than 10^2 GPa/s will promote category (iii) type, and lower values of dP/dt will promote category (ii) type solidification. This distinction is however arbitrary and the cut off value of dP/dt will be determined by comparing the nucleation rates at the high pressure to the actual pressurization rate. Figure 2 compares the interface velocities from both processes for different

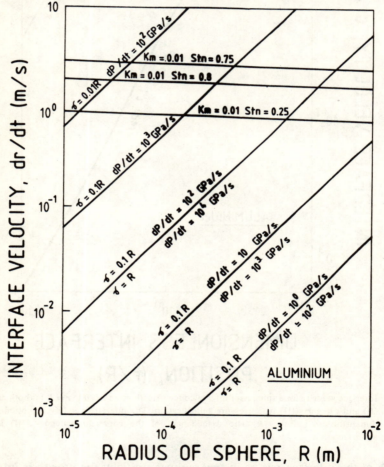

Fig. 2. A comparison of the interface velocities possible with category (II) and category (III) type solidification with increasing size of the sphere being pressurized. For category (III) type solidification, only the initial high solidification velocity is shown. The solidification velocity under category (III) is independent of the radius of the sphere as long as it is interface controlled.

conditions. The category (iii) solidification is not dependent on the radius of the sphere and only the initial high velocity values (i.e. proportional to the undercooling) are shown in the figure. The category (ii) type solidification is dependent on the radius and rate of pressure application. We notice that even for a $dP/dt = 10\,\mathrm{GPa/s}$ the interface velocities in category (ii) are much lower than in category (iii). The velocities become comparable only after the radius exceeds 1m.

From the theoretical development it seems evident that of all the methods available for pressure assisted solidification, high interface velocities are available only from category (iii) type of solidification. As shown below it is also fortuitous that this is a practical method of rapidly applying high pressure and causing rapid solidification of large volumes.

3. EXPERIMENTAL TECHNIQUES

Experiments were performed in the moderate and high pressure regimes. The rate of pressure application was varied from 100 ms to the order of minutes. The various experimental techniques used are given here.

3.1. Direct moderate pressure application

Experiments on various alloys were carried out where pressure was directly applied on the liquid metal or alloy in the following fashion. Aluminium alloys for example were charged into graphite crucibles and melted under protective covering agents to prevent oxidation and dissolution of gases in liquid metal. The die assembly comprising the die, top punch and bottom punch were coated with a graphite coating (approximate thickness smaller than 100 μm) on the internal surfaces with the help of a paint brush. The die assembly was heated to a temperature of 473K in an electric furnace. The liquid metal was poured in the die at predetermined temperature. The schematic representation of the process is shown in Fig. 3. A detachable type of top-punch was placed on the liquid metal surface and pressure was applied by advancing the ram of the 100 tonnes hydraulic press. The pressure developed by the technique was in the order 100–300 MPa. The die was provided with an internal taper to ensure that the sample could be ejected from the die with ease.

The sample (60 mm diameter, 120 mm long) was vertically cut into two halves. The vertical section was polished and etched for microstructure. The chemical composition of aluminium alloys studied in this pressure regime are given in Table 2.

3.2. Cooling curves

In order to investigate the cooling rates during solidification of alloys when pressurized as stated above a special bush and thermocouple assembly was designed and developed as shown in Fig. 3. Two thermocouples were placed

90

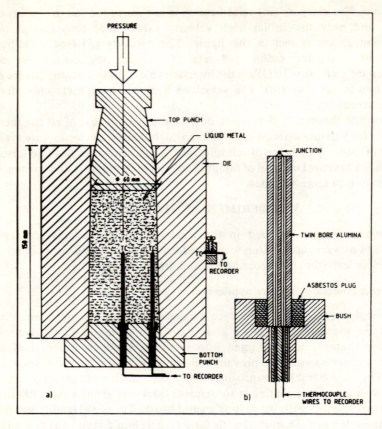

Fig. 3. (a) Arrangement used for recording cooling curves of liquid metals and alloys during pressure solidification and during solidification at atmospheric pressure. The figure is drawn to scale.
(b) Exploded view of thermocouple bush (schematic).

Table 2. Chemical composition of aluminium alloys

Metal or Alloy	Weight percent						
	Cu	Zn	Mg	Mn	Fe	MM	Si
Al (99.75%)	–	–	–	–	0.12	–	0.10
Al (99.99%)	–	–	–	–	0.04	–	0.04
Al-2.2% Cu	2.2	–	–	–	0.15	0.13	–
Al-4.4% Cu	4.4	–	–	–	0.15	–	0.13
Al-Zn-Mg-Cu	1.5	5.6	2.6	0.08	0.13	–	0.10
Al-25% Cu	–	–	–	24.93	0.11	–	0.10

inside the die such that one thermocouple measured the temperature in the centre region while the other thermocouple measured the temperature in the edge region of the 60 mm diameter 120 mm long casting. A third thermocouple was clamped on the external surface of the die. All the three thermocouples were exactly at the same level of height equal to 60 mm from the bottom of the component. Chromel/alumel thermocouple wires of different sizes 0.3 mm and 0.22 mm diameter and 0.05 mm diameter was used. Twin bore recrystallized alumina sheath of 2.5 mm outer diameter, 75 mm long was used for the centre and edge thermocouples. The thermocouples were connected to a six pen quick response (0.18 second for full-scale deflection) and high speed, 120 cm/minute chart recorder. The liquid metal was poured into the preheated die and the pressure was applied. The cooling curve was recorded until the solidification was complete.

3.3. High pressure solidification

3.3.1. Slow pressure application

High pressure slow solidification experiments were carried out on Mg-Zn and Pb-Sb alloy systems. High pressures roughly about 2.0 G.Pa were applied slowly (over 10s) by a hydraulic press. The same die assembly (Fig. 4)

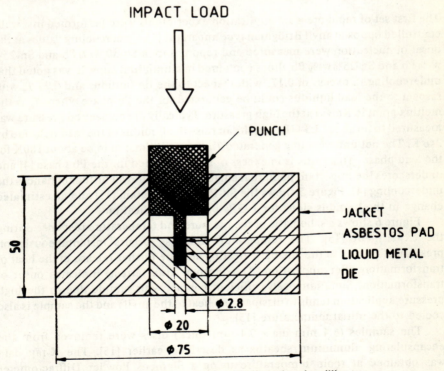

Fig. 4. Die and punch assembly, with dimensions in millimeters.

was also used for RPA experiments, to be described in the next section. The die and punch were heated to a temperature above the liquids so that the solid charge material was now in liquid state. The die cavity was then positioned on the plattens of the hydraulic press. An asbestos plug was initially placed on the surface of the liquid alloy for two reasons: (a) To ensure that the top punch did not quench or cool the liquid alloy when both come in contact (b) prevent oxidation of the alloy contained in the cavity.

The top punch was placed on the asbestos plug and pressurization was commenced slowly such that the pressure was built-up in 10 to 15 seconds. Immediately after the completion of pressurization, the die assembly was quenched in water. Marageing (Mar M-200) was used for the punch. The steel punch was machined in solution treated condition.

3.3.2. Rapid pressure application (RPA)

Rapid pressure application as described by Reddy et al. [3] was conducted by the quick application of a forge hammer to the die assembly shown in Fig. 4.

4. RESULTS

4.1. Pb-Sn alloys

The first set of rapid pressure application experiments were performed in a well-controlled opposed anvil Bridgman type apparatus [1]. Undercooling values at the onset of nucleation were measured and reported for a Sn-30 wt% Pb and Sn-25% wt% Pb and Sn-25% wt% Pb alloys contained in aluminium cups. It was noted that undercooling in excess of $0.1T_m$ with respect to the tin liquidus and $0.25\ T_m$ with respect to the lead liquidus could be generated for the Pb phase where T_m is the melting point in Kelvin at the high pressure. Typically the nucleation pressure was measured to be about 3.3 GPa which can raise the liquidus of the lead rich side by 246 K. The net undercooling generated Tu then is estimated to be about 180K for the lead phase. This value is in excess of that calculated for the Pb phase [4] and underscores the important point that adiabatic heat removal further enhances the undercooling [4]. Figure 5 shows the Sn-Pb phase diagram [15] and an estimated change in the liquidus slope for 1 GPa pressure application.

Figure 6 shows a schematic of the pressure and temperature response during the experiment [15]. The temperature increase occurs instantaneously on pressurization due to the adiabatic heating and the further release of the heat of transformation after nucleation. The thermal kink (arrest) indicates onset of transformation. Simultaneously, efficient contact with sample due to the high pressure application tends to dissipate the heat to the anvils and the sample is also cooled to the initial temperature [15].

The samples (6.4 mm dia × 0.4 mm thick) discs were removed from the encapsulating aluminium sheaths as described earlier [15]. The X-ray data was obtained at room temperature using a Siemens Powder Diffractometer

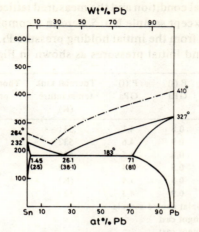

Fig. 5. The lead-tin phase diagram (from M. Hansen, Constitution of binary alloys, McGraw-Hill, New York, 1958) and estimated liquidus at 1 GPa pressure.

(Cu-K$_\alpha$ radiation was used). The diffraction data were utilized for calculating the lattice parameters and also for a qualitative interpretation of intensity variations. Accurate parameter values were obtained by using graphical extrapolation (Bradley-Jay correction method [16]. Only reflections above 60° were considered for the graphical plots.

Results for the lattice parameter values of the lead phase are shown in Table 3 along with the experimental condition during pressurization. Values obtained for a splat quenched sample, a conventionally cooled sample and pure

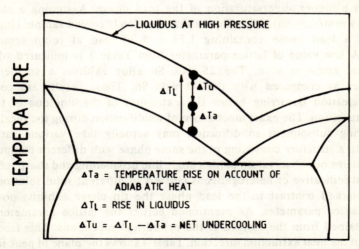

Fig. 6. Schematic illustration of the net undercooling temperature ΔTu; ΔTa is the adiabatic heat release and ΔT_l is the rise in liquidus respectively when pressure is applied.

Table 3. Experimental condition and the measured lattice parameter values. All pressurized samples except sample No. 5 were in a completely liquid condition prior to pressurization from the initial holding pressure Pi. Pf and Pi are the final and initial pressures as shown in Fig.

Sample	Composition wt% Pb	P (i) GPa	P (f) GPa	Thermal kink temperature (K)	Thermal kink pressure GPa	Lattice parameter A°
1.	30.0	0.8	4.1	531	3.2	4.944
2.	30.0	0.9	4.6	532	3.5	4.944
3.	25.0	0.8	4.5	529	(-)	4.941
4.	25.0	0.5	3.4	522	(-)	4.9425
5.	25.0	0.5	3.4	(N)	(N)	4.9445
6.	30.0	0.9	4.3	527	3.4	4.945
7.	25.0	Splat cooled sample				4.944
8.	25.0	Ingot cast				4.9475
9.	99.99	Ingot cast				4.9500

(-) Indicates that the measurement was not made
(N) Indicates that the kink was not observed

lead (99.99 wt%) samples are also shown. A thermal kink (arrest) indicating onset of nucleation at a well-defined kink pressure is noted when the samples were pressurized from a semi-solid state which allowed for epitaxial solidification [15]. Results from a conventionally cooled sample and pure lead (99.99 wt%) sample are also shown. The work of Tyzack et al. [17] on Pb-Sn alloys indicate that it is difficult to retain more than 5 wt% Sn in lead on rapid quenching from temperature of 483 K even though the equilibrium solubility at the eutectic temperature is about 29 at.% (19 wt%). Tin reduces the lattice parameter value indicates a higher supersaturation of the lead phase. Assuming a change of (0.000139 mm/at.% Sn) [15] indicates that the slowly cooled sample (Ingot cast) exhibits a lead phase containing 1.79 at.% of lead at room temperature 298 K. A low value of lattice parameter from Table 3 is indicated whenever a thermal arrest is seen. The 25 wt% Sn alloy exhibits a smaller lattice parameter as compared with the 30 wt% Sn. These results are consistent with nucleation occurring below the extension of the liquidus of the lead rich composition. The exact mechanism of solidification during the recalescence and during subsequent solidification may actually take various paths [18], leading to a structure consisting of the same phase with different compositions. This explains our observation of high angle line broadening and sharp reflections of a form indicative of heterogeneous clusters of different solid solutions.

In marked contrast to the lead phase the tin phase exhibits no change in the lattice parameter. As mentioned earlier the lattice parameter values were obtained from the surface of the cylindrical specimens. This face is also normal to the heat extraction direction. Table 4 shows the plane of peak intensity for the specimens listed in Table 3. We observe that samples showing the thermal

Table 4. Planes of maximum intensities for the Pb and Sn phases. The sample numbers correspond to those in Table 3

Sample	Plane of peak intensity (Sn)	Plane of peak intensity (Pb)
1	(200)	(111)
2	(200)	(111)
3	(211), (200)	(111)
4	(101), (200)	(111)
5	(101)	(111)
6	(101), (200)	(111)
7	(200)	(111)
Pure (99.99 wt%)	(101)	(111)

arrest (kink in Fig. 7) display a (200) plane as the plane of maximum intensity whereas the rest of the samples except the splat cooled display the (101) plane to be the highest intensity plane.

Microstructural examination of the samples were conducted and the micrographs were presented [1]. From the micrographs the splat cooled and the RPA samples displayed similar distribution and morphology of the lead phase. However the volume fraction of the lead phase was higher in the RPA processed

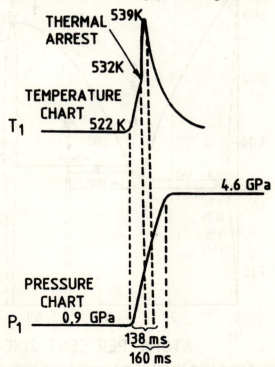

Fig. 7. Schematic of a pressure and temperature versus time response from [1].

samples. (Similar microstructures are shown subsequently in the explosive experiments described in the following section.)

4.2. Mg-Zn alloy

The experiments described above were well characterized and were conducted in a controlled Bridgman opposed anvil apparatus. The pressure-induced undercooling was measured and reported. However the apparatus could not be used beyond 573 K for fear of permanently damaging the delicate press. In this section we report on some unique phase selection during RPA-induced solidification in Mg-Zn alloy. The experiments were conducted on a more routine forge hammer and hydraulic press and the maximum pressures, though completely reproducible, are limited by yielding of the die and punch.

The equilibrium phase diagram of the Mg-Zn system [19] is shown in Fig. 8. The domains for the existence of several reported phases in many instances

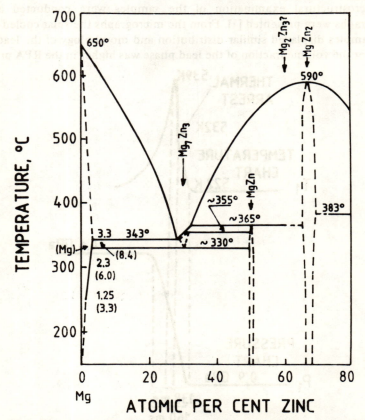

Fig. 8. Mg-Zn phase diagram (from M. Hansen and Constitution of binary alloys, McGraw-Hill, New York, 1958).

are not adequately established [20]. The phases Mg_7Zn_3, $MgZn$, Mg_2Zn_3 and $MgZn_2$ have all been reported [19] to exist in the range of the composition of interest. Some of these phases have been reported as being in equilibrium with the Mg-rich primary solid solution [19] and to be stable down to room temperature. Glass formation has also been achieved in the composition range 48 to 57 wt% Zn [21]. There have been reports that the Mg_7Zn_3 may decompose into $\alpha Mg + Mg_2Zn_3$ or $\alpha Mg + MgZn$ [19] through a eutectoid reaction and that a metastable moidification of MgZn [22] is possible.

4.2.1. Experiments and results

A Mg-35 at% Zn alloy was made from metals of 99.99% purity, by melting under a flux in a stainless steel crucible and chill-casting into ingots 50 mm diameter by 30 mm in height. This composition is hypereutectic with respect to αMg at normal pressures with a eutectic temperature at 343°C (616 K) as shown in Fig. 8.

The liquid metal was pressurized at 723 K some 50 K above the 1 bar $MgZn_2$ liquidus as discussed below. Two types of pressure experiments were performed (i) where pressure was applied up to 2 GPa by slow application (over 10 to 15 s) in a hydraulic press and (ii) where a dynamic pressure of 3.5 GPa was applied over an estimated time of less than 200 ms. The hardened and tempered low alloy steel die and a maraging steel punch used as shown in Fig. 4 was identical for both of the experiments and both die and punch were maintained at a temperature of 723 K prior to pressurization. Some 15 seconds after pressurization the whole assembly was purged with water while still under pressure. The dies were then machined open as they tended to seize after the high pressure application. The various experiments performed during this investigation are detailed in Table 5. Each experiment was repeated at least three times and the results were found to be reproducible. We have ascertained that no major melting or segregation losses occurred during any of these experiments, and final analyses from experiments I to V gave zinc contents of 34 ± 2 at.% not markedly different from the original composition. Resulting microstructures are shown in Figs. 9 and 10 and are summarized in Table 5. Phase identifications were confirmed by X-ray diffraction (Table 6) and wavelength dispersive microprobe analysis (Table 7). Cooling rates during solidification (Table 5) were measured with *in situ* thermocouples for experiments I and II and estimated assuming a contact heat transfer coefficient of $10^3 W/m^2$ K and Newtonian conditions for experiment III.

4.2.2. Discussion

The results show the Mg_7Zn_3 was present under all conditions studied with αMg as the other phase under conditions of lower cooling rate and no or slow pressurization (experiments I, II and VI) and MgZn as the other phase either at increased cooling rate with no pressurization (experiment III) or after rapid pressurization (experiments IV and V). At the lowest cooling rate with no

98

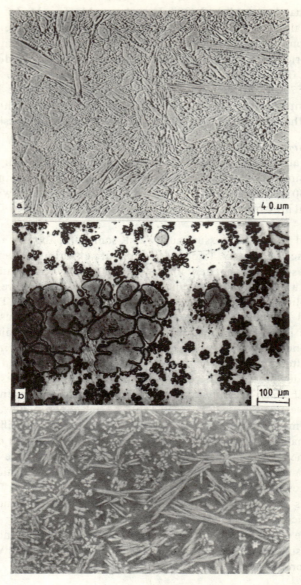

Fig. 9. (a) Scanning electron micrograph of structure resulting from experiment I. The dark phase in the eutectic is αMg and the more voluminous lighter phase (primary solidification and also one phase of the eutectic) is Mg_7Zn_3.

(b) Optical micrograph of structure resulting from experiment II. The flowery and globular regions are αMg. The matrix is Mg_7Zn_3.

(c) Scanning electron micrograph of structure resulting from experiment III. The platelets are MgZn. The matrix is Mg_7Zn_3.

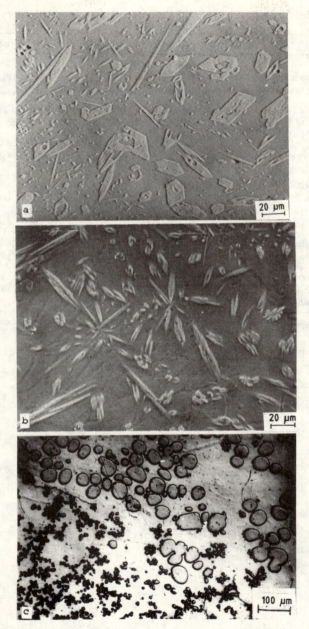

Fig. 10. (a) Scanning electron micrograph of structure resulting from R.P.A. experiment IV. The platelets are MgZn. The matrix is Mg_7Zn_3.

(b) Scanning electron micrograph of structure resulting from experiment V. The platelets are MgZn. The matrix is Mg_7Zn_3.

(c) Optical micrograph of structure resulting from the slow pressure experiment VI. The flowery and globular forms are αMg. The matrix is Mg_7Zn_3.

Table 5. Experiments and their resulting microstructures

	Experiment	Microstructure
I.	Slow cool ($\dot{T} = 0.25\ K/s$) at 1 bar Air cooled from 723 K in hot die punch assembly in contact with forge hammer platens. Sample dimensions 2.8 mm dia × 6 mm long	Plates of Mg_7An_3 in Mg/Mg_7Zn_3 eutectic matrix (Fig. 9a)
II.	Chill cast ($\dot{T} = 1\ K/s$) at 1 bar Cast from 723 K in a cold mould internal dia 25 mm × height 10 mm	Primary α Mg in a matrix of Mg_7Zn_3, (Fig. 9b)
III.	"Splat-cooled" ($\dot{T} \sim 5 \times 10^4\ K/s$) at 1 bar Teemed at 723 K on to liquid N_2 chilled copper block. Resulting "splat" thickness 0.5 mm	Platelets of MgZn in a matrix of Mg_7Zn_3 (Fig. 9c)
IV.	Rapid pressurization ($\dot{P} \sim 10^2$ GPa/s) and slow cool pressurization in < 200 ms at 723 K in die assembly by forge hammer application with asbestos pad between die assembly and platens	Platelets of MgZn in a matrix of Mg_7Zn_3 (Fig. 10a)
V.	Rapid pressurization ($\dot{P} \sim 10^2$ GPa/s) and water quench As IV but cold water quenched 15 s after pressurization	Platelets of MgZn in a matrix of Mg_7Zn_3 (Fig. 10b)
VI.	Slow pressurization ($\dot{P} \sim 0.25$ GPa/s) and water quench Pressurization in 10 to 15 s at 723 K in a hydraulic press and die assembly water quenched as in V	Primary αMg in a matrix of Mg_7Zn_3 (Fig. 10c)

T and $\dot{P}$ are respectively cooling rate and pressurization rate associated with solidification

Table 6. 'd' spacing (nm) obtained from X-ray diffractograms using Cu K_a radiation and solid samples

Expt. 1	Expt. II	Expt. IV	Expt. V	Expt. VI
—⁺	0.572⁺	—⁺	0.575⁺	—⁺
—	(0.520)	—	(0.528)	—
—	0.501⁺	—ˣ	0.501ˣ	—
—⁺	0.453⁺	—⁺	—⁺	—⁺
0.440⁺	0.441⁺	0.439⁺ˣ	0.441⁺ˣ	—⁺
—	(0.431)	—	—	—
—	0.421⁺	—ˣ	0.426ˣ	—
0.399⁺	0.409⁺	—⁺ˣ	—⁺ˣ	—⁺
—	0.385⁺	0.380ˣ	0.385ˣ	—
0.379⁺	—⁺	—⁺	—⁺	0.377⁺
(0.375)	(0.375)	(0.375)	(0.374)	—
(0.349)	(0.348)	—	(0.348)	—
—⁺	—⁺	0.324⁺	—⁺	—⁺
—⁺	—⁺	—⁺	—⁺	0.318⁺
—⁺	—⁺	—⁺	0.308⁺	0.304⁺
—	—	—	(0.289)	—
—⁺	—⁺	—⁺	0.277⁺	—⁺
0.276*	—*	—	—	—*
(0.274)	(0.274)	—	(0.274)	—

Table 6 (*Contd.*)

Expt. 1	Expt. II	Expt. IV	Expt. V	Expt. VI
— $^{+}$	— $^{+}$	— $^{+\times}$	0.267$^{+\times}$	0.267^{+}
—	—	—	—	—
0.256^{+}	0.257^{+}	— $^{+\times}$	0.256$^{+\times}$	0.255^{+}
0.253^{+}	— $^{+}$	— $^{+}$	0.253^{+}	— $^{+}$
— $^{+}$	— $^{+}$	0.247$^{+\times}$	— $^{+\times}$	— $^{+}$
0.245*	— *	—	—	— *
0.243^{+}	0.243^{+}	0.242$^{+\times}$	0.242$^{+\times}$	0.244^{+}
0.241^{+}	0.239^{+}	— $^{+}$	— $^{+}$	0.241^{+}
0.233^{+}	0.233^{+}	0.233^{+}	0.233^{+}	0.232^{+}
0.231^{+}	0.231^{+}	— $^{\times}$	— $^{\times}$	—
0.228^{+}	0.228^{+}	0.229^{+}	0.228^{+}	0.228^{+}
0.222^{+}	0.222^{+}	0.221$^{+\times}$	0.222$^{+\times}$	— $^{+}$
0.217^{+}	0.216^{+}	0.215$^{+\times}$	0.216$^{+\times}$	0.215^{+}
0.214^{+}	— $^{+}$	— $^{+\times}$	0.212$^{+\times}$	0.210^{+}
0.199^{+}	0.200^{+}	0.201$^{+\times}$	0.199$^{+\times}$	0.196^{+}
0.196^{+}	— $^{+}$	— $^{+}$	— $^{+}$	— $^{+}$
0.193^{+}	—	— $^{\times}$	— $^{\times}$	—
0.191^{*+}	0.191^{*+}	0.191^{+}	0.191^{+}	0.191*
0.189^{+}	0.189^{+}	— $^{+}$	— $^{+}$	0.189^{+}
0.184^{+}	0.185^{+}	0.184$^{+\times}$	0.184$^{+\times}$	— $^{+}$
0.178^{+}	— $^{+}$	— $^{+\times}$	— $^{+\times}$	— $^{+}$
— $^{+}$	0.175^{+}	0.176$^{+\times}$	0.175$^{+\times}$	— $^{+}$
— $^{+}$	0.173^{+}	— $^{+\times}$	— $^{+\times}$	—
0.170^{+}	—	— $^{\times}$	— $^{\times}$	—
— $^{+}$	0.166^{+}	0.168$^{+\times}$	0.168$^{+\times}$	0.167^{+}
0.164^{+}	— $^{+}$	— $^{+\times}$	0.164$^{+\times}$	0.163^{+}
— *	0.160*	— $^{\times}$	— $^{\times}$	— $^{\times}$
—	0.152	0.152$^{\times}$	— $^{.\times}$	—
— *	0.146*	0.148$^{\times}$	0.148$^{\times}$	0.146*
0.145	—	0.144$^{\times}$	0.145$^{\times}$	0.145
0.143	0.143	—	—	0.143
0.142	0.142	0.142^{+}	0.142*	0.142
0.140	—	— $^{\times}$	0.140$^{\times}$	—
0.139*	— *	— $^{\times}$	0.138$^{\times}$	0.139*
0.137*	0.137*	— $^{\times}$	— $^{\times}$	0.137*
—	0.136	0.130$^{\times}$	0.136$^{\times}$	0.136
0.131	—	— $^{\times}$	— $^{\times}$	—
—	0.124	0.125$^{\times}$	— $^{\times}$	—
0.122	0.122	0.121	0.121	0.121
—	—	—	0.120	—
0.111	0.119	—	—	—
—	0.115	—	—	0.108
—	—	—	—	0.105

* αMg JCPDS Card No. 4-770; × MgZn JCPDS Card No. 8-206; + MgZn, JCPDS Card No. 8-269. Parentheses indicate unidentified reflections possibly due to Zn-rich phases e.g. $MgZn_2$ and Mg_2Zn_{11}.

Table 7. Results of wavelength dispersive microprobe analysis and microhardness measurements of the phases shown in Figs. 9 and 10

Experiment	Phase	Analysis Mg	at. % Zn	Phase Identification	Microhardness* Kg/mm^2
I	Plate	72.7	27.2	Mg_7Zn_3	320 ± 15
	Eutectic second phase	96.5	3.5	αMg	–
II	Primary phase	97.2	2.8	αMg	128 ± 2
	Matrix	72.3	27.7	Mg_7Zn_3	322 ± 7
III	Platelets	50.0	50.0	$MgZn$	290 ± 4
	Matrix	71.1	28.9	Mg_7Zn_3	298 ± 9
IV	Platelets	45.4	54.6	$MgZn$	314 ± 16
	Matrix	73.9	26.1	Mg_7Zn_3	$316 \pm 7^+$
V	Platelets	48.5	51.5	$MgZn$	316 ± 3
	Matrix	72.0	28.0	Mg_7Zn_3	322 ± 5
VI	Primary phase	95.4	4.6	αMg	116 ± 9
	Matrix	68.7	31.3	Mg_7Zn_3	322 ± 8

25 g load except for [+] 15 g load

pressurization (experiment I), the microstructure consists of plates of Mg_7Zn_3 in a matrix of coupled $\alpha Mg/Mg_7Zn_3$ eutectic (Fig. 9a). Increasing the cooling rate by a factor of 4, to 1K/s, in experiment II apparently gives sufficient undercooling to form αMg as the primary phase in a matrix of Mg_7Zn_3 (Fig. 9b). There was no evidence of formation of the MgZn phase in these specimens but unidentified X-ray reflections in Table 6 could possibly be due to the formation of Zn-rich phases $MgZn_{11}$ which would be necessary to account for the average composition of the alloy richer in Zn than both αMg and Mg_7Zn_3 unless the composition range of the Mg_7Zn_3 phase had been extended to higher Zn content in the experiments. The microstructure and constitution obtained on slow pressurization and quenching (experiment VI) is similar to those of experiment II. Increasing cooling rate during solidification to $\sim 5 \times 10^4$K/s in the absence of pressurization in experiment III has the notable effect of replacing primary Mg with MgZn as well as substantially refining the scale of the microstructure (Fig. 9c). Rapid pressure application in experiments IV and V has the same effect in forming MgZn but with a lesser degree of microstructural refinement (Figs. 10a and 10b). This suggests that rapid pressurization is sufficient to generate the undercooling required to form MgZn. The increased refinement of this MgZn in experiment II is attributable to the high cooling rate applied during solidification, which also generates sufficient undercooling to form MgZn.

The rapid pressure application technique may be employed to produce metastable microstructures similar to those obtained during rapid solidification. RPA experiments performed on a Mg-35 at.% Zn alloy indicate that an undercooling process may cause new phase selections different from the alloy solidified by slow pressure application.

4.3. Fe-B alloys

Alloys in the range of 12.27 at.% boron are known to be glass forming (reviewed by Frank et al. [23]). Further the decomposition products on heating the glass to induce crystallization are also adequately known [23]. We also note that in several Fe-B amorphous alloys the decomposition products on heating initially leads to metastable Fe_3B in the tetragonal crystal structure which changes to Fe_3B in the orthogonal crystal structure on further heating [23]. We report here preliminary results of an investigation on a few Fe-B alloys subjected to RPA (listed in Table 8).

The RPA samples remain at a high temperature, 1300 K, for about 10 seconds before quenching the bulk die assembly in cold water (at 300 K). As the temperature of the experiment is high we expect the RPA processed microstructure if formed after a undercooling process to contain products of the metastable state; namely Fe_3B [24]. Table 8 lists the 'd' spacings from as cast $Fe_{69}B_{31}$ and RPA processed $Fe_{80}B_{20}$ and $Fe_{69}B_{31}$ alloys. This datum was obtained on a Debye Scherrer Camera with Fe K radiation. The RPA processed samples show the existence of Fe_3B possibly in the tetragonal form with $a = 0.857$ nm and $c = 0.444$ nm, 'd' spacings corresponding to this phase are marked in Table 8. Compared to the $\alpha-$Fe and Fe_2B lines the Fe_3B lines are broad and diffuse.

4.4. Ni-Cu alloys

Nickel alloys containing copper are known to display reduced dendrite spacing with increasing undercooling [25]. Further, at undercooling values exceeding 174K nucleation induced cavitation leading to pressure pulses are known to drastically reduce the equiaxied dendrite size [26, 27]. RPA experiments were performed on a Ni–42.3 at.% Cu alloy. Figures 11(a) and (b) compare the microstructures for the *as* cast and RPA sample. RPA experiments were carried out at 1723K. The *as* cast sample refers to a specimen melted in an identical die as used for the RPA experiment and quenched identically as the die after the RPA experiments, the only difference being that pressure is not applied. The *as*-cast microstructure is similar in scale, morphology and segregation to those reported earlier [25]. The RPA processed sample however shows an extremely small equiaxed dendrite size of 12.1 ± 2.1 μm. For the cast sample in an identical die, the dendrite arm spacings, varied from 90–133 μm, at the centre of the casting. The average dendrite arm spacing of the cast sample at the edge was 55.6 μm. The RPA processed sample has a uniform grain size throughout the sample and the dendrite arm spacing is smaller than those obtained by conventional undercooling techniques [25, 26].

Table 8. 'd' spacings in nm of Fe-B alloys processed by RPA and casting. The symbols *, +, @, £, refer to α-Fe, Fe_2B, Fe_3B and not-observed respectively. RPA processing for $Fe_{69}B_{31}$ and $Fe_{80}B_{20}$ were carried out at 1503 K and 1473 K respectively

As cast $Fe_{69}B_{31}$		RPA $Fe_{69}B_{31}$		RPA $Fe_{80}B_{20}$	
0.359	+	0.360	+	£	
0.254	+	0.255	+	0.254	+
£		0.224	@	0.2220	@
0.212	+	0.212	@	0.2142	@
0.201	*	0.202	*	0.202	*
£		£		0.196	@
0.184	+	0.184	+	£	
0.182	+	0.163	+	0.164	+
0.162	+	0.163	+	0.164	+
0.161	+	0.161	+	0.161	+
£		£		0.148	@
0.143	*	0.143	@*	0.143	@*
0.128	+	0.128	+	0.129	+
0.126	+	0.126	+	£	
0.1204	+	0.1203	+	0.120	+
0.1189	+	0.1189	+	0.118	+
0.1169	*	0.1169	*	0.117	*
0.1095	+	0.1109	+	0.109	+
0.1049	+	0.1048	+	0.1047	+
0.1015	*	0.1016	*	0.1014	*
0.0980	+	0.0982	+	£	

4.5. Fe-Ni-B-Si alloy

The composition of $Fe_{41}Ni_{41}Si_5B_{13}$ was cast in an experiment similar to the Ni-Cu alloys and was subjected to RPA as well. Figures 12(a) and (b) compared the two microstructures. Segregation of various elements is seen in the *as* cast sample while no such seen in a RPA processed sample. This indicates rapid growth of the solid/liquid interface bringing about changes in the partition coefficient and leading to partitionless solidification. The rapid interface growth occurs as the solid-liquid interface recalasces towards its equilibrium state. Evidence for a mostly single phase material is also provided by X-ray examination which showed the material to be fcc phase with lattice parameter, $a = 0.3568$ nm as measured with FeKα radiation. Extremely weak diffuse lines at 0.226 nm and 0.215 nm were also observed again, suggesting a possibility of Fe_3B type phase as noted above with the RPA experiment on Fe-B alloys.

4.6. Al-Mn alloy [5]

Rapid pressurization of the $Al_{80}Mn_{20}$ alloy was carried out in a simple piston

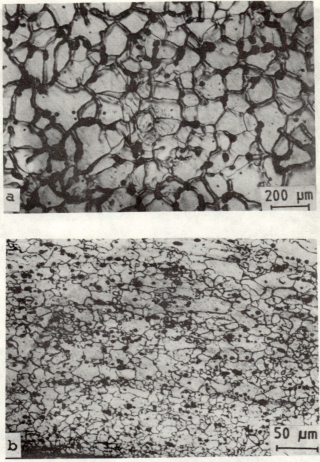

Fig. 11. Equiaxed dendrite size in Ni-42.3 at.% Cu alloy
(a) as cast
(b) R.P.A. processed.

cylinder apparatus made of maraging steel. The alloy was melted in air at 1245 K in a 3 mm diameter (4 mm height) cylindrical cavity and was rapidly pressurized by forge hammer application on the cold piston. The piston and the liquid alloy were thermally isolated by a hot 1 mm asbestos cloth placed on the liquid metal. We estimate that the sample was pressurized to about 2 GPA in less than 200 ms [1, 4].

The product of the rapid pressurization experiment was found to be magnetic. A vibrating sample magnetometer measurement of the sample showed the presence of a ferromagnetic phase with a curie temperature of 800 K. The sample was then analysed by the X-ray powder diffraction technique. The observed d-spacings with their assignment to various phases are shown in Table 9. The

106

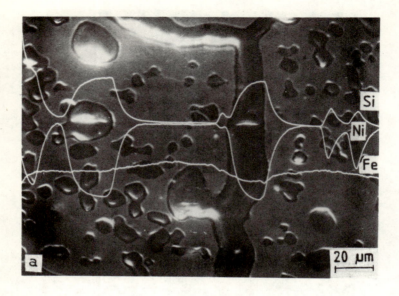

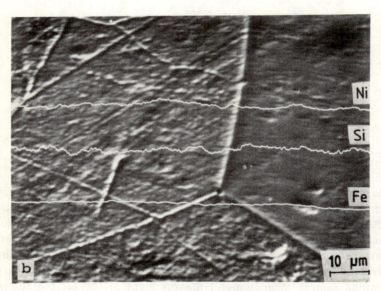

Fig. 12. Segregation profile of Ni, Fe and Si in a Fe-41% Ni-5% Si-13% B alloy as measured by wavelength dispersive analysis in a microprobe (a) as cast (b) R.P.A. processed. The microprobe beam diameter is estimated to be 1 μm and the beam spread ~ 3 μm. R.P.A. experiments were carried out at 1723 K and the female die used was made of mild steel for this alloy.

Table 9. The table shows the observed d-spacings and intensities from $Al_{80}Mn_{20}$ rapidly pressurised alloy. A long exposure (12 hrs) Debye-Scherrer photograph obtained with FeK_α radiation was used. The icosahedral phase was indexed as explained in the text, the indices for Al_6Mn and α-Al_2O_3 were obtained from the JCPDS powder data file (Card No. 06–665 & 10–173) and the d-spacings of the μ phase were obtained as described in the text

St: Strong, m: medium, W: weak, V: very

Iobs	d/nm	Icosahedral indices	Observed d/d (100000)	Expected d/d (100000)	μ phase (d/nm)	Intensity	Al₆Mn (hkl)	Cubic phase (hkl)	Al₂O₃ (hkl)
(1)	(2)	(3)	(4)	(5)	(6)	(7)	(8)	(9)	(10)
W	0.638				0.636	mW			
VW	0.610				0.607	VW			
VW	0.555				0.549	VW			
VW	0.505				0.505	VW			
W	0.487				0.492	VW	(110)	(111)	
VW	0.438						(002)		
VW	0.413				0.412	VW			
W	0.374	(110001)	0.58	0.56	0.371	w	(020)		
VW	0.331	(111010)	0.65	0.65	0.334	w			
VW	0.295				0.299	mW		(220)	
VVW	0.286						(022)		
W	0.268				0.269	W			
W	0.261						(202)		
St	0.252				0.249	W	(113)	(311)	(104)
W	0.244				0.241	m		(222)	(110)
m	0.232				0.231	mW	(130)		
St	0.226				0.225	mW	(131)		
St	0.215	(100000)	1	1	0.215	m			
St	0.209				0.208	VSt	(310)	(400)	(113)
St	0.204	(110000)	1.058	1.051	0.203	St	(311)		
St	0.201				0.201	m			
mW	0.188				0.189	W	(223)		
VW	0.184						(133)		
VW	0.177				0.177	VW			
VW	0.173						(042)	(422)	(024)
VW	0.169								
VVW	0.164						(224)		
VW	0.161							(333)	
VW	0.160								(116)
VVW	0.153				0.152		(242)		
VVW	0.151								
m	0.148	(111000)	1.45	1.45	0.148	VW	(006)	(440)	
VW	0.147								
VW	0.145	(111100)	1.49	1.49	0.145	VW			
m	0.143				0.143	VW	(243)		
VW	0.140				0.140	m		(531)	(124)
VW	0.137								(030)
W	0.131				0.1306	W			
					0.1296	St			

108

Table 9. (*Contd.*)

(1)	(2)	(3)	(4)	(5)	(6)	(7)	(8)	(9)	(10)
m	0.1272	(101000)	1.69	1.70	0.1272	St			
m	0.1258				0.1260	W			
					0.1254	W			
W	0.1248				0.1248	VW			
W	0.1238				0.1236	m			
W	0.1226				0.1228	VW			
VW	0.1169	(211000)	1.84	1.82					
VVW	0.1140								
VVW	0.1123				0.1121	mW			
VW	0.1117				0.1115	VW			
VW	0.1093	(110010)	1.97	1.97					
m	0.1070	(200000)	2.01	2.00	0.1070	mW			
					0.1067	mW			
VVW	0.1056				0.1056	W			
VVW	0.1035				0.1036	mW			(226)
VVW	0.1013	(220000)	2.13	2.10	0.1010	VW			
VVW	0.0996								(1210)

St: Strong, m: medium, W: weak, V: very

diffraction pattern from the icosahedral phase (we have retained the icosahedral indexing over the cubic indexing by Pauling [28]) was indexed on the basis of six independent vectors $(1, \pm \tau, 0), (0, 1, \pm \tau)$ and $(\pm \tau, 0, 1)$, pointing to the vertices of an icosahedran as done by Bancel et al. [29], where, $\tau = 1.618$, the golden mean. In addition to the icosahedral phase, the X-ray powder pattern shows the presence of a cubic phase ($a = 0.838$ nm), the μ phase of the Al-Mn system [30], small amounts of Al_6Mn and a little contamination of Al_2O_3. The X-ray powder pattern after magnetic separation of a part of the material showed a clear reduction in intensity of the lines from the cubic phase. The material when heated at 1180 K for three hours in an evacuated quartz ampule (at $\sim 10^{-6}$ torr) was no longer ferromagnetic and X-ray photographs showed a complete transformation of the material to the phase (with a little Al_2O_3), showing conclusively that the cubic phase is a new metastable ferromagnetic phase formed by the rapid pressurization process in the Al-Mn system. The transformed phase is the same as that obtained on heating a rapidly quenched $Al_{78}Mn_{22}$ alloy at 700°C for three hours. A comparison of the X-ray photographs from the rapidly pressurized and rapidly cooled alloys after transformation shows a one to one correspondence between the X-ray lines, but for the presence of a few weak lines of Al_2O_3 contaminant in the pressurized sample. Figure 13 shows the typical microstructure of the alloy. The presence of four distinct phases marked A, B, C and D was observed after initially etching mildly with a (FeCl₃ + Ethanol + HCl) solution followed by a deep etch with Kellers Reagent (lml HF + 1.5 ml HNO₃ + 2.5 ml HCl + 95 ml H₂O). The compositions of these phases were determined using an electron microprobe. Results obtained from the microprobe were corrected for flourescence, beam absorption and atomic number against pure standards (99.99 wt% pure) and are

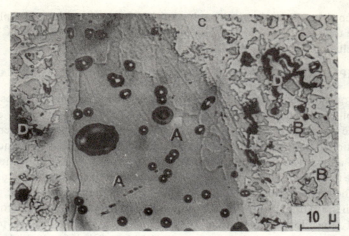

Fig. 13. Microstructure of the rapidly pressurized sample. The compositions of the phases A, B, C and D are given in Table 4.

Table 10. Chemical composition of the various phases shown in the micrograph of Fig. 13 (Composition in atomic percents)

	Al	Mn	Fe	O
Phase A	74.99	24.81	0.20	—
Phase B	78.53	20.58	0.89	—
Phase C	82.41	17.20	0.39	—
Phase D	32.70	1.01	0.09	66.20

the steel dies had contaminated the material. The dark chain marked D was found to be an Al_2O_3 stringer left over from the melting experiment.

Since the extreme brittleness of the sample precluded transmission electron microscopy, our evidence for correlating the phase to the microstructure is as follows. Phase A could be pitted completely with the $FeCl_3$ etch when etched for a long time. The rest of the phases were relatively stable to this etch. We melt spun alloys with compositions from $Al_{75}Mn_{25}$ through $Al_{84}Mn_{16}$ but could not find any phase that was similarly etched. However, as expected, ,we found the icosahedral phases in these ribbons and those with compositions from $Al_{84}Mn_{16}$ to $Al_{80}Mn_{20}$ etched similar to phase C. The pressurization experiments were repeated with a 1 mm die cavity where a higher pressure then that applied earlier could be achieved. The sample thus obtained showed larger areas that would etch with $FeCl_3$ and also showed higher X-ray intensities of the cubic phase. μ phase X-ray lines were also then much weaker. We conclude that phase A is the magnetic cubic phase. The μ phase from the equilibrium phase diagram of the Al-Mn system [30] contains approximately 22 at.% Mn and of the phases marked B and C in Fig. 13, the phase

110

B must correspond to the μ phase because of its composition. The phase C is therefore the icosahedral phase. The measured composition of phase C supports this conclusion.

5. EXPLOSIVE TECHNIQUES FOR RAPID PRESSURE APPLICATION

As the forge hammer technique (Fig. 4) was destructive in nature and restricted to small volumes (typical maximum dimensions about 10 mm) an explosive technique for larger volume was also developed. Figure 14 shows a schematic of a unidirectional explosive pressure wave experiment performed on a liquid Sn-25 wt% Pb alloy. The die assembly was kept hot at 473K and the contact explosive used was 150 g trimonite. Pressures generated are estimated to be about 2 GPa. The metal formed a 150 mm long and 20 mm diameter cylinder. After the explosion the cylinder was vertically sectioned and examined. The microstructure obtained (Fig. 15) was identical to that reported in [1]. Lattice parameter calculations gave the lead phase to be cubic with $a = 0.49435$ nm which compares very favourably with the supersaturated lattice parameter shown in Table 3. This technique is presently being pursued further.

6. NUCLEATION AND TRANSFORMATION STUDIES AT MODERATE PRESSURES

Phase transformations can be studied at moderate cooling rates ($10^2–10^3$ K/s) with *in situ* thermocouples during moderate pressure solidification in metal moulds. Unlike high pressure solidification described above the pressures applied for low to moderate pressure solidification are of the order of only 200 MPa which leads to an efficient thermal contact of the liquid metal which the die. Cooling rates in the liquid and during solidification can be obtained by this technique and

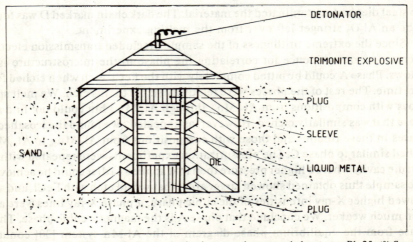

Fig. 14. Schematic of the explosive pressurization experiment carried out on a Pb-25wt% Sn alloy.

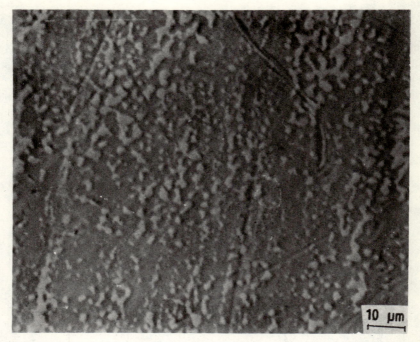

Fig. 15. Microstructure of the explosively pressurized alloy of Figure 14. Please note similarity with microstructure shown in reference [1].

shown in Table 10. Microprobe results indicate that a small amount of iron from TTT (time transformation temperatures) can be rigorously studied. The method has been applied to several aluminium alloys and results are reported here.

6.1. Transformation: (Aluminium)

As shown in Fig. 16(b) pressure solidification led to the nucleation of aluminium at 865.5 K, i.e., with a 65.5 K undercooling (ΔT). The normalized undercooling parameter ($\Delta T / T_1$) is 0.07 when $T_i = 931$ K (T_m and T_1 are used interchangeably to imply melting point or liquidus temperature). The curve also showed a recalescence of 23.5 K. The high cooling rate leading to undercooling is attributed to the improved heat transfer coefficient at the liquid/die interface on account of pressurization causing intimate contact [7] of the liquid with the die wall.

The transformation temperature (considered the nucleation temperature, T_N) was recorded at various cooling rates (for over 20 experiments) and is plotted in Fig. 17 as a undercooling versus cooling rate diagram. In the range of measurements the undercooling is seen to increase with cooling rate. The undercooling parameter versus cooling rate curve for the case of pure aluminium showed a large scatter when the 0.05 mm thick foil was used. The rest of the values for the round bead thermocouples indicate an ascending curve. The line shown in

112

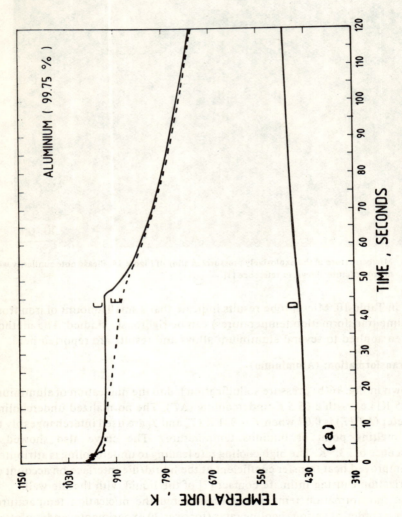

Fig. 16. (a) Cooling curve of aluminium solidified at normal atmospheric pressure.

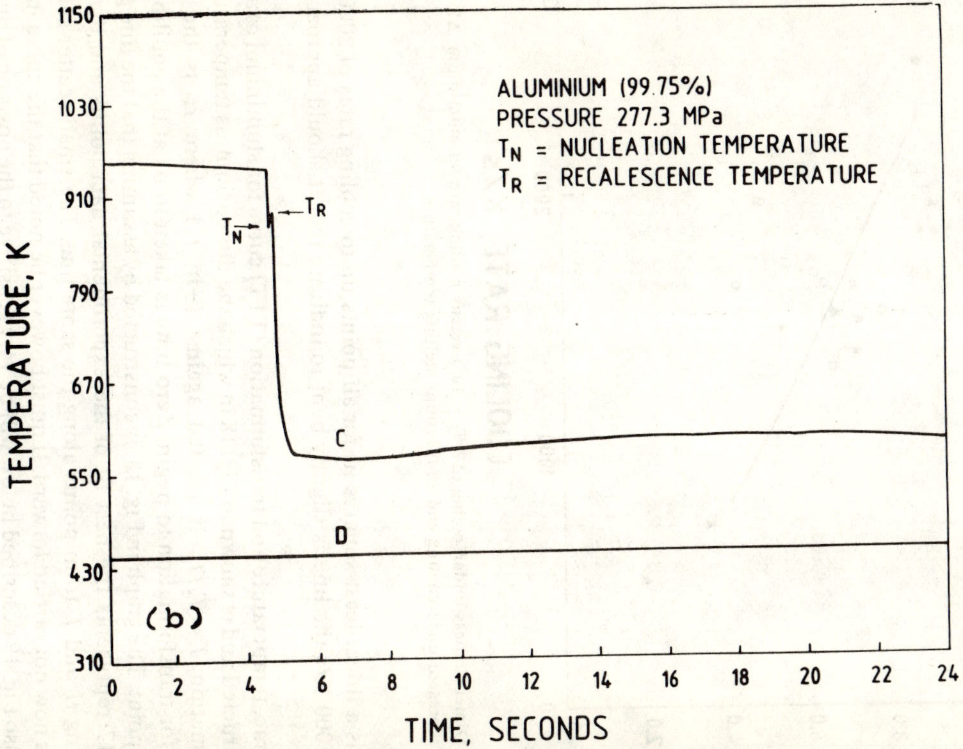

Fig. 16. (b) Cooling curve of aluminium solidified under 277.3 MPa pressure.

114

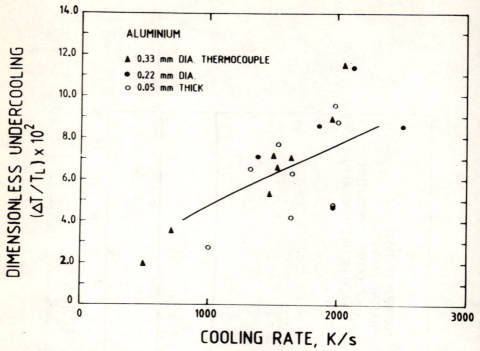

Fig. 17. Dimensionless undercooling ($\Delta T/T_l \times 10^2$) versus cooling rate for aluminium. ΔT and T_l represent undercooling and equilibrium melting temperature.

Fig. 17 is a linear-least-squares fit for all points up to cooling rates of 2000 K/s. Below 1000 K/s the line is slightly bent to indicate that it should approach the origin.

A time, temperature and transformation (TTT) curve for aluminium can now be constructed and is shown in Fig. 18 in which the dimensionless temperature of transformation $T_r = T_N/T_M$ is plotted against $(-\ln t_N)$ where t_N is the time elapsed for transformation to begin. Zero time is taken to be at the equilibrium melting point. The graph in Fig. 18. is constructed by assuming that the line given in Fig. 17 represents the best fit of the experimental data points and then by calculating t_N and T_r from points along the same line. The small scatter seen in Fig. 17 is now not carried forward. It must however be noted that the curve in Fig. 18 and also Fig. 19 described below will be very sensitive to the position of the line in Fig. 17. As described above for the purpose of discussion in this paper we have chosen a least squares fit for the data points in Fig. 17.

The form of the TTT curve thus obtained is similar to that postulated by Turnbull [31] and Boettinger and Perepezko [32]. The dynamic TTT curve was calculated by them by integrating the expression for the steady state nucleation

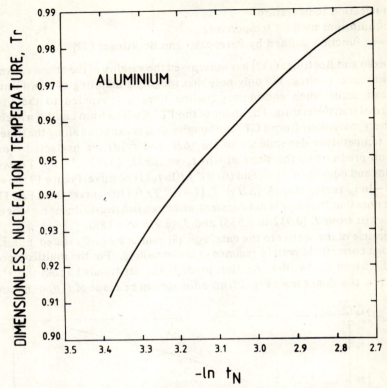

Fig. 18. $(-\ln t_N)$ versus dimensionless nucleation temperature, Tr, for aluminium. t_N and T_r represent the time for the first nucleus to form and dimensionless nucleation temperature respectively.

rate (J) and working out the condition when the first nucleus forms. This led to an expression of type

$$-\ln t_N = -\ln (a\,(\Omega a)\,Z(T_N) + 16\pi r^3\,f(\phi)/3K\,((\Delta H_v)^2\,T_m\,(1-T_r)^2\,Tr \qquad (8)$$

where

t_N = time for the first nucleus to form
a = heterogeneous catalytic surface area
Ωa = pre-exponential factor in the steady state nucleation rate expression. This term contains the jump frequency term of atoms from liquid to solid as well as the number of atoms in contact with the liquid.
r = solid-liquid surface energy.
$f(\phi) = (2-3\,\mathrm{Cos}\,\phi + \mathrm{Cos}^3\,\phi)/4$; where, ϕ is the solid-liquid contact angle
K = Boltzman constant
ΔH_v = heat of fusion
T_r = dimensionless nucleation temperature = T_N/T_M

116

T_N = nucleation temperature
T_M = equilibrium melting temperature
$Z(T_N)$ = a function defined by Perepezko and Boettinger [32]

Perepezko and Boettinger [32] have discussed the validity of the above expression for continuous cooling. We only note that roughly a doubling of transformation time can occur when continuous cooling time is compared to the time of isothermal transformation. The shape of the TTT derived from eqn. [8] will change extremely marginally from a CTT, although a shift is expected along the time axis. If the temperature dependence of Ωa, ΔH_v and $f\,(\theta)$ are neglected then the equation predicts: (i) the slope of $-\ln t_N$ vs $16\pi 3K\,(\Delta H_v)^2\,(1-T_r)^2\,Tr$, will be constant and equal to $v^3 f(\theta)$, and (ii) $d\,T_r/d(\ln t_N)$ is positive. Figure 19 shows the plot of $-\ln t_N$ versus $16\pi/3K\,(\Delta Hv)^2\,T_m\,(1-T_r)^2\,Tr$ for the curves from Fig. 17 and 18. We note that the slope is not constant, although two nearly linear regions of the slopes exist from T_r (0.912 to 0.958) and T_r (0.978 to 0.988).

Because of the scatter in the data, eqn. [8] cannot be easily tested, but is seen to predict correctly in certain regimes of undercooling. (For the multicomponent alloy discussed below, the equation predicts correct trends.) From the $r^3\,f(\theta)$ values (i.e. the slopes from Fig. 19) an estimate can be made of $f(\theta)$ or θ (contact

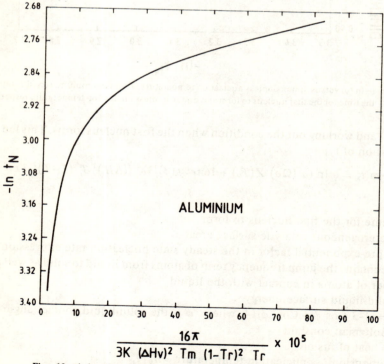

Fig. 19. $(-\ln t_N)$ versus $16\pi/3K\,(\Delta H_v)^2\,Tm\,(1-Tr)^2$ for aluminium.

Table 11. Contact angle: Aluminium

Region	Tr	Contact Angle, Ø
I	0.908–0.9575	10.56–17.4
II	0.978–0.988	5.52

angle) for aluminium being nucleated on the chromel-alumel thermocouple. Table 11 gives the value of θ at different undercoolings. It is noted that the published values of θ [33] for heterogeneous aluminium nucleation on Ni$_3$Al are nearly the same ($\sim 18°$) as the angle for the higher undercooling regime. Material constants for aluminium are taken from Kurz and Fisher [34] and Mondolfo [35].

A more appropriate method of calculating dT_r/d (ln t_N) is by calculating N (number of nuclei formed per second) and has been given by Reiss and Katz [36]

$$N = (1/\xi) \int_{T_o}^{T_N} J_N \, dT \qquad (9)$$

(ξ, is the imposed cooling rate). Reiss and Katz [36] show that transient nucleation for metals is only a very small contributor to t_N. (A magnitude of these transient terms and the time to form nuclei in sluggish systems has been discussed by Kelton and Greer [37].) However using the steady state nucleation rate term (J) is still a drawback in solving eqn. [9] and there is no clear evidence to suggest that it is adequate. Although appearing simple, eqn. [9] is cumbersome to solve as pointed out by Reiss and Katz and a fully generalized solution is not immediately available to predict the TTT curve. We have therefore tested the experimental data only with equation [8].

Cooling curves were also recorded for a multi-component liquid aluminium alloy (Al-5.6% Zn-2.6% Mg-1.5% Cu) while solidifying under a pressure of 277.3 MPa. A typical curve is shown in Fig. 20. The nucleation temperature is indicated. (please note that this is not at the bottom of the blip but where the cooling curve deviates from linearity.) The undercooling parameter ($\Delta T/T_L$) versus the cooling rate is shown in Fig. 21. The curve is nearly similar to the one obtained for pure aluminium. The significant point to be observed with this alloy is that the dimensionless undercooling parameter ranges from 0.1262 to 0.3612, whereas the maximum value was only 0.115 for pure aluminium.

The microstructural changes with increasing undercooling were investigated and are given below. The TTT curve was determined from the nucleation temperature and cooling rate data and is given in Fig. 22. The $-\ln t_N$ versus $16\pi/3K$ $(\Delta H_v)^2$ T_M $(1-T_r)^2$ Tr curve is given in Fig. 23. From the slope of this curve the contact angle θ was determined and is given in Table 12. The thermophysical values assumed are for pure aluminium [34, 35]. The θ value is 32.4° when the curve is considered as straight line parallel to the central nearby

118

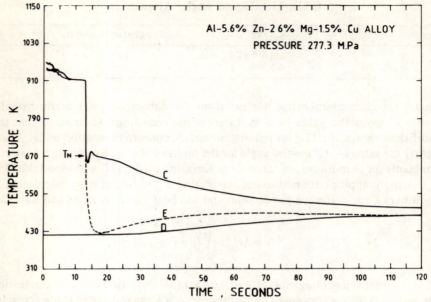

Fig. 20. Cooling curve of Al-5.6% Zn-2.6% Mg-1.5% Cu alloy solidified under 277.3 MPa pressure.

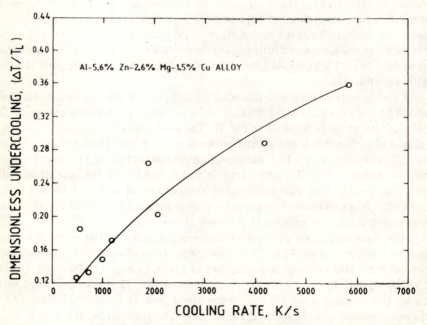

Fig. 21. Dimensionless undercooling ($\Delta T/T_l$) versus cooling rate for Al-5.6% Zn-2.6% Mg-1.5% Cu alloys.

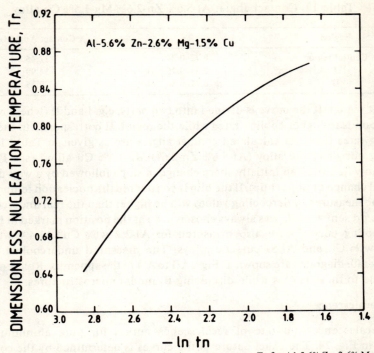

Fig. 22. $(-\ln t_N)$ versus dimensionless nucleation temperature, Tr, for Al-5.6% Zn-2.6% Mg-1.5% Cu alloy.

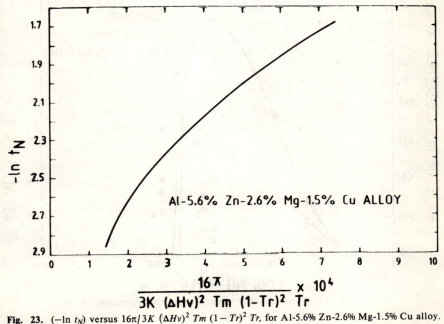

Fig. 23. $(-\ln t_N)$ versus $16\pi/3K \ (\Delta Hv)^2 \ Tm \ (1 - Tr)^2 \ Tr$, for Al-5.6% Zn-2.6% Mg-1.5% Cu alloy.

Table 12. Contact angle: Al-5.6% Zn-2.6% Mg-1.5% Cu alloy

Region	Tr	Contact Angle Ø
Central region	0.72–0.84	32.45
I	0.64–0.75	34.99
II	0.81–0.86	29.35

straight region. If the curve is divided into two parts, e.g. I and II (where region I corresponds to higher cooling rates while the region II corresponds to the lower cooling rates then the calculated contact angles are as given in Table 12. The cooling curves of the alloy (Al-5.6% Zn-2.6% Mg-1.5% Cu alloy) (see Fig. 20) invariably indicated an initially sharp change in slope followed by a well-defined gradual change in slope (blip). If the blip is considered the nucleation temperature then the measured undercooling values will be higher than the real undercooling. For the present work T_N was always chosen to be at the position marked in Fig. 20.

Cooling curves were also measured for Al-2.2 wt% Cu, Al-4.4 wt% Cu, Al-33 wt% Cu and Al-Si eutectic alloys. The measured undercooling versus cooling rate diagrams are shown as Figs. A 1 to A 4 in the appendix. Reference will be made to these curves while discussing Bi modal microstructures.

6.2. Recalescence

The recalescence and rate of recalescence curves for pure aluminium are shown in Fig. 24. The exact nature of the curves is determined by the complex

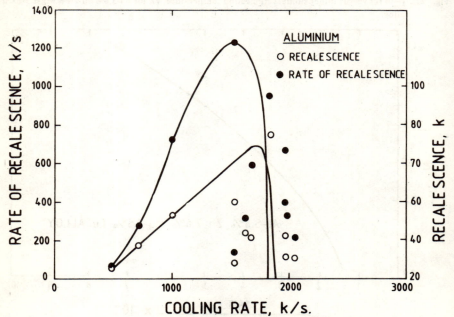

Fig. 24. Recalescence versus cooling rate for aluminum.

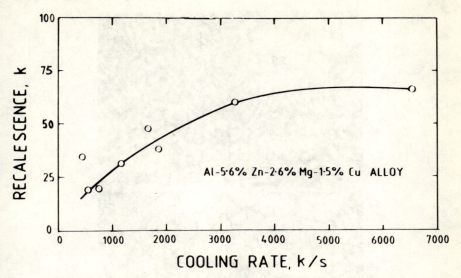

Fig. 25. Recalescence versus cooling rate for Al-5.6%Zn-2.6%Mg-1.5%Cu alloy.

heat flow (including fluid motion near the thermocouple tip). As expected, both the recalescence and rate of recalescence initially increase with cooling rate. The curves then indicate that as external cooling rate increases the amount and rate of heat extracted by the external boundaries can offset the heat generation by rapid growth from an undercooled nucleus. This is a surprising result since the rate of growth of the undercooled nucleus is expected to release heat much faster than the heat removal rate by the surrounding liquid [13, 14]. At the present time we attribute this phenomenon to the detachment of the growing nucleus from the thermocouple tip on account of cavitation bursts causing rapid fluid flow to provide the momentum for the detachment. The recalescence measurement for the Al-5.6%Zn-2.6%Mg-1.5%Cu alloy is shown in Fig. 25. The amount of recalescence noted for this case is seen to reach a steady state in the range of cooling rates recorded in our experiments.

6.3 Microstructure

6.3.1. Bi-modal dendritic structure

All the alloy systems solidified under pressure exhibited two distinct regions of fine and coarse dendritic structure (Bi-modal dendritic structure) as shown in the microstructure in Figs. 26 a, b, c, (taken from different regions of the same casting) for an Al-5.6% Zn-2.6% Mg-1.5% Cu alloy solidified under a pressure of 277.3 MPa.

Kattamis and Flemings [25] and Kattamis [38] have indicated that a substantial quantity of liquid always remains after recalescence is completed.

122

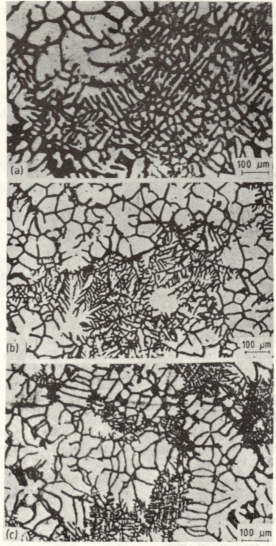

Fig. 26. a, b and c; Bi-modal dendritic structure of Al-5.6%Zn-2.6%Mg-1.5%Cu alloy solidified under 277.3 MPa pressure.

Dendrite coarsening occurs during this period, and the dendrite arm spacing at each point within the ingot depends on the time required for complete solidification at this point. They also point out that in cases where the bulk quantity of liquid metal is involved, the final ingot spacing appears to be not greatly different from that of undercooled melts. The dendrite arm spacing (DAS)

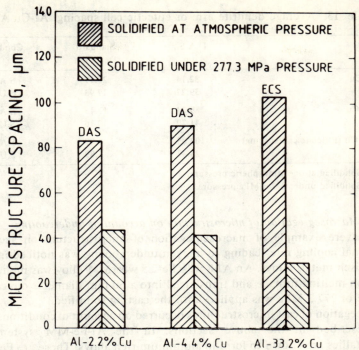

Fig. 27. Dendrite arm spacing (DAS) of Al-2.2%Cu and Al-4.4%Cu alloys solidified under both the conditions of normal atmospheric pressure and 277.3 MPa pressure. Eutectic cell spacing (ECS) of Al-33.3%Cu alloy is also shown for both conditions.

for Al-2.2 wt% Cu, Al-4.4 wt% Cu and Al-33.2 wt% Cu alloy are shown in Fig. 27. More than one hundred measurements for each of these three alloys were made from optical micrographs. DAS values for these alloys when solidified in the same die assembly with punch on top of liquid metal surface but at usual atmospheric pressure are also shown in Fig. 27. In the case of Al-Cu eutectic alloy (with silicon as impurity) the cell spacing (containing the eutectic) was measured. A significant point of observation is the spread in the microstructure spacing as reflected in the standard deviation given in Table 13. This deviation remains roughly the same with increased cooling rate. As the mean values of spacings lie within the same order of magnitude, a similar spread reflects much higher undercooling (with cooling rate) at the lower values of spacing. Such a behaviour is reflected in Fig. A 1 and A 2. In the case of both Al-2.2% Cu and Al-4.4% Cu alloys the DAS values varied from as low as 13 μm for pressure solidified castings. Such a small value of DAS is indicative of the substantial local undercooling and high growth rate during recalescence. Significant deviation from the $t^{1/3}$ law was observed for microstructure of castings solidified under pressure.

124

Table 13. Average dendrite arm or eutectic cell spacing: Al-Cu Alloys

Alloy	D.A.S., um	Standard Deviation	Condition
Al-2.2% Cu	82.14	25.17	S.A.P.
	39.27	17.93	S.U.P.
Al-4.4% Cu	89.24	13.15	S.A.P.
	41.79	10.16	S.U.P.
Al-33.2% Cu (Eutectic cell spacing)	102.05	6.09	S.A.P.
	28.42	5.89	S.U.P.

S.A.P.: Solidified at one atmospheric pressure.
S.U.P.: Solidified under 277.3 MPa pressure.

6.3.2. Macrosegregation of microstructure on account of undercooling

A severe example of macrosegregation of microstructure on account of differential cooling rate leading to uneven undercooling was noticed in Al-Fe-MM (misch metal) alloys. An Al-8 wt% Fe-2.5 wt% MM alloy was melted in an induction melting furnace and transferred into a 60 mm diameter die cavity. A pressure of 277.3 MPa was applied until the casting solidified.

Segregation of cast microstructures occurred on account of conditions of heat flow described in the above sections. In the Al-Fe-MM system three intermetallics are known to form in aluminium rich alloys. These are FeAl and FeAl$_6$ and Al$_8$CeFe$_4$ [35]. FeAl is monoclinic and is the more stable phase while the FeAl$_6$ is orthorhombic and is found only when the alloy is rapidly cooled from the liquid state. Al$_8$CeFe$_4$ is tetragonal and found in the inter-dendritic regions. In the present work FeAl occurred in various morphological forms when solidified under pressure of 277.3 MPa; while the FeAl was seen only as flakes when solidified at usual atmospheric pressure (0.1 MPa).

The various forms of FeAl$_3$ which occurred during the solidification under pressure could be classified into six types. The rod-like, ten-point prism type, ribbon-shaped plates with branching, ten-point stars with prism base, plate-like and blocky star-like are illustrated respectively in Figs. 28a, b, c, d, e and f. Phase identification to verify composition was carried out using an electron probe micro analyser. The six morphological forms were noted for all compositions ranging from 4-10 wt% Fe and 2.5-4.5 wt% MM. Although several reports (39, 40) have shown a maximum of two forms of FeAl$_3$ in any specific experiment, the present work shows six different forms of FeAl$_3$ in the same casting when the alloys were solidified under pressure.

6.3.3. Undercooling versus cooling rate

The cooling rates versus undercooling for several aluminium alloys are shown in Figs. A1-A4.

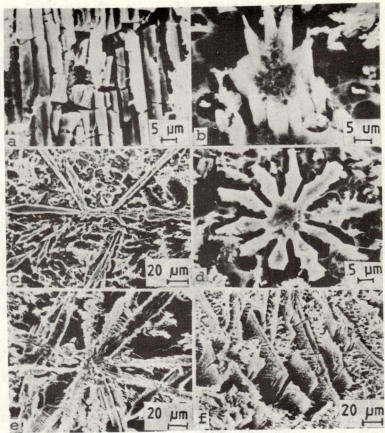

Fig. 28. Scanning electron micrographs (SEM) of Al-Fe-Misch metal alloy solidified under 277.3 MPa pressure. The various forms of FeAl₃ intermetallic phase are shown. (a) the rod like, (b) ten-point prism type, (c) ribbon shaped platelets with branching, (d) ten-point stars with prism base, (e) blocky-star, and (f) plate-like.

6.4. Transformation: Al-14%Mn alloy

Experiments were performed on an Al-14 at% Mn alloy [50] to study the formation of phases at moderate cooling rates. The experimental procedure involved application of 200 MPa pressure to a liquid contained in a die of 30 mm diameter. Figure 29 shows the cooling curves recorded by a chromal-alumel thermocouple placed at the centre of the die cavity for two different experiments. Two types of experiments were performed; (i) where pressure was applied at or below the liquidus, Figure 29(a) corresponding to experiment A (i.e. quenched from near the liquidus) and (ii) where pressure was applied in liquid state, Figure 29(b), corresponding to experiment B (i.e. quenched from the liquid state). The

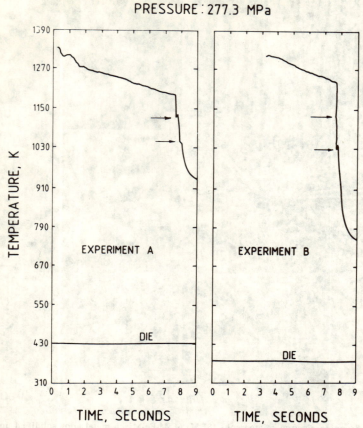

Fig. 29. Cooling curves for two different low pressure solidification experiments on an Al-14at. % Mn alloy.

liquidus temperature of the alloy is 1191 K. Four experiments were performed and labelled A, B, C, D, i.e., where pressurization was commenced when the liquid temperature was at 1192 K, 1236 K, 1573 K and 1111 K respectively. The measured cooling rates in the liquid prior to the first arrest were 3540 K/s, 1787 K/s, 184 K/s, 700 K/s respectively for experiments A to D.

6.4.1. *Transformation on cooling*

In all instances on slicing the cylindrical sample a casing of Al_6Mn was observed as shown schematically in Fig. 30. An inner core of diameter 16 mm was also observed. The Al_6Mn nucleates at the wall of the die and grows inwards as cooling proceeds. However the growth of the phase is slow. Figure 31(a) is a SEM photograph showing the Al_6Mn platelets in the casing part of the cast cylinder.

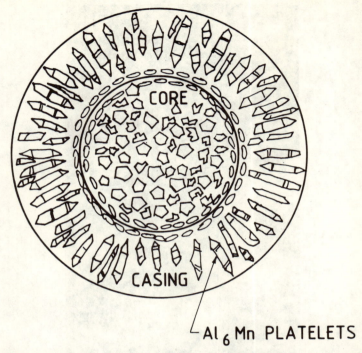

CORE

CASING

└ Al₆ Mn PLATELETS

Fig. 30. Schematic of a typical microstructure observed on slicing the cylindrical sample obtained from low pressure solidification of Al-14at. % Mn alloys. The Al_6Mn platelets grow as long as a strong positive gradient exists. The core undercools to yield I and T phase while suppressing Al_4Mn formation with increasing cooling rate.

Figures 31(b) and (c) show SEM photographs taken from the central part of the samples of experiments A and D respectively. Figure 32 is a plot of the X-ray diffraction pattern taken from the central regions of experiments A to D. Also shown are the diffraction patterns of pure Al, Al_4Mn, T phase, icosahedral phases and the Al_6Mn phase in a similar fashion to that given by Schaefer et al. [41]. The T phase is a decagonal phase identified by Chattopadhya et al. [42] and Bendersky et al. [43] and which also shows streaking in the reciprocal space of the transmission electron microscope (TEM). This phase forms sometimes in preference to and/or epitaxically on the icosahedral phase [41]. Icosahedral phases were observed in experiments A, B and D, with a larger amount present in experiment A as compared to experiment B.

This is related to the higher cooling rate in experiment A (3540K/s). The cooling curves are interpreted as follows:

A higher cooling rate (3540K/s) (experiment A) leads to the icosahedral (I) phase forming at 1121 K and growing up to 1049 K, where the T phase nucleates and grows. The T phase nucleates when the average cooling rate in the liquid drops to 411 K/s. The formation temperature of the icosahedral phase is identical

128

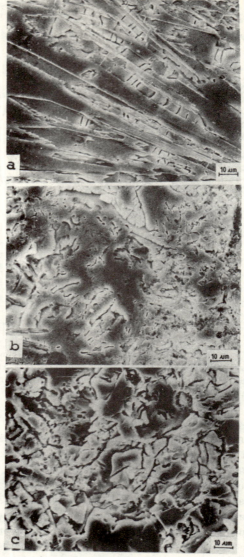

Fig. 31. (a) SEM photograph showing Al$_6$Mn platelets in the casing part of the cast cylinder.
(b) and (c) SEM photographs taken from the central part of the samples of experiments A and D respectively.

to that calculated by Schaejar et al. [41]. At a lower cooling rate (1787 K/s) (experiment B) although a slight disturbance in the cooling curve is noticed at exactly 1121 K the melt actually cools to 1028 K where nucleation of the T phase occurs. In Sample D the first arrest is noticed at 1027K. X-ray evidence indicates

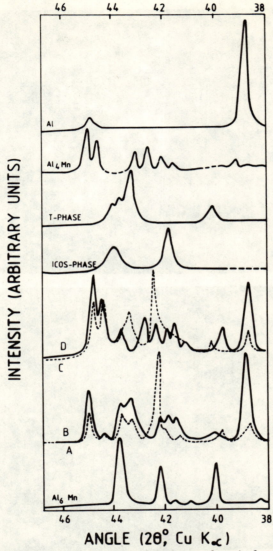

Fig. 32. A plot of the diffraction patterns obtained from the core of samples from experiments A to D. Also shown are the standard diffraction patterns from the I, T, Al_6Mn and pure aluminium. This composite figure and the Al_4Mn lines follows a style first given in reference [41].

the presence of Al_4Mn in experiment B whereas the Al_4Mn lines are absent in experiment A (where a greater amount of the icosahedral phase was indicated). A streaked electron diffraction pattern of the T phase was observed in experiment C. X-ray data from the centre of experiment D indicated that in addition to the Al_4Mn, T and Al phases, the presence of the icosahedral phase I [44, 29] and the

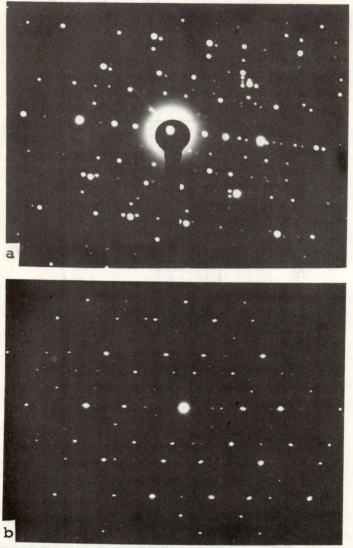

Fig. 33. (a) TEM selected area diffraction pattern showing the icosahedral symmetry from experiment D. The 5 fold symmetry pattern is seen amongst spots from adjacent T and equilibrium phases.
(b) TEM photograph showing the streaked pattern of the T phase from experiment A.

$(Al_{10}Mn_3)$ [45, 46] phase. TEM investigations indicated that the icosahedral phase was finely distributed between many other phases. From the diffraction pattern shown in Fig. 33(a), the five-fold character of the icosahedral phase could be identified even in sample D, although the presence of the other phases somewhat masked the diffraction patterns. Fig. 33 (b) shows a streaked pattern of the T phase from sample A. In this sample T phase overgrowth on the I phase made it difficult to locate the I phase in the transmission electron microscope.

The formation of the T and icosahedral phases by this technique reflects that the liquid could undercool to the liquidus of these metastable phases primarily because Al_4Mn could not form while the liquid was cooling. The microstructure therefore shows the casing of Al_6Mn which grew radially inward in the presence of a strong initial radial temperature gradient and a central region of supercooled liquid in which the T and icosahedral phases formed. The icosahedral and the T phases have a competetive existence [41]. The formation of icosahedral phase in experiment D (i.e. at low cooling rate) is on account of the liquid composition shifting to a lower manganese contents because of the formation of $(Al_{10}Mn_3)$. A lower manganese content in the liquid has been found to be unfavourable for the T phase formation [41, 47].

From these measurements a TTT diagram can be constructed for the phases and is shown as Fig. 34. Implicit in the diagram is that although the T phase is known to grow epitaxically on the icosahedral phase, our results indicate that the T phase may independently nucleate (unconnected grains of the I and T phase have been previously shown earlier [48]. Even while growing on the I phase, epitaxy of the T phase may begin with a nucleation event. The TTT diagram indicates that the liquidus temperature of the icosahedral phase is a weak function

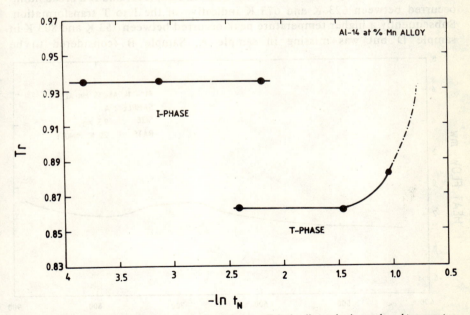

Fig. 34. The measured TTT curves for the I and T phase. Tr is the dimensionless reduced temperature given by ratio of the arrest temperature divided by the liquidus temperature (1191K). t_N is the time for the arrest with zero time taken as when the melt reaches the liquidus temperature.

132

of the cooling rate in the range of measurement. The T phase is progressively easier to nucleate at low undercooling (along the T phase TTT) obtained at a low cooling rate. At low cooling rates the icosahedral phase is unable to grow effectively and the T phase dominates unless the composition of the liquidus has shifted to lower manganese contents from which there is some difficulty for the T phase to form. The dominance of the T phase over the I phase at low average cooling rates during solidification has also been recently reported [49].

6.4.2. *Transformation on heating*

Transformation on behaviour of the I and T phases has been reported [47, 48]. The transformation peaks in a DSC seem to depend on the initial state of the sample. For the composition $Al_{86}Mn_{14}$ if no T phase is present and if only the I phase is present then a transformation peak is seen at 743 K. If the T phase is present then a broad hump is seen between 623 K and 723 K indicating a transformation of the I to the T phase, and the equilibrium orthorhombic Al_6Mn forms with a higher energy release (12.9cal/g). The I to T transformation gives out (3.6 cal/g) (47). It is not clear whether the T phase formation between 623 K and 723 K can also occur with the simultaneous transformation to the equilibrium phase.

Figure 35 shows the DSC transformation curve for sample A. DSC measurements on samples A–D indicated that in samples A and D a broad hump occurred between 623 K and 673 K indicative of the I to T transformation. Subsequently a higher temperature peak occurred between 753 K and 803 K in sample D but was missing in sample A. Sample B (considered to be

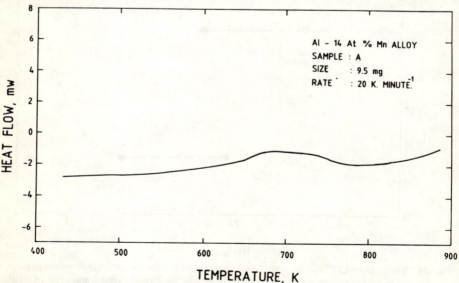

Fig. 35. The DSC curve obtained on reheating the central part of the casting corresponding to sample A.

predominantly T phase) shows a broad peak between 633 K and 753 K. Sample C (mostly considered to contain the equilibrium phases along with some T phase) shows a very shallow and broad peak between 753 and 853 K. We note from the above results that the low pressure solidification technique can be tailored to modify the cooling rates such that the initial state of the sample can be varied considerably.

7. CONCLUSION

In this paper we have discussed the various ways in which pressure affects solidification. Pressure solidification techniques have been characterized into medium and high ranges. Depending on the pressure level and the time scales in which pressure can be applied, several different solidified phases can be obtained and have been discussed. A broad spectrum of experiments have been presented which highlight the versatility of pressure induced solidification.

ACKNOWLEDGEMENTS

The authors thank Dr. P. Rama Rao, Director, DMRL, for his encouragement and permission to publish this paper. Discussions with Professor P. Ramchandra Rao and Professor R. Trivedi are gratefully acknowledged.

REFERENCES

1. Sekhar, J.A., M. Mohan, C. Divakar and A.K. Singh. *Scripta. Metall.* **18**, 1327 (1984).
2. Sekhar, J.A., G.S. Reddy and P.V. Rao. *Proceedings of the 1st ASM conference on Rapidly Solidifed Materials.* P.W. Lee and R.S. Carbonara (eds.), p. 83–90 (1986).
3. Reddy, G.S., P.V. Rao and J.A. Sekhar, *Int. J. Rapid Solidification* (1986).
4. Sekhar, J.A., *Scripta Met.*, **19** (1985).
5. Sekhar, J.A. and T. Rajasekharan. *Nature*, Vol. 320, p. 153, March 13 (1986).
6. Plyatskii, V.M. *Extrusion Casting*, Primary Sources, New York (1965).
7. Sekhar, J.A., G.J. Abbaschian and R. Mehrabian. *Mater. Sci. and Engineering,* **40**, 105 (1979).
8. Chatterjee, S. and A.A. Das. *Brit. Found.* **69**, 274 (1972).
9. Nishida, Y. and H. Matsubara, *Brit. Found.* **69**, 274 (1976).
10. Jayaraman, A., W. Klement Jr., R.C. Newton and G.C. Kennedy. *J. Phy. Chem. Solids*, **24**, 7 (1963).
11. International Critical Tables of Numerical Data, Physics, Chemistry & Technology, E.W. Washburn (ed.-in-chief) **1**, 102 (1982).
12. Smithells, C.J., *Metals Reference Book,* Butterworths, Washington, 3rd Edition, Volume 2, 618 (1962).
13. Sekhar, J.A. and T. Chande. *Transactions Ind. Inst. of Metals*, **37**, 67 (1984).
14. Levi, C.G. and R. Mehrabian. *Metall. Trans.* **13A**, 221 (1982).
15. Rao, P.V. and J.A. Sekhar. *Mat. Letters*, 3, 216–218 (1985).
16. Cullity, B.D. *Elements of X-ray diffraction.* Adison Wesley Readings, 334 (1956).
17. Tyzac, C and G.V. Raynor. *Acta. Cryst.* 7, 505–510 (1954).
18. Shiohara, Y, M.G. Chu, D.G. Macisaac and M.C. Flemings. In: *Rapidly solidified amorphous and crystalline alloys,* Materials Research Society Symposia, Proceedings, B.H. Kear, B.C. Giessen and M. Cohen (eds.), North Holland, Amsterdam, 65 (1982).
19. Hansen, M. and K. Anderko. *Constitution of Binary Alloys,* McGraw Hill, New York, pp. 927–928 (1958).

134

20. Elliot, R.P. *Constitution of Binary Alloys,* First suplement, McGraw Hill, New York, pp. 602–603 (1965).
21. Calka, A., M. Madhava, D.E. Polk and B.C. Giessen. *Scripta Met,* 11, 65–70 (1977).
22. Clark, J.B. and F.N. Rhines. *Trans. AIME,* 209, 425–430 (1957).
23. Frank, H, V. Herold, U. Koster and M. Rosenberg, Rapidly Quenched Metals II, B. Cantor (ed.), 1, The Metals Society, 155–162 (1978).
24. Koster, U. and U. Herold. *Topics in Applied Physics, Glassy Metals,* 1, H.J. Gunthrhodt and H. Beck (eds.), Springer-Verlag, Berlin Heidelharg, New York, 46, 225 (1981).
25. Kattamis, T.Z. and M.C. Flemings. *Trans. Met. Soc. of AIME,* 236, 1523–1531 (1966).
26. Walker, J.L. *Physical Chemistry of Process Metallurgy,* Interscience, New York, 845 (1961).
27. Horvay, G. *Proc. 4th National Congress of Appl. Mech. ASME,* 1315–1325 (1962).
28. Bursil, L.A. and Peng, Ju Lin. *Nature* 316, 50 (1985).
29. Bancel, P.A. et al. *Phy. Rev. Letts.* 54, 2422 (1985).
30. Taylor, M.A. *Acta Met.* 8, 256 (1960).
31. Turnbull. D. *Contemp. Phys.* 10 (1969) 473.
32. Perepezko, J.H. and W.J. Boettinger. *Mater. Res. Soc. Symp. Proc.* 19, 223 (1983).
33. Cantor, B. and R.D. Doherty. *Acta Met.* 26, 33 (1979) data from R.T. Southin and G.A. Chadwick, *Acta Met,* 26, 223 (1978).
34. Kurz, W. and D.J. Fisher. *Fundamentals of Solidification,* Trans. Tech. Publications, Aedermannsdorf Switzerland, p. 23 (1984).
35. Mondolfo, L.F. *Aluminium Alloy-Structure and Properties,* Butterworths, London (1976).
36. Reiss, H. and J.L. Katz. *Rapid Solidification Processing, Principles and Technology,* R. Mehrbian, B.H. Keal and M. Cohen (eds.), Claitors Publishing Division, Bapton Rouge (1978).
37. Kelton, K.F. and A.L. Greer. *J. of Non-Crystalline Solids,* 79, 295–309 (1986).
38. Kattamis, T.Z. *Trans. Met. Soc. of AIME,* 251, 1401, (1980).
39. Biloni, H. and G.F. Bolling. *Trans. Met Soc. AIME,* 227, 1351 (1963).
40. Louis, E., R. Mora and J. Pastor. *Metal Science,* 591–593 (1980).
41. Schaefer, R.J., L.A. Bendersky, D. Schechtman, W.J. Boettinger and F.S. Biancaniell. *Met. Trans* (1986).
42. Chattopadhya, K., S. Lele, S. Ranganathan, G.N. Subbanna and N. Thangaraj. *Current Science,* 54, 895 (1985).
43. Bendersky, L., R.J. Schaefer, F.S. Biancaniello, W.J. Boettinger, M.J. Kauman and D. Shechtman. *Scripta Met.* 19, 969 (1985).
44. Schechtman, D, I. Blech, D. Gratias and J.W. Cahn. *Physical Review Letters,* 53, 1951 (1985).
45. Pearson, W.B. (ed.) *Handbook of Lattice Spacings and Structures of Metals and Alloys,* Pergamon Press, New York, 2, 574 (1967).
46. Moffatt, W.G. *Binaly Phase Diagrams Handbook,* (Al-Mn), Gen. Electrical Company, USA (1977).
47. Kimura, K.T. Hashimoto, K. Suzuki, K. Nagayama, H. Ino and Shin Takenchi. Technical Report of ISSP, published by the Institute of Solid State Physics, University of Tokyo, Japan (1985).
48. Sekhar, J.A. *Current Science,* 54, 904 (1985).
49. Schaefer, R.J. and L. Bendersky. *Scripta Met.* 745 (1986).
50. Reddy, G.S., J.A. Sekhar and P.V. Rao. *Scripta Met.,* 21, 13–18 (1987).

APPENDIX

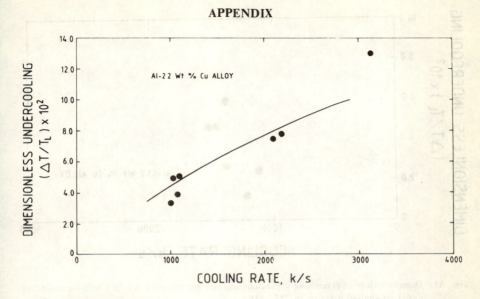

Fig. A1. Dimensionless undercooling versus cooling rate for the alloy. Al–2.2 wt% Cu solidified under an applied pressure of 277.3 MPa.

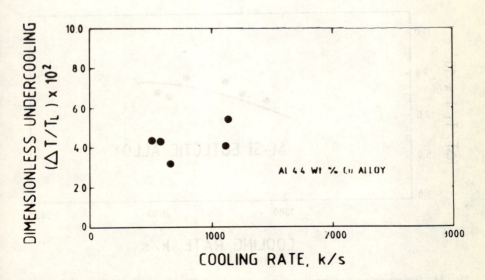

Fig. A2. Dimensionless undercooling versus cooling rate for the alloy Al-4.4 wt% Cu solidified under an applied pressure off 277.3 MPa.

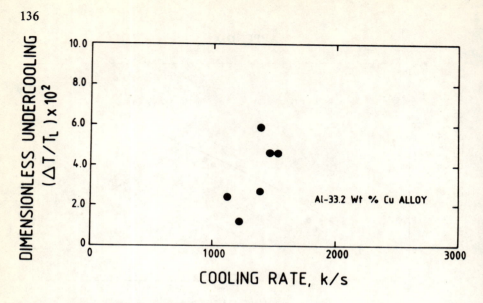

Fig. A3. Dimensionless undercooling versus cooling rate for the alloy A1-33.2 wt% Cu solidified under an applied pressure of 277.3 MPa.

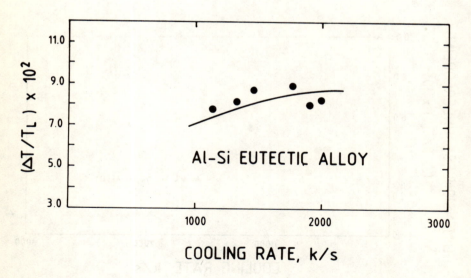

Fig. A4. Dimensionless undercooling versus cooling rate for the alloy Al-11.3 wt% Si (eutectic) solidified under an applied pressure of 277.3 MPa.

NOMENCLATURE

A = Area
C_p = Specific heat
h = Heat transfer coefficient
H = Heat of fusion
K = Thermal conductivity
K_m = Kinetic parameter relating interface velocity and undercooling in the linear law
P = Pressure
P_t = Pressure required for complete solidification
R = Radius of sphere
r_* = Radial direction
r = Dimensionless radius = r/R
t = Time
T_* = Temperature
T = Dimensionless temperature T/T_f
ΔT_u = Net undercooling
T_{mo} = Melting point at atmospheric pressure
T_f = Ambient temperature
β = Coefficient of volume expansion
l = Density
F_o = Dimensionless time = $Kt/C_p R^2$
ϵ = Epsilon number = $HK_m R/K$
Ai = Adiabatic number = $(R^2 \beta/K)\, dP/dt$
Bi = Biot number = hR/K
f = Dimensionless solid fraction formed
Stn = Dimensionless undercooling = $Cp\,(T_m - T_o)/H$
St = Stefan number = $Cp\,(T_{mo} - T_f)/H$
1 = Subscript indicating liquid
s = Subscript indicating solid
ΔV = Volume contraction on solidification
V = Volume

6

Nonequilibrium Processing with Lasers

J. MAZUMDAR AND A. KAR

Laser Aided Materials Processing Laboratory, Department of Mechanical and Industrial Engineering, University of Illinois at Urbana-Champaign, 1206 West Green Street, Urbana, IL 61801, U.S.A.

ABSTRACT

Inherent rapid cooling in laser processing often produces novel materials due to nonequilibrium partitioning during solidification. Such nonequilibrium processing has drawn a lot of interest in both theoretical and experimental areas of materials science. This paper reports theoretical and experimental investigations on *in situ* alloy formation with extended solid solution.

A mathematical model for determining the composition of metastable alloys and the effects of various process parameters on the composition of extended solid solution is presented. Initially, dilute solution theory was applied for the nonequilibrium partition coefficient. To model the nonequilibrium partitioning of solute in concentrated solution, rate equations and conservation of mass equations are used and an expression for nonequilibrium partition coefficients is derived for concentrated solution, which is more realistic for laser alloying and cladding. Effect of rapid cooling on solute segregation is presented by deriving an expression for nonequilibrium partition coefficient.

INTRODUCTION

Laser technology has drawn considerable interest in its use for materials processing and manufacturing. It not only makes manufacturing processes simpler and more economical but also provides a unique way of modifying the surface chemistry of materials. Inherent rapid heating and cooling rate in laser alloying and cladding processes can be utilized to obtain novel materials without being restricted by equilibrium phase diagram. High cooling rate results in extended solid solution to produce metastable alloys. Recently, the laser cladding technique has been applied to increase the solid solubility of reactive elements such as Hf in nickel superalloy [1] for improved high temperature properties. The mechanism of extended solubility encompasses the energy, momentum, and mass transport of solute atoms. The heating and cooling rates are determined by the heat transfer equation while the extent of mixing and the redistribution of solute atoms in the molten cladding pool are obtained from the momentum and mass transport equation. This paper presents a summary of our theoretical studies on nonequilibrium solidification and extended solid solubility during laser cladding. We will discuss infinite dimensional (the substrate and the cladding are semi-infinite) and finite dimensional (the substrate and the cladding are finite) mathematical models for such nonequilibrium phenomena. In the infinite dimensional model, the freezing temperature of the cladding melt has been considered independent of the concentration of solute. This assumption has been relaxed in the finite dimensional model and consequently nonequilibrium phase diagram of a given system can be obtained from its equilibrium phase diagram by using this model.

There is very little information in the available literature on mathematical modeling of laser surface cladding and alloying with rapid cooling rate. As mentioned earlier, the composition of a rapidly cooled metastable alloy depends greatly on the heat transfer rate during the solidification process. Three-dimensional heat transfer models for various material processing with CW laser were presented by several authors [2–5]. It was found experimentally that surface tension arising due to very high temperature gradient in laser melted pools causes convection. This aspect was modeled by Chan, et al. [6]. They studied the effects of surface tension on the cooling rate, surface velocity, surface temperature, and the pool shape. Chande and Mazumder [7] examined the distribution of solute by diffusion and convection in a laser melted pool after it was delivered to such a pool.

The problem of extended solid solution due to rapid cooling has been studied by various investigators using thermodynamic variables such as free energy and chemical potential. The thermodynamics of nonequilibrium solidification has been examined very well by Baker and Cahn [8]. Boettinger and Perepezko [9] discussed the process of rapid solidification from the thermodynamics point of view. Boettinger, et al. [10], also used the response function approach of Baker and Cahn [8] and stability analysis for microsegregation-free solidification. Further details on rapidly solidified materials can also be found [11, 12].

Mathematical model

Process physics

In laser cladding processes the substrate material to be clad is moved at a constant rate while the cladding powder is poured onto it using a pneumatic powder delivery system and melted simultaneously by a laser beam (see Fig. 1a). The uniform speed of the substrate and the constant feed rate of the cladding powder ensure uniform cladding thickness. The molten pool of the cladding material which is formed just below the laser beam solidifies by dissipating heat to the surrounding air, adjacent cladding,

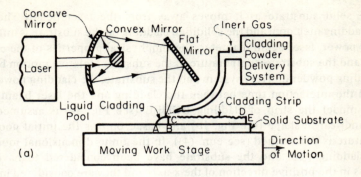

(a)

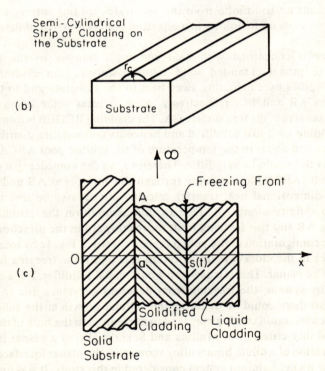

(b)

(c)

Fig. 1.(a) Schematic diagram of laser surface cladding.
    (b) Three-dimensional view of the cladding and the substrate.
    (c) Geometric configuration used in the present model. The model substrate and the solidification of the model cladding have been shown after rotating the pool ABCA (see Fig. 1a) by 90° clockwise.

and the solid substrate as it moves away from the laser beam. The shape of the cladding melt pool and the solidified cladding on the substrate is influenced by laser power, laser beam diameter, thermophysical properties of the cladding powder and the substrate, temperature of the substrate, relative motion between the cladding powder delivery system and the substrate, the cladding powder feed rate and the interaction time between the cladding and the laser beam. In the present model the strip of cladding BCDEB (see Fig. 1a) is assumed to be semicylindrical in shape (see Fig. 1b) and based on this the initial pool mean temperature is determined (see eqn. (4)). In the finite dimensional model, the liquid cladding pool and the substrate have been considered to have finite thickness in the positive direction of the x-axis and they are considered infinitely large in the other two directions (see Fig. 1c). On the other hand, they are considered semi-infinite in the infinite dimensional model where the cladding melt pool extends up to infinite from the substrate-cladding interface and the substrate also extends up to infinity in a direction opposite to that of the cladding pool.

This paper is concerned with heat and mass transfer in the cladding melt ABCA to model extended solid solution during laser cladding. This liquid pool solidifies by conducting away heat to the substrate and to the solid cladding across AB and BC, respectively. Also, it loses some energy to the ambient inert gas across the free surface CA. The cladding BCDEB is formed from the liquid cladding melt just solidified and hence its temperature distribution is almost uniform and equal to the temperature of the molten pool ABCA. So the heat loss across BC would be very little. Therefore, we can consider that the pool of cladding melt (ABCA) loses energy through the surfaces at AB and CA. To carry out one-dimensional heat transfer calculations we assume that the pool extends up to infinity along AB. Due to this assumption the cladding melt solidifies along AB and this freezing front moves upward in the direction of BC. The geometric configuration of the pool has been shown in Fig. 1c by rotating the pool 90 degrees in the clockwise direction. In this figure the freezing front has been shown to be planar. This is true when a pure metal solidifies. For a cladding melt of an alloy system, the freezing front develops curvature due to surface tension and also there could be dendrite and cellular growth at the solid-liquid interface. Moreover, rapid cooling may lead to nucleation in the bulk of the liquid phase. The stability criterion of Mullins and Sekerka [13] of a planar interface during solidification of a dilute binary alloy shows that the planar interface will be unstable for the nickel-hafnium system considered in this study. It was found [14] that there are dendrites which are very small in size. Nevertheless, the dendrites will affect the diffusion of solute atoms at the freezing front. However, to simplify the mass transfer analysis, the freezing front has been assumed to be planar due to the small size of dendrites. Moreover, the trend predicted by the model is more important than the exact numerical values due to the paucity of high temperature materials data and simplifying assumptions. Besides this the present model has been developed based on a few more assumptions listed as follows:

1) The thermal conductivity and the thermal diffusivity for a mixture is the sum of the volume-averaged value of the respective transport properties of each element of the mixture,

2) The mass diffusivity of each element in the liquid phase is the average value of self-diffusivity over the room temperature and the initial temperature with modified activation energy for the mixture,

3) The cladding pool and the substrate are in thermally perfect contact,

4) There is no mass diffusion in the solid phase.

5) The solute segregated at the solid-liquid interface moves into the liquid phase by diffusion only. This is because a boundary layer is formed near the interface where diffusion of solute atoms is dominant,

6) The cladding melt forms a uniform solution of composition equal to that of the cladding powder mixture before its solidification begins,

7) Only 50% of the laser energy is absorbed by the cladding material. Studies [12] show that the amount of laser energy absorbed by different materials is 37 to 60%.

With these assumptions, the one-dimensional heat conduction equations for the substrate, solidified cladding, and the liquid cladding regions are solved. The solutions of these heat transfer equations are used to obtain the velocity of the solid-liquid interface and then the mass transfer equation is solved to determine the distribution of solute atoms in the liquid phase. From this, the concentration of the solute atoms in the solid phase is computed by using an expression for nonequilibrium partition coefficient.

Mathematical formulations

In this section the governing heat conduction equations for energy transport in the substrate, solidified cladding, liquid cladding regions, and the mass transfer equation for diffusion of solute atoms in the liquid cladding pool are presented along with boundary and initial conditions.

The governing equations for energy transport are

i) Solid substrate region

$$\frac{\partial^2 T_1'}{\partial x^2} = \alpha_1 \frac{\partial T_1'}{\partial t}; \; t \geq 0 \tag{1}$$

This equation is applicable in the regions $-\infty < x \leq 0$ and $0 \leq x \leq a$ for the infinite and the finite dimensional models, respectively.

ii) Solidified cladding region

$$\frac{\partial^2 T_2'}{\partial x^2} = \alpha_2 \frac{\partial T_2'}{\partial t}; \; t \geq 0, 0 \leq x \leq S(t); \tag{2}$$

iii) Liquid cladding region

$$\frac{\partial^2 T_3'}{\partial x^2} = a_3 \frac{\partial T_3'}{\partial t}; \quad t \geq 0; \qquad (3)$$

This equation holds good in the regions $S(t) \leq x < \infty$ and $S(t) \leq x \leq b$ for the infinite and the finite dimensional models, respectively.

The auxiliary conditions are

$$K_1 \frac{\partial T_1'}{\partial x} = h_1 (T_1' - T_r) \text{ at } x = 0 \text{ and } t \geq 0,$$

for the finite dimensional model.

$T_1' = T_r$ at $x = -\infty$ and $t \geq 0$ for the infinite dimensional model.

$$T_1'(\bar{a}, t), = T_2'(\bar{a}, t),$$

$$K_1 \frac{\partial T_1'}{\partial x} = K_2 \frac{\partial T_2'}{\partial x} \text{ at } x = a \text{ and } t \geq 0,$$

In these two conditions $\bar{a} = 0$ and $\bar{a} = a$ for the infinite and the finite dimensional models, respectively.

$$T_2'(S(t), t) = T_3'(S(t)t) = T_I(C_I, \dot{S}(t)),$$

The solid-liquid interface temperature T_I is considered constant and equal to the melting temperature T_m of the cladding powder mixture in infinite dimensional model and it is considered a function of C_I and $\dot{S}(t)$ in the finite dimensional model.

$$-K_3 \frac{\partial T_3'}{\partial x} = h_3 (T_3' - T_r) \text{ at } x = b \text{ and } t \geq 0$$

for the finite dimensional model.

$T_3' = T_r$ at $x = \infty$ and $t \geq 0$

for the infinite dimensional model.

$$K_2 \frac{\partial T_2'}{\partial x} - K_3 \frac{\partial T_3'}{\partial x} \varrho L \frac{dS}{dt} \text{ at } x = S(t) \text{ and } t \geq 0,$$

and

$$S(0) = d'.$$

where $d' = 0$ and $d' = a$ for the infinite and the finite dimensional models, respectively. The initial conditions for the above problems are obtained from the following considerations.

The cladding has been assumed to melt almost instantaneously as soon as it is exposed to the laser beam to reach a uniform temperature $\bar{T}_2$. $\bar{T}_2$ is calculated by taking a lumped-parameter energy balance of the laser energy imparted to the cladding and the energy required to melt the cladding powder and raise its temperature to $\bar{T}_2$. This yields the following expression for $\bar{T}_2$:

$$\bar{T}_2 = T_r + \frac{1}{C_p}\left[\frac{2pf}{\pi r_c^2 vp} - L\right] \tag{4}$$

This expression is derived by considering the cladding to be formed in the shape of a semi-cylindrical strip (see Fig. 1b). In the finite dimensional model, we obtain the initial condition for the solid substrate by constructing a linear temperature profile between the temperature $\bar{T}_2$ at $x = a$ and T_r at $x = 0$, whereas the solid substrate is considered to be at the room temperature in the infinite dimensional model. Thus the initial conditions are

$$T_1'(x,0) = (\bar{T}_2 - T_r)\frac{x}{a} + T_r \text{ for the finite dimensional model.}$$

$$\bar{T}_1(x,0) = T_r \text{ for the infinite dimensional model.}$$

and

$$T_3'(x,0) = \bar{T}_2.$$

Now we will consider the diffusion of solute atoms in the cladding melt pool. This phenomenon is governed by the following mass transfer equation

$$\frac{\partial^2 C}{\partial x^2} = \frac{1}{\bar{D}}\frac{\partial C}{\partial t} \tag{5}$$

subject to the following initial and boundary conditions

$$C(x,0) = C_o,$$

$$-\bar{D}\frac{\partial C}{\partial x} = 0 \text{ at } x = b \text{ for the finite dimensional model,}$$

$$C(\infty, t) = C_o \text{ for the infinite dimensional model, and}$$

$$-\bar{D}\frac{\partial C}{\partial x} = \dot{S}(t)\,[k-1]\,C(x,t) \text{ at } x = S(t).$$

The nonequilibrium partition coefficient k is taken to be same as the one derived by Aziz [15].

Method of solution

We will now discuss the solution for the infinite and the finite dimensional models as described above. Since the procedure of solving the above equations is very long, we will resort to briefly outline the solution technique used in our study.

a) Infinite dimensional model

The method of solution for this model has been described in details elsewhere [14]. The multi-region heat transfer problem is solved by the method used by Carslaw and jaeger [16]. It gives the following expressions for the location and the speed of the freezing front.

$$S(t) = 2\gamma (a_e t)^{1/2} \tag{6}$$

$$\dot{S}(t- = \gamma (a_3/t)^{1/2} \tag{7}$$

Here k_2 and α_2 are taken to be equal to k_3 and α_3 respectively and γ is obtained from the following transcendental equation:

$$\frac{T_m - T_r}{1 + \mathrm{Erf}(\gamma)} + \frac{T_m - \bar{T}_2}{1 - \mathrm{Erf}(\gamma)} = \sqrt{\pi}\ \rho\frac{\alpha_3}{k_3}\ L\gamma e^{\gamma^2} \tag{8}$$

After knowing the rate of formation of the solid phase the distribution of solute in the liquid phase can be obtained by solving the mass diffusion problem described by eqn. (5) and the associated boundary and initial conditions. An exact analytical solution for the above problem has been obtained by using similarity variables which are determined using the Lie group theory. Details of this theory and its application for the solution of differential equations can be found in [17, 18]. Using this group theoretic approach the similarity variables are found to be

$$y = \frac{x}{\sqrt{t}}\ \text{and}\ \psi_n = \frac{C(\sqrt{t})^n}{A_n},\ n = 0, 1, 2, \ldots, \infty \tag{13}$$

and the distribution of solute in the cladding pool is given by

$$C = \sum_{n=0}^{\infty} A_n \psi_n \frac{1}{(\sqrt{t})^n} \tag{14}$$

where

$$A_n = \left(\frac{\lambda}{D^*}\ r\sqrt{\alpha_3}\right)^n, \ n = 0, 1, 1, \ldots\ldots, \infty$$

$$r = (y + 2\gamma\sqrt{\alpha_3})\ (2\bar{D})^{1/2}$$

For this problem four terms of the series (14) have been computed by using (13) in the governing equation (5) and by satisfying the boundary conditions. The expressions for ψ_0, ψ_1, ψ_2 and ψ_3 are given by

$$\psi_o(r) = C_o \left[1 - \frac{G\,\mathrm{Erfc}\,(\sqrt{2}\,r)}{(\sqrt{8/\pi})\,\mathrm{Exp}(-4\gamma^2\,\alpha_3/\bar{D}) + G\,\mathrm{Erfc}[2\gamma(\alpha_3/\bar{D})^{1/2}]}\right]$$

$$\psi_1(r) = -\frac{G\,\psi_o(r_0)}{r_o + G}\,e^{-(r^2 - r_o^2)/2}$$

$$\psi_2(r) = -\frac{G\,[\psi_o(r_o) - \psi_1(r_o)]}{r_o G + (r_o^2 - 1)}\cdot r e^{-(r^2 - r_o^2)/2}$$

$$\psi_3(r) = -\frac{G\,[\psi_o(r_o) + \psi_1(r_o) + \psi_2(r_o)]}{(r_o^2 - 1)\,G + r_o(3 - r_o^2)}\,(r^2 - 1)\,e^{-(r^2 - r_o^2)/2}$$

where

$$G = \gamma\sqrt{\alpha_3}(k_e - 1)\,(2/\bar{D})^{1/2} \text{ and } r_0 = \gamma(2\alpha_3/\bar{D})^{1/2}$$

Using these expressions for ψ_0, ψ_1, ψ_2 and ψ_3 the concentration of solute in the liquid phase is computed using eqn. (14) and then the corresponding concentration in the solid phase (C_s) is obtained by using the following expression derived by Aziz [15] for nonequilibrium partition coefficient.

$$\frac{C_s}{C_l} = \frac{\dot{S}(t)\lambda/D^* + k_e}{\dot{S}(t)\lambda/D^* + 1} \tag{15}$$

In the present model the interdiffusivity D^* is taken to be equal to the mass diffusivity and the interatomic spacing λ is taken to be 4Å which is the average diameter of an atom.

b) Finite dimensional model

This model differs from the above-mentioned infinite dimensional model in two notable aspects, namely, the substrate and the cladding pool are considered to have finite thickness in this model as opposed to the semi-infinite dimension assumed in the infinite dimensional model. The second aspect is that the solid-liquid interface temperature is not constant in this model, whereas it is considered constant in the infinite dimensional model. Besides this, the governing equations have been solved in the infinite dimensional model by using similarity variables and in the finite dimensional model by using integral transform and the Runge-Kutta methods. The following expression for T_l as derived by Boettinger and Perepezko [9] has been used in this model.

$$T_l = T_m^* + \bar{m}_l C_l^* + \frac{\bar{m}_l C_l^*}{1 - k_e}\left[k_e - k\left(1 - ln\frac{k}{k_e}\right)\right] + \frac{\bar{m}_l}{1 - k_e}\,\frac{\dot{S}(t)}{v_o}.$$

Here v_o represents the velocity of sound in the liquid phase. The velocity of sound in a medium depends on its temperature and composition. In this study v_o has been taken to be equal to the velocity of sound in the pure solvent. T_m^* and m_l are determined from the equilibrium phase diagram by taking a segment of the liquidus around the point of interest. This segment is considered linear whose slope and intercept with the temperature axis are $\bar{m}_l$ and T_m^* respectively. C_l^* is

148

obtained by evaluating $C(x, t)$ at $x = S(t)$ and by expressing it in terms of mole fraction.

Nonequilibrium partition coefficient for concentrated solution
    Nonequilibrium partition coefficient for concentrated solution is of great interest for rapid solidification of metallic alloys. Recently, Aziz [22] reported an expression. However, one of the problems is too many unknowns involved in such derivation. In this work we will present another relatively simple expression with an attempt to minimize the number of unknown parameters.
    By using the procedure discussed in ref. [23], we obtain the following expression for nonequilibrium partition coefficient for concentrated solutions:

$$k = \frac{k_e}{k_e + (1 - k_e)\ e^{-(\bar{\beta} - \bar{\delta})}} \qquad (16)$$

$$\bar{\beta} = \beta_c\ e^{-\alpha A\ \varrho m_i^*\ k_e}$$

$$\bar{\delta} = \sum_{n=1}^{\infty} \frac{\{\alpha_A\ \varrho m_i^*(1 - k_e)\}^n - \{\alpha_A\ \varrho m_i\ (1 - k)\}^n}{n \cdot n!}$$

$$\beta_c = ud/D_i$$

The diffusion coefficient of A in B is given by

$$D_{AB} = D_A(T)\ e^{\alpha_A \varrho_A} \qquad (17)$$

where
k = nonequilibrium partition coefficient,
k_e = equilibrium partition coefficient,
ϱ_A = concentration of A,
ϱ = density of the alloy
m_i^* = mass fraction of solute at the solid-liquid interface in the liquid phase under equilibrium condition,
m_i = mass fraction of solute at the solid-liquid interface in the liquid phase under nonequilibrium condition,
u = speed of the solid-liquid interface,
d = length of the diffusion zone,
D_i = $D_A(T)$
$ln\{D_A(T)\}$ = intercept of the lnD_{AB} versus ϱ_A graph,
α_A = slope of the lnD_{AB} versus ϱ_A graph

Results and discussion

Parametric results
    The above model was used to study the effect of various important process

parameters such as laser power, laser-cladding interaction time and cladding powder delivery rate on the composition of hanfium and aluminum in nickel matrix. Results were obtained for ambient temperature $T_r = 293°K$ and the heat transfer coefficients $h_1 = 1$ watt/cm²−°K and $h_2 = 0.01$ watt/cm²−°K. Only convective heat loss from the cladding and the substrate to the ambient inert gas has been considered in this model because the heat loss due to conduction is usually very small and the radiative loss is usually about 5% [2] of the total energy of the melt pool. The convection heat transfer coefficients are usually 25 to 250 watt/m²−°K and 50 to 2×10^4 watt/m² − °K for the processes of forced convection by gases and liquids, respectively [21]. The heat conduction within the cladding melt is mainly unidirectional for reasons mentioned earlier and its heat loss to the ambient gas has been taken to be due to forced convection by the surrounding inert gas. For this reason, the above value of h_2 has been selected. The heat conduction within the substrate has been considered to be unidirectional even though there would be conduction in the other two directions. Consequently, the heat loss due to conduction in these two directions has not been accounted for in the one-dimensional model. This is why a rather high convective heat loss from the substrate has been considered by choosing the above value of h_1. The laser power was fixed at 5, 6, and 7 kW. For each power of laser, the speed of the workpiece was varied as 40, 50, and 60 inches/min. Since the transport properties depend on temperature, average values, which are calculated according to formulae of Kar and Mazumdar [14], were used in this model.

Results obtained from this study have been presented in Figs. 2 through 12. Figure 2 is based on the results of infinite dimensional model and Figs. 3 through

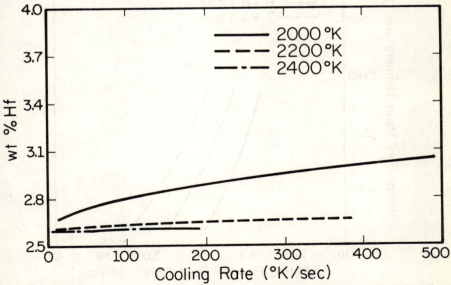

Fig. 2. Effect of cooling rate on the composition of hafnium in the solid phase at the freezing front.

150

12 are based on the results of the finite dimensional model. Figure 2 shows the extension of solubility of Hf in the α phase of Ni-Hf alloy as a function of cooling rate for initial pool mean temperatures 2000, 2200, and 2400°K. It can be seen from this figure that the higher the cooling rate the higher is the composition of hafnium in nickel matrix. However it should be noted that same cooling rate for different initial pool mean temperatures does not result in the same composition of hafnium in the solid solution. This is because the higher the initial pool mean temperature the longer it takes to first form the solid phase at the interface of the substrate and the liquid cladding material and during this time the solute atoms diffuse away from the interface causing a formation of alloy lean with solute atoms. This is true if there is no adsorption between the solute atoms and the solid substrate. Figures 3 and 4 are concerned with the variation of initial pool mean temperature and solute concentration in the solid phase with the cladding powder feed rate. The cladding powder feed rate is determined by using the expression $\frac{1}{2}\pi r_c^2 v\varrho$ and assuming that the cladding takes the shape of a semicylindrical strip (see Fig. 1b). Cladding powder feed rate is a very important process parameter since it is related to cladding thickness and velocity of the workpiece as well as the initial pool mean temperature. Thus if the powder delivery rate is known, then $\bar{T}_2$ can be obtained from Fig. 3 and the concentration of solute can be obtained from the Fig. 4 and then the velocity of the workpiece can be selected for a desired cladding thickness. As can be seen from Fig. 4, the concentration of hafnium

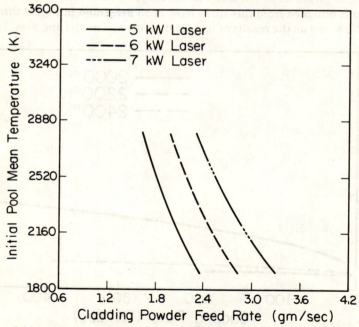

Fig. 3. Initial pool mean temperature of nickel-hafnium versus cladding powder feed rate.

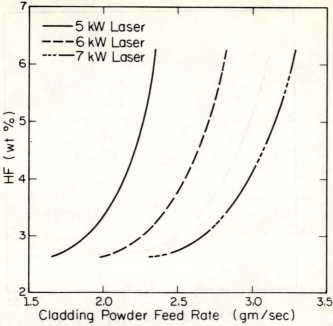

Fig. 4. Concentration of hafnium in the extended solid solution of nickel-hafnium versus cladding powder feed rate.

increases as cladding powder feed rate increases for a given laser power. This is due to the fact that the initial pool mean temperature, $\bar{T}_2$, varies inversely with the cladding powder feed rate as evident from Fig. 3. The closer is $\bar{T}_2$ to the melting point of the cladding powder the higher would be the cooling rate and this will result in a higher concentration of solute in the solid phase. This was discussed in detail by Kar and Mazumder [14]. However, it would be erroneous to conclude that one can enrich the solid phase with solute to any concentration by increasing the cladding thickness indefinitely for a given laser power and a given speed of the workpiece. There is a critical value of the cladding thickness at which T_2 becomes equal to the melting point of the cladding powder for a given workpiece speed and laser power. This critical cladding thickness can be obtained from eqn. (4). If the cladding thickness is greater than this critical value, then there will be some unmelted powder between the substrate and the cladding melt and hence the substrate will not be clad. Thus for a given workpiece speed and laser power, the concentration of solute in the solid phase increases with cladding thickness as long as the thickness is less than the critical value. Figure 5 indicates the effect of specific laser energy on solute concentration in the solid phase. Specific laser energy is defined as laser power required to produce a cladding of unit mass per unit time. It is determined by the relation $P/(\frac{1}{2}\pi r_c^2 v\varrho)$. The importance of selecting specific laser energy as a parameter is that it allows representation of the

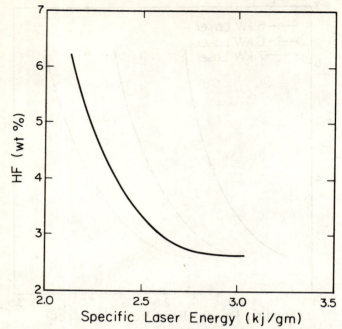

Fig. 5. Concentration of hafnium in the extended solid solution of nickel-hafnium versus specific laser energy.

solute concentration data for any combination of laser power, workpiece speed, and cladding thickness parameters by a single curve. This is because the initial pool mean temperature is proportional to specific laser energy. Thus the initial pool mean temperature and hence the thermal characteristic of the cladding melt pool will not be different as long as the laser specific energy remains the same for any choice of the process parameters.

The effect of another important process parameter, laser-materials interaction time, on the solute concentration and the initial pool mean temperature is shown in Fig. 6 for different values of laser power. Interaction time is defined as the ratio of laser beam diameter to the speed of the workpiece. It can be observed that the trends of the graphs with respect to laser-material interaction time are opposite to those observed with respect to cladding powder feed rate. This is because lowering of cladding powder feed rate leads to higher initial pool mean temperature than what is obtained as interaction time decreases provided other process parameters are kept unchanged. We can deduce from Figs. 3 and 4 that the solute concentration in the solid phase increases as the initial pool mean temperature decreases. This explains why the solute concentration curves with respect to the interaction time should be opposite to those which are plotted with respect to the cladding powder feed rate.

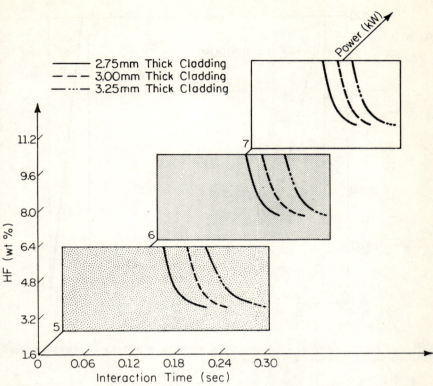

Fig. 6. Concentration of hafnium in the extended solid solution of nickel-hafnium versus laser-cladding interaction time.

Figure 7 shows the nonequilibrium phase diagram for Ni-Hf alloy computed by using the finite dimensional model. The characteristic parameters for this figure have been presented in Table 1. Deviation of nonequilibrium phase diagram from the equilibrium one shows the extension of solid solubility that can be obtained during laser cladding. The nonequilibrium phase diagram has been plotted in the neighborhood of the melting point of the cladding powder because the value of k_e used in the expression for nonequilibrium partition coefficient (eqn. (15)) corresponds to this melting point. It can be seen from the Fig. 7 that the width of the solid-liquid region between the equilibrium solidus and liquidus has reduced considerably. The nonequilibrium solidus line of this figure shows the extension of Hf concentration in Ni that can be obtained due to the rapid cooling in laser cladding. This phenomenon can be understood from the fact that the equilibrium phase diagram of Ni-Hf has a negative slope at the point which corresponds to the nominal composition of the cladding powder. Due to this, Hf is rejected from the Ni matrix when the solution of Ni-Hf solidifies and this results in increasing the concentration of Hf in the liquid phase.

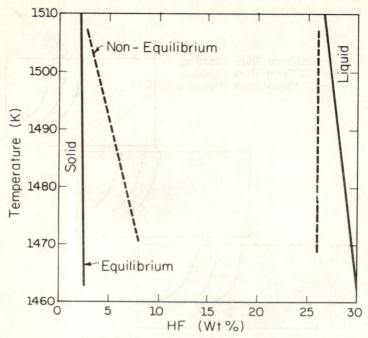

Fig. 7. Comparison of nonequilibrium phase diagram of the extended solid solution of nickel-hafnium with its equilibrium phase diagram.

But since the solid phase retains more Hf than its equilibrium composition, the liquid phase will have less Hf than the equilibrium value and hence the extended solid solution (that is, the nonequilibrium) phase diagram will shrink.

Figures 8 through 12 show results for Ni-Al alloy based on the finite dimensional model. The effects of cladding powder feed rate on the initial pool mean temperature and the concentration of aluminum in the solid phase have been shown in Figs. 8 and 9, respectively. Figures 10 and 11 show, respectively, the effects of specific laser energy and laser-cladding interaction time on the concentration of aluminum in the solid phase. Finally, the nonequilibrium phase diagram for the Ni-Al system has been presented in Fig. 12. The characteristic parameters for this figures have been given in Table 1.

It should be observed that the trends of the graphs for Ni-Al system as shown in Figs. 8 through 12 are opposite to those of the graphs for Ni-Hf system shown in Figs. 3 through 7. This is because the slope of the equilibrium phase diagram of Ni-Al alloy is positive at the nominal composition of the cladding powder. This means that decrease in temperature $(\bar{T}_2)$ will cause reduction in the concentration of aluminum in the Ni-Al solid solution. This is what we deduce from Figs. 8 and 9. On the contrary, the slope of

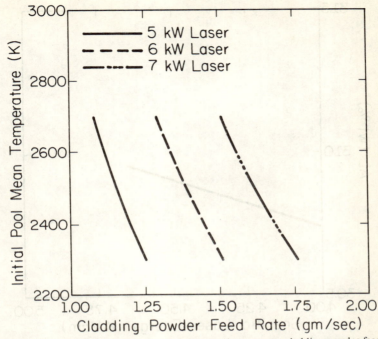

Fig. 8. Initial pool mean temperature of nickel-aluminum versus cladding powder feed rate.

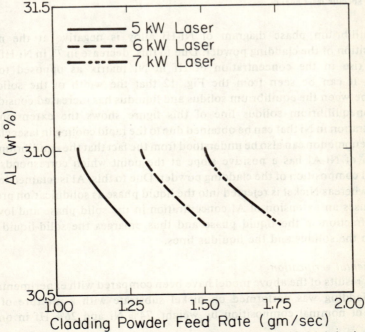

Fig. 9. Concentration of aluminum in the extended solid solution of nickel-aluminum versus cladding powder feed rate.

156

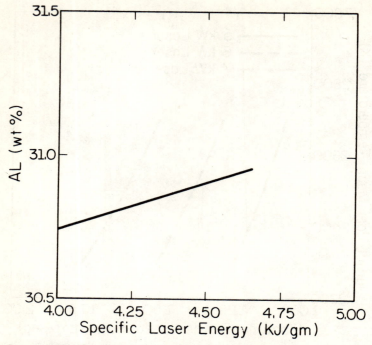

Fig. 10. Concentration of aluminum in the extended solid solution of nickel-aluminum versus specific laser energy.

the equilibrium phase diagram of Ni-Hf alloy is negative at the nominal composition of the cladding powder. Due to this, decrease in $\bar{T}_2$ in Ni-Hf system causes rise in the concentration of Hf in Ni matrix as opposed to Ni-Al system. It can be seen from the Fig. 12 that the width of the solid-liquid region between the equilibrium solidus and liquidus has increased considerably. The nonequilibrium solidus line of this figure shows the extension of Al concentration in Ni that can be obtained due to the rapid cooling in laser cladding. This phenomenon can also be understood from the fact that the equilibrium phase diagram of Ni-Al has a positive slope at the point which corresponds to the nominal composition of the cladding powder. Due to this, Al is retained in the Ni matrix whereas Nickel is rejected into the liquid phase as solidification proceeds. This causes an extension of Al concentration in the solid phase and lowers its weight fraction in the liquid phase and thus enlarges the solid-liquid region between the solidus and the liquidus lines.

Experimental verification

The results of the above model have been compared with experimental data. Laser cladding was performed on nickel substrate with a mixture of Ni-Hf powder of nominal composition by weight, 74% Ni and 26% Hf in one case

157

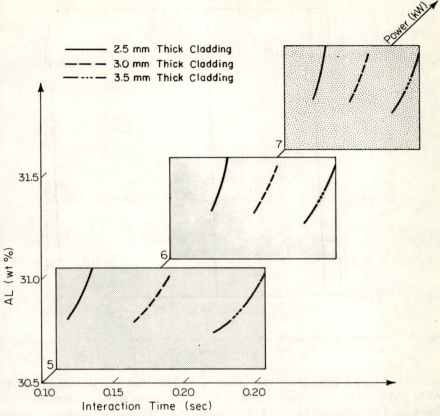

Fig. 11. Concentration of aluminum in the extended solid solution of nickel-aluminum versus laser-cladding interaction time.

Table 1. Solidus composition of Hf, solidus/liquidus temperature at the substrate-cladding interface and the speed of solidification at this interface

Solute Element	Composition of Hf (Wt%)	Solidus/Liquidus Temperature (K)	Interface Speed (cm/sec)
	3.19	1505	3.26
	4.83	1499	2.88
Hafnium	6.23	1489	2.74
	7.15	1480	2.68
	9.58	1453	2.55
	30.83	1864	6.1
Aluminum	30.74	1863	5.6
	30.56	1862	4.5

158

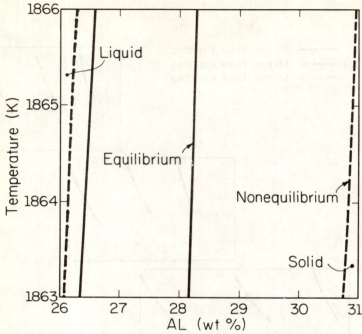

Fig. 12. Comparison of nonequilibrium phase diagram of the extended solid solution of nickel-aluminum with its equilibrium phase diagram.

and 74% Ni and 26% Al in the other case. The cladding is formed almost in the shape of a semi-cylindrical strip of metal on the substrate (see Fig. 1b). Scanning Transmission Electron Microscope (STEM) analysis of these samples shows the concentration of Hf in the α phase and Al in the martensitic solid solution regions in excess of that predicted by the equilibrium phase diagram. The percent model predicts the composition of the extended solid solution quite well. These results have been presented in Table 2.

Table 2. Comparison of theoretical results with experimental results of extended solid solution

Nominal composition of the cladding, wt. %	74% Ni, 26% Hf	74% Ni, 26% Al
Laser power, (Kw)	5	5
Laser beam diameter, mm	3	3
Workpiece speed, in/min	50	50
Initial pool mean temperature $(\bar{T}_2)$, °K	1862	2420
Composition in the solid solution, wt.%		
Theoretical results		
(a) Infinite dimensional model	3.05 Hf	27.3 Al
(b) Finite dimensional model	7.15 Hf	30.76 Al
Experimental results	6.5 Hf	29 Al

A small fraction of the Ni-Hf alloy was found experimentally to contain 9.3% (by wt.) Hf in the Ni matrix. There is a little discrepancy between the theoretical and experimental results which may be due to some of the assumptions of this model. For example, it has been assumed that all of the solute dissolves in the solvent before the cladding melt starts solidifying. But the time it takes to melt and freeze the cladding powder may not be sufficient for dissolution of all solute atoms to occur. Also, the present model utilizes an expression, eqn. (15), for the nonequilibrium partition coefficient which is applicable to dilute solution. Apart from these conditions, the presence of a two-phase zone between the solidus and the liquidus lines and the surface tension driven flow causing convection in the liquid pool will affect the mixing of solute in the liquid phase and thus alter its composition in the solid phase. Paucity of high temperature liquid metal data also contributes to the numerical error. However, the objective of this study is to determine the trend of, and thus understand, the underlying process physics.

CONCLUSION

This work examines the extension of solid solubility based on the transport of energy and mass. Solute transport has been considered to take place only in the liquid phase while the energy transport has been considered in both solid and liquid phases. The effect of nonequilibrium cooling rate on solute segregation at the freezing front has been taken into account by considering a nonequilibrium partition coefficient. Using this, the mass transfer problem has been solved for solute distribution in the liquid phase and the heat transfer problem has been solved to obtain the velocity of the solid-liquid interface, its location, and the freezing temperature of the interface. These mathematical solutions have been utilized to study the effect of various process parameters on the concentration of solute in an alloy.

It is found that the same composition of solute is obtained in the alloy for different cladding powder feed rate by varying the laser power. This is because the initial pool mean temperature of the cladding material is the same even though other parameters are different. Thus it can be concluded that the choice of the initial pool mean temperature determines the composition of the alloy. Since initial pool mean temperature is directly proportional to specific laser energy, when the initial temperature of the cladding melt is fixed or the specific laser energy is selected other process parameters can be determined by selecting any two of the three parameters, r_c, v, and P and the third one will then have to be determined from eqn. (4). The composition of solute in the solid phase can be represented by a single curve for all possible values of the process parameters if it is plotted against the specific laser energy or the initial pool mean temperature. The effect of laser-cladding interaction time on solute concentration in the alloy is found to be opposite to that of cladding powder feed rate. Also, it has been found that if the equilibrium liquidus has a positive slope at

160

the nominal composition of the cladding powder, then the nonequilibrium phase diagram shrinks from the equilibrium phase diagram. If the corresponding slope is positive the nonequilibrium phase diagram shows an enlarged solid-liquid region between the solidus and the liquidus compared to that of the equilibrium phase diagram.

ACKNOWLEDGMENT

This work was made possible by a grant (AFOSR : 85-0333) from the U.S. Air Force Office of Scientific Research. Continued encouragement from the program managers Dr. A. Rosenstein in also appreciated. The authors also like to acknowledge the help provided by Dr. J. Singh and Dr. S. Sircar for the experimental data, and Mrs. June Kempka and her co-workers at the Department of Mechanical and Industrial Engineering Publications Office for their help in preparation of this manuscript.

REFERENCES

1. Singh, J. and J. Mazumder. *Acta Metall.* (accepted for publication).
2. Mazumder, J. and W.M. Steen. *J. Appl. Phys.* **51**, 941–947 (1980).
3. Cline, E. and T.R. Anthony. *J. Appl. Phys.* **48**, 3895–3900 (1977).
4. Kou, S., S.C. Hsu and R. Mehrabian. *Met. Trans. B.* **12B**, 33–45, (1981).
5. Ashby, M.F. and K.E. Easterling. *Acta Met.* **32**, 1935–1948 (1984).
6. Chan, C., J. Mazumder and M.M. Chen. *Met. Trans. A.* **15A**, 2175–2183, 1984.
7. Chande, T. and J. Mazumder. *J. Appl. Phys.* **57**, 2226–2232, 1985.
8. Baker, J.C. and J.W. Cahn. Thermodynamics of Solidification. In: *Solidification, ASM,* Metals Park, OH, pp. 23–58, (1971).
9. Boettinger, W.J. and J.H. Perepezko. *Proc., Rapidly Solidified Crystalline Alloys, TMS-AIME,* New Jersey, May 2–3, 1985.
10. Boettinger, W.J., S.R. Coriell and R.F. Sekerka. *Mat. Sci., Eng.* **65**, 27–36, 1984.
11. Kear, B.H., B.C. Giessen and M. Ghen (eds.) *Proc. MRS* **8**, Boston, Nov. 1981.
12. Li, L.J. and J. Mazumder. In: Laser Processing of Materials (eds.) K. Mukherjee, and J. Mazumder, *Proc. Metal. Soc. AIME,* Los Angeles, pp. 35–50, (1981).
13. Mullins, W.W. and R.F. Sekerka. *J. Appl. Phys.* **35**, 444–451 (1964).
14. Kar, A. and J. Mazumder. *J. Appl. Phys.* **61** (7), 2645–2655.
15. Aziz, M.J. *J. Appl. Phys.* **53**, 1158–1168) (1982).
16. Carslaw, H.S. and J.C. Jaeger. *Conduction of Heat in Solids,* 2nd Ed., Clarendon Press, London, pp. 282–296 (1959).
17. Bluman, G.W. and G.D. Cole. *Similarity Methods for Differential Equations,* Springer-Verlag, New York (1974).
18. Ovsiannikov, L.V. *Group Analysis of Differential Equations,* Translation edited by W.F. Ames, Ist ed., Academic Press, New York (1982).
19. Kar, A. and J. Mazumder. *Acta Metallurgica* (in Press).
20. Kar, A. and J. Mazumder. Metal. Trans. A (submitted).
21. Incropera, F.P. and D.P. Dewitt. *Fundamentals of Heat and Mass Transfer,* 2nd Ed., John Wiley and Sons, New York, p. 8 (1985).
22. Aziz, M.J. In: *Science and Technology of Rapidly Quenched Alloys,* (ed.) M. Tenhover, W.L. Johnson and L.E. Tanner, Materials Research Society, p. 25 (1987).
23. Kar, A. and J. Mazumder. *Phys. Rev. B.* (submitted).

NOMENCLATURE

a	Width of the substrate
b	Width of the substrate and the cladding melt
B	Laser beam diameter
C	Concentration of solute in the liquid phase
C_l	Concentration of solute in the liquid phase at the solid-liquid interface
C_p	Average specific heat of the cladding material
C_s	Concentration of solute in the solid phase at the solid-liquid interface
C_l^*	Concentration (mole fraction) of solute in the liquid phase at the solid-liquid interface
$\bar{D}$	Average mass diffusivity of solute in the liquid phase
D^*	Interdiffusivity of solute at the interface
D_{AB}	Diffusion coefficient of species A into species B at the solid-liquid interface
$D_A(T_l)$	Intercept of the graph lnD_{AB} versus C
D_l	Value of $D_A(T_l)$ in the liquid phase at the sollid-liquid interface
f	Fraction of laser energy absorbed by the cladding material
h_1	Heat transfer coefficient at the substrate boundary
h_2	Heat transfer coefficient at the cladding surface boundary
k_e	Equilibrium partition coefficient
k	Nonequilibrium partition coefficient
k_i	Thermal conductivity of the i-th region
L	Latent heat of fusion of the cladding material
m_l	Mass fraction of solute in the liquid phase under nonequilibrium cooling condition
m_l^*	Mass fraction of solute in the liquid phase under equilibrium cooling condition
$\bar{m}_l$	Liquidus slope (°K/mole fraction of solute)
P	Laser power
q	Rate of cladding powder delivery
r_c	Radius of the semi-cylindrical strip of cladding
s	Position of the solid-liquid interface
$\bar{T}_2$	Initial pool mean temperature
T_i'	Temperature of the i-th region
T_l	Temperature at the solid-liquid interface
T_m	Melting temperature of the cladding powder mixture
T_m^*	Intercept of liquidus line on temperature axis
T_r	Ambient temperature
v	Speed of the workpiece
v_o	Speed of sound in the cladding melt
α	Slope of the graph lnD_{AB} versus C
α_i	Average thermal diffusivity in the i-th region

162

β A non-dimensional parameter defined as $\dot{S}(t)\lambda/D^*$

β_c A non-dimensional parameter defined as $\dot{S}(t)\lambda e^{-\alpha\varrho/2}/D_i$

λ Interatomic distance

ϱ Average density of the cladding material

Subscripts

i 1, solid substrate; 2, solidified cladding; 3, liquid cladding.

7

Laser Surface Cladding

W.M. STEEN

Materials Department
Imperial College of Science and Technology, London SW7 2BP
Now at Mechanical Engineering Department,
The University of Liverpool, Liverpool, U.K.

ABSTRACT

A review is made of the various ways in which the laser can be used for surface cladding. The operating conditions for successful cladding by the blown powder technique to produce low dilution fusion bonded clad layers are described. In particular it is shown that there is likely to be a fast resolidification at the interface region which ensures, for the correct operating conditions, that the level of dilution can be very low. A method of rapid alloy scanning based on this process and relying upon this low dilution is illustrated with two examples in the development of hardfacing alloys.

INTRODUCTION

The laser has some unique properties for surface heating. The electromagnetic radiation in a laser beam is absorbed within the first few atomic layers for opaque materials. Such as metals, and there are no associated hot gas jets, eddy currents or even radiation spillage outside the optically defined beam area. In fact the applied energy can be placed precisely on the surface only where it is needed. Thus it is a true surface heater and a unique tool for surface engineering.

Many surfacing techniques have been suggested for the laser. They include surface transformation hardening, surface melting, surface alloying, surface cladding and surface shock hardening. Among the cladding routes are those which melt preplaced powder [1], or blown powder [2], those which decomposed vapour (Laser chemical vapour deposition, LCVD) by pyrolysis [3, 4] or photolysis [5], those which are based upon local vapourization as in laser physical vapour deposition or sputtering and those based on enhanced electroplating or cementation [6]. Most of these processes are still at the laboratory stage. Only

164

transformation hardening [7], and cladding [8] are used in production at present. The versatility of optical engineering will ensure that this is not always going to be the case.

This paper discusses the blown powder process from two angles. The first is the operating conditions for successful cladding to produce low dilution, fusion bonded clad layers. The second is the use of such layers, made with a mixed powder feed, to scan alloy compositions and thus rapidly to locate interesting alloys for specific uses.

Blown powder cladding

The process is described in oreater detail elsewhere [2]. The general experimental arrangement is illustrated for a mixed feed system in Fig. 1. The main reason for the current interest in this process is that it is one of the few cladding techniques which has a fusion bond with low dilution and is adaptable to automatic processing. It thus has the low dilution associated with the forge bonded processes such as jetkote, D-gun, Gator Gard or Flare, but the good interface strength and low porosity associated with the welding processes such as TIG, flame, or plasma. The covering rate for laser powers greater than 5kW is also attractive and when consideration of powder costs and after machining costs are taken into account then the process becomes economically comparable with other processes for covering large areas.

It stands almost alone in its ability to cover very small areas and in particular areas near to thin walls which might be thermally sensitive, i.e. melt in the heat from a plasma torch or other hot jet system.

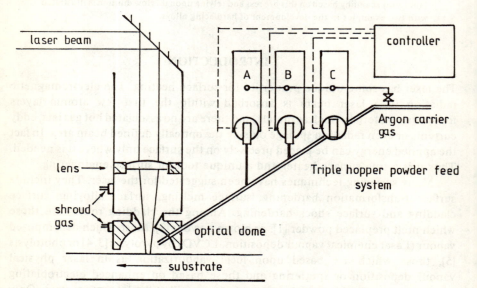

Fig. 1. Experimental arrangement with mixed powder feed.

Fig. 2. Examples of track transverse sections made under different processing conditions.
  a) Aspect ratio incorrect for overlapping.
  b) Some dilution appearing.
  c) Good low dilution fusion bonded track.

In cladding with blown powder, the processing is essentially conducted over a small melt pool area which travels over the surface of the substrate. This has the great advantage that thermal penetration is minimal thus reducing the distortion and heat affected zone (HAZ) problems-though not eliminating them. However it does mean that if area coverage is to be achieved then overlapping of clad tracks is required. It is found [9] that there are three basic cross sections of such clad tracks as shown in Fig. 2. These sections illustrate the limits of successful low dilution fusion bonded cladding. For example, the angle marked α in Fig. 2a must be acute for successful overlapping without porosity. This angle is essentially defined by the aspect ratio of the clad track. In Fig. 2b some dilution is apparent. This is principally due to excessive energy delivery which is not absorbed in melting powder. It is thus defined by an energy balance. The third limit occurs when there is a lack of fusion bonding. A discontinuous clad or 'balling' of the powder results. This will also be determined from an energy balance.

Thus we have three limits to low dilution fusion bonded cladding:
* Onset of porosity   (aspect ratio)
* Onset of significant dilution   (upper energy balance)
* Loss of clad continuity   (lower energy balance).

These limits are discussed in more detail elsewhere [9]. From this work operating curves of the form shown in Fig. 3 were calculated.

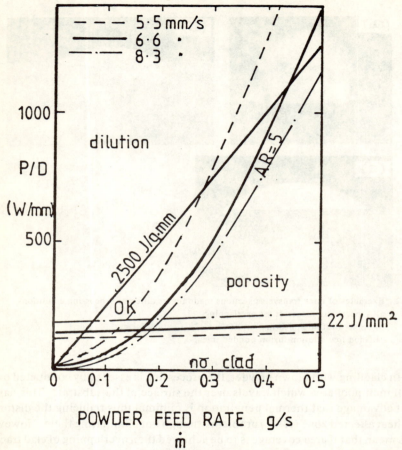

Fig. 3. Approximate operating curves suggested from [9]. 'OK' signifies low dilution fusion bonded clad tracks.

The limits marked on these operating charts are defined by:

A dilution parameter $\qquad (P/\dot{m}D) \qquad$ J/kg m

where P = Incident absorbed energy, W: $\dot{m}$ = Powder feed rate. kg/s: D = Incident beam diameter. m:

A porosity parameter $\qquad (PV\rangle\dot{m}^2). \qquad$ Js/m kg^2

where V = the traverse speed. m/s:

A fusion parameter $\qquad P(1-r-s)\rangle DV. \qquad$ J/m^2

where r = reflectivity: s = shadow effect of particles in screening the substrate surface.

For example, it is found that an aspect ratio (width/height) of greater than 5 is required for successful overlapping of tracks. The dilution parameter should be less than around 2500J/g.mm for cladding cobalt based hardfacing

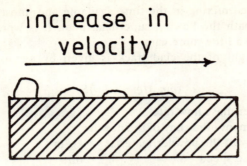

Fig. 4. Effect of speed on track section.

alloys such as Wall Colmonov Wallex 6PC with 50% overlap with no preheat: and more than 10J/mm^2 is required for a continuous clad track in this material.

It is seen from Fig. 3 that the operating region is larger for higher traverse speeds due to the lack of dependence of the dilution parameter on speed. The effect of varying speed on track section is shown in Fig. 4. Height and width change but the level of dilution does not.

The low dilution attainable with this technique opens up possibilities not available before. In particular clad alloys can be prepared in-situ by blowing mixed powders without contamination from the substrate [10]. The reason for the low dilution may be seen in Fig. 5 showing longitudinal sections through a clad track which has been made over a substrate in which a copper marker wire was embedded.

The resulting EDAX scan revealed the extent of the melt pool, delineated in Fig. 5. The most interesting aspects are the extent of the pool, the implied high velocity of backward flow from the front end of the clad required to

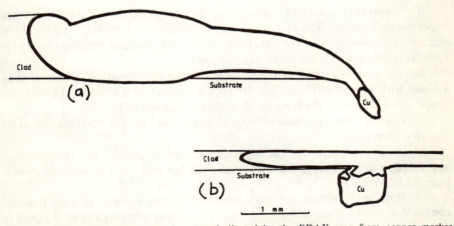

Fig. 5. Outline of the melt pool shapes as indicated by the EDAX scan from copper marker experiments [10], a) 4 mm/s b) 12 mm/s.

create the observed mixing in the time available and, most interestingly, the tranquil gap beneath the flowing superficial pool. It is worth examining this tranquil area with a little more care.

The final height of the clad track is given by a mass balance of the form:

$$(\dot{m}\eta_m)/\varrho_r = (vhw)S \tag{1}$$

Where $\dot{m}$ = powder mass flow rate   kg/s
     η_m = catchment efficiency of the powder
     ϱ_p = density of the powder material   kg/m^3
     V = traverse speed   m/s
     h = final height of clad track   m
     w = width of track   m
     S = cross sectional shape factor.

From equation (1) we get $h = \dot{m}\eta_m/(\varrho_r vwS)$.   (2)

and hence for a uniformly growing clad, the height at any distance x, from the leading edge h_x, is:

$$h_x = h \, x/L$$

Where L = Length of the melt pool, m.
Surface tension effects would almost certainly make the molten pool thicken faster than this. For example a convex pool shape might be described by:

$$h_x = h\sqrt{x/L}$$

Thus, in general, we have: $h_x = h(x/L)^n$ (3)
Where n = a constant probably between 0.5 and 1.0.
The significance of this will be discussed later.

If we consider a section through the molten pool, then Takeda's marker experiments suggest a flow pattern similar to that shown in the Fig. 6. The order of size of the surface convection flow driven by surface tension forces is approximately found from these experiments.

For the slow traverse experiment (Fig. 5a) the pool length was measured as 4.6 mm with a melt distance into the copper marker of 0.23 mm. The traverse speed was 4 mm/s so the time for flow was 0.23/4 = 0.057s.

This gives the minimum speed near the copper marker of 4.6/0.0576 = 80 mm/s.

A similar calculation on the higher speed run assuming that the pool had again been traversed after 0.23 mm of the marker had been penetrated gave a velocity of 73 mm/s.

Both these values are high compared to the velocity of heat flow by conduction (of the order of 2 mm/s). Thus the surface temperature of the pool is dependent upon the temperature of the convected layer, not directly upon the incident laser power.

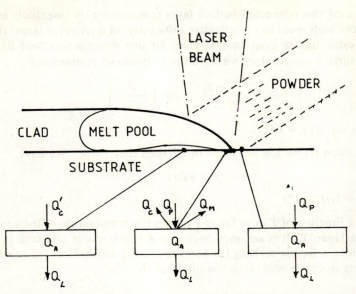

Fig. 6. Diagram of the major heat flow systems operating at the leading edge of a clad track.

This layer will have a temperature, T, decided by the heat balance on it:
Laser power absorbed = heat to melt arriving powder + convected energy
+ conduction losses.

$$\frac{4P(1-r)\,wdx}{\pi D^2} = \dot{m}\eta_m(C_p T + L_m)\,wdx + v_c w\delta h\varrho\, C_p\frac{dT}{dx}\,dx + q_c wdx \qquad (4)$$

Where symbols are listed at the end of the paper.
Hence the surface layer temperature, T, is given approximately by a relationship
of the form:

$$T = \frac{\pi D_b\left[\dfrac{4P(1-r)}{\pi D^2} - q_c - V_c\,\delta h\varrho C_p\dfrac{dT}{dx}\right] - L_m}{\dfrac{\dot{m}\eta_m}{C_p}} \qquad (5)$$

Now $v_c \propto dT/dx$: Thus:

$$T = AP - B(dt/dx)^2 - C$$

Since (dT/dx) is positively correlated to P this would signify that the convected
layer temperature, T, is not a very strong function of P.

Assuming this convected surface layer temperature is essentially constant and that the melt pool has an effective diffusivity of α (A value larger than the material value due to convection). Then for one dimensional heat flow with constant surface temperature we have the well-known relationship:

$$T^* = \frac{T - T_0}{T_1 - T_0} = erf \left(\frac{y^2}{4\alpha t} \right) \tag{7}$$

If $y = h_x = \dot{m}\eta_n x / v\varrho_p w\, LS)$ (8)

From equation [7] for a given value of T^* such that $T = T_m$. We have:

$$y = A\sqrt{\alpha t} \tag{9}$$

where: $A = 2erf^{-1}(T^*)$

A is only a function of P in so far as P affects the temperature of the superficial convective layer, T_1. It is assumed here that A is essentially constant.

The time. t, for the melting isotherm to reach a depth, h_x, is (x/v) and hence the melting isotherm will reach the interface if:

$$y = \frac{\dot{m}\,\eta_m\, x^n}{V\varrho_p\, w\, L^n\, S} \leq A\sqrt{\alpha \frac{x}{V}} \tag{10}$$

From which we can derive an expression for the distance. x. at which the interface is no longer molten:

$$x = \left[\frac{SA\sqrt{\alpha v}\; L^n\, w\varrho_p}{\dot{m}\;\eta_m} \right]^{\frac{2}{2n-1}} \tag{11}$$

If $n \leq 0.5$ then $x \to \infty$. i.e. melt interface never reaches the substrate surface except at clad initiation point.

Substituting values from the slow traverse result of Takeda's gives a value of $x = 0.6$ mm. If:

$\varrho_p = 7000$ kg/m³ : $w = 2 \times 10^{-3}$m: $L = 4.6 \times 10^{-3}$ m: $V = 4 \times 10^{-3}$ m/s
$\alpha = 1.5 \times 10^{-5}$ m²/s: $\dot{m} = 0.2 \times 10^{-3}$ kg/s: $\eta_m = 0.8$: $A = 0.5$: $n = 1$: $S = 0.5$

The value observed from the Takeda trace is ~ 0.6 mm.

For a satisfactory understanding of the factors affecting the value of x. it is necessary to know how L behaves with the laser parameters.

A heat balance on the melt pool, which is in a quasi-steady state yields a relationship of the form:

$$\eta_c P = k\, L\, w\, (dT/dZ)_{av} + \dot{m}\, \eta_m\, H_m \tag{12}$$

So

$$L = (\eta_c P - \dot{m}\eta_m\, H_m)/(kw\, (dT/dz)_{av}) \tag{13}$$

Where η_c = Optical coupling coefficient: k = Thermal conductivity of substrate W/mk; $H_m = (\varrho_p C_p T_m + L_m)$ = heat capacity of melted particles. J/kg

Now the temperature gradient at the melt interface is functionally described in Rosenthal's solution for a moving point source [12].

$$T - T_0 = \frac{P'/V}{2\pi k} \exp\left(\frac{-V\xi}{2\alpha}\right) \exp\left(\frac{-Vr}{2\alpha}\right)/r \tag{14}$$

From which we get:

$$\left(\frac{dT}{dz}\right)_{av} = (T_m - T_o)\left[\frac{2VZ_m}{2\alpha T_m} - \frac{1}{r_m^2}\right] \tag{15}$$

i.e. $\left(\dfrac{dT}{dz}\right)_{av} \alpha \ (aV + b)$ \hfill (16)

since at $x = y = 0$ $z \simeq r$. Note $b \simeq 1/h_x^2$ which is ignored.
Thus it is expected that $L = \eta_i F'/kwBv$
Where B is a constant, and $P' = (P - \dot{m} \ \eta_m \mu_m/\eta_c)$

$$\text{And so } x = \left[\frac{A\sqrt{\alpha} \ V^{(0.5-n)} \ W^{(1-n)} \ \eta_c^n \ P'^n \ \varrho_p}{k^n \ B^n \ S \ \dot{m} \ \eta_m}\right]^{\frac{2}{2n-1}} \tag{17}$$

This gives a clue to the parameter best describing the extent of dilution. That it to say the smaller, x, the less the level of dilution. Assuming $n = 1$, this signifies that $P^2/V\dot{m}^2$ should be smaller than a certain value, and the more convex the pool shape, i.e. $n \rightarrow 0.5$, the less significant in V.

The most significant point in this qualitative calculation is that it is highly probable that the tranquil zone observed by Takeda is in actual fact a resolidification almost immediately after forming the fusion bond. It thus explains the remarkably good results previously observed concerning the low dilution levels found with laser cladding.

The secondary observation from this analysis is that the shape of the leading edge of the clad affects the probability of dilution. The analysis also includes the functional effects of the operating parameters.

This surprising, and convenient, conclusion is supported by the fact that successful cladding can be achieved on scaled mild steel, the oxide scaling making no difference due to the fast flow of superficial material away from the clad interface: also EDAX analysis of heavily dilute traces shows the same tendency, as in Fig 7, where the step in the composition is indicative of some non-uniform mixing and therefore discontinuous solidification.

172

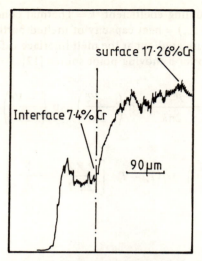

Fig. 7. Crystal spectrometer scan of Cr across the thickness of a highly diluted clad layer.

RAPID ALLOY SCANNING

One new way of using this unique process is to feed a multiple powder feed from several feed hoppers at the same time as shown in Fig. 1. While a clad track is being laid down one or more of the hopper feed screws (Quantum Laser Proprietary System) can be altered thus laying down a clad with a varying composition from one end to the other.

For example, this was done on the system Ni-Si-Cr-B with three powder streams of composition shown in Table 1. and the experimental conditions given in Table 2.

Twelve separate runs were each repeated in triplicate with the composition varying to cover the alloy range shown in the shaded area of Fig. 8. From these runs one of the triplicate tracks was surface ground and hardness values obtained along the track length. These values were plotted against the estimated composition as shown in Fig. 9 [9] producing a pseudo-ternary diagram with a good indication of the location of eutectics. Specific composition locations on the track were chosen for detailed analysis by WDX, EDX and SIMS in order to check the estimated compositions. All the tracks were examined by an Ardrox magnetic ink particle crack detector. In which it was found that the severity of the cracking rose with hardness, being particularly severe for values over 750 VHN. The microstructures (Figs. 10, 11) show a refinement of the dendrites as the eutectic is approached. This is unexpected since the thermal gradient would be reduced, if anything, with a lowering of the freezing point as the eutectic is approached. The solidification rate is unlikely to change significantly being dependent upon the traverse speed and dendrite growth direction. The effect is probably due to the reduced concentration difference in the liquid phase between the dendrites as the

Table 1. Composition of the powders used: (composition established by absorption and emission spectroscopy, carbon combustion and wet chemistry)

Alloy system 1: (Ni-Cr-Si-C-B)

	Powder		
Element	1 Ni-Si-Cr atomized 50–150 μm wt%	2 Cr-C crushed 50–150 μm wt%	3 Ni-B crushed 50–150 μm wt%
Ni	bal	—	bal
Cr	1.60	bal	trace
Si	5.60	0.06	0.15
C	trace	10.54	0.16
Fe	trace	0.29	0.8
B	trace	—	15.4
Al	trace	trace	trace
S	trace	trace	trace

Alloy system 2: (Mo-Ni-Cr-C-Si-B)

Element	Powder Mo based mixture atomized + crushed 50–150 μm wt%	Ni based alloy no 5 atomized 50–150 μm wt%
Mo	bal	—
Ni	14.19	bal
Cr	8.91	11.71
Si	0.03	3.43
Fe	0.17	5.12
C	1.08	0.60
B	2.62	2.61
Al	trace	trace
S	trace	trace

Table 2. Experimental conditions used

Laser power	1.6	kW
Spot size	5	mm
Traverse speed	6.6	mm/s
Mode structure	approx	TEMO1*
Powder delivery speed	1.6	m/s
Powder feed rates:		
Ni-Cr-Si	10	g/min
Cr-C	0–2.5	g/min
Ni-B	0–6	g/min
Mo based alloy	0–10	g/min
Ni based alloy	0–10	g/min
Initial temperature of substrate	20	C

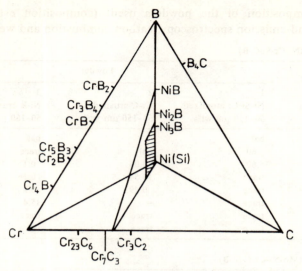

Fig. 8. Area of alloy variation examined for system 1 (Ni-Cr-Si-C-B).

eutectic represents a concentration limit. Thus from dendritic solidification theory it can be shown [12] that:

$$\Delta C_L \,(\text{max}) - GRl^2/2m_L D_L$$

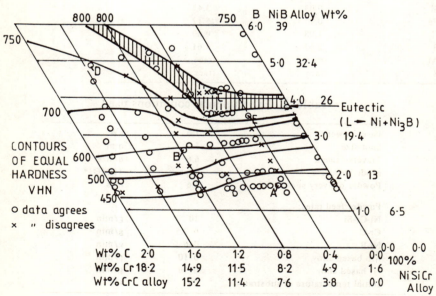

Fig. 9. Hardness vs composition for system 1 (Ni-Cr-Si-C-B).

Fig. 10. Micrograph of the clad structure of sample A (see Fig. 8 for composition)

Fig. 11. Micrograph of clad structure of sample D (see Fig 8 for composition).

176

where:

ΔC_L(max) = largest concentration difference in liquid phase.
G = thermal gradient C/m
R = rate of solidification m/s
l = half width of the liquid space between dendrite walls, m
m_L = slope of the solidus C kg/m^3
D_L = diffusivity of diffusing species, m^2/s

At the eutectic composition ΔC_L (max)$\rightarrow 0$ and thus $L\rightarrow 0$ creating the interdendritic phase and fine structure typical of a eutectic. As the composition is varied ΔC_L (max) would be essentially constant dependent only on maintaining constitutional supercooling to a very low value, probably similar to the variation in melting point due to the effect of curvature on the liquid/solid interface around the dendrites. If significant constitutional supercooling were to occur then branching would be expected and this would reduce ΔC_L (max), in fact altering the dendrite arm spacing, as we are discussing. However when the eutectic is sufficiently close then ΔC_L (max) would be expected to be restricted and hence the dendrite secondary arm spacing would alter, but separation of dendrites would increase due to increased eutectic phase. More work is required on this subject since it has serious implications on the interpretation of cooling rates which have been calculated in many papers via the dendrite arm spacing. Fortunately most previous work is based on constant composition alloy systems and hence after some calibration the results would be expected to be reasonably accurate.

A second alloy system was examined since the first system suffered from cracking with high hardness alloys when cladding with no preheat. The second system chosen included a Molybdenum alloy as given in Table 1.

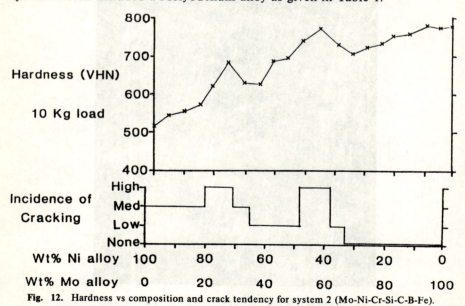

Fig. 12. Hardness vs composition and crack tendency for system 2 (Mo-Ni-Cr-Si-C-B-Fe).

The results of a rapid alloy scan are shown in Fig. 12 together with the cracking data.

CONCLUSIONS

The laser-cladding process using blown powder is capable of generating clad tracks of very low dilution. The reason is probably that there is rapid resolidification at the melt interface prior to convective stirring having any significant effect. The operating regions for producing such clads have been identified.

This property of generating low dilution clad tracks has been used to develop a novel technique for rapid alloy scanning. Two examples of the use of this technique have been shown.

ACKNOWLEDGEMENTS

The author would like to acknowledge the assistance of the SERC, Quantum Laser Corp and Control Laser Ltd. in the equipping of the laboratory in which this work was done. Also to Wall Colmonoy Ltd. for their collaboration in the alloy scanning work through an SERC/CASE award with P. Monson.

REFERENCES

1. Powell, J. Proc. Conf on Surface Engineering with Lasers, paper 17, Publ. Metal Society, London, (1985).
2. Weerasinghe, V.M. and W.M. Steen. *Proc. 4th Int. Conf. on Lasers in Material Processing*, Los Angeles Jan. (1983), (ed.) E.A. Metzbower, ASM Ohio, USA, pp. 166-175 (1984).
3. Steen, W.M. Proc. Conf. Advances in Coating Techniques, W.I., Cambridge. UK, pp. 175-187 (1978).
4. Jervis, T.R. In: *Laser Surface Treatment of Metals*, (eds.) C.W. Draper and P. Mazzoldi, *Proc. NATO ASI.* San Miniato. Italy, Martinus Nijhoff, Dordrecht, Netherlands. pp. 567-576 (1986).
5. Tardieu de Maleissye, J. Ibid., pp. 555-566.
6. Roos, J.R., J.P. Celis and W. Van Vooren. In: Ibid, pp. 577-590.
7. Hawkes, I.C. In: *Proc. 4th Int. Conf. on Lasers in Manufacture* (LIM4), Birmingham, U.K., pp. 19-32 (1987).
8. McIntyre, M. *Proc. 2nd Int: Conf on Applications of Lasers in Material Processing*, Los Angeles, USA (1983).
9. Steen, W.M., V.M. Weerasinghe and P. Monson. *Proc. SPIE Conf.* Innsbruck. Austria April 1986, SPIE. PO Box 10. Bellingham, Washington, USA, vol. 650, pp. 226-234 (1986).
10. Takeda, T., W.M. Steen and D.R.F. West. *Proc. LIM 2 conf.*, Birmingham, UK (ed.) M.K. Kimmitt, IFS (publications) Ltd., Kempston, UK, pp. 85-96 (1985).
11. Flemings, M.C. *Solidification Processing*, McGraw Hill, New York, USA, pp. 83 (1974).
12. Rosenthal, D. *Trans ASME* **68**, 849 (1946).

178

NOMENCLATURE

a, A	Constants
b, B	Constants
c	Constant
C_p	Specific heat J/kg C
ΔC_L	Largest concentration difference in liquid phase. kg/m^3
D	Incident beam diameter m
D_b	Diameter of powder stream
D_L	Diffusivity of diffusing species in liquid melt m^2/s
G	Thermal gradient 'C/m
H_m	Heat capacity of melted particles J/kg
ϱh	Approximate thickness of surface converted layer m
h	Depth of clad m
h_x	Depth of clad a distance, x, from leading edge
k	Effective thermal conductivity of melt W/mK
l	Half width of liquid space between dendrite walls—m
L	Length of clad melt pool—m
L_m	Latent head of fusion of clad particles J/kg
$\dot{m}$	Powder feed rate kg/s
m_L	Slope of the solidus C/kg/m^3
Ω	Cläd leading edge shape constant
P	Incident laser power W
R	Rate of solidification m/s
r	Reflectivity
r_m	Radius of melt isotherm
s	Shadow factor
S	Shape factor
t	Time
T	Temperature; T_m—melting point, T_o—initial, T_1—final,
T^*	$(T-T_o)/(T_1-T_o)$ dimensionless temperature
v_c	Convection velocity of surface layer m/s
V	Traverse speed m/s
w	Clad width m
x	Distance in clad direction m
y	Distance to side of clad centre m
z	Depth m
Z_m	Depth of melt isotherm
α	Effective thermal diffusivity m^2/s
η_c	Optical coupling coefficient
η_m	Powder catchment efficiency
ϱ	Pool density kg/m^3
ϱ_p	Particle density kg/m^3

8

Fundamental Analysis of the Continuous Casting Process for Quality Improvements

J.K. BRIMACOMBE AND I.V. SAMARASEKERA

The Centre for Metallurgical Process Engineering, The University of British Columbia, Vancouver, B.C. Canada V6T 1W5

ABSTRACT

The continuous-casting process for the production of semi-finished steel shapes has been subjected to detailed analyses, particularly with respect to aspects of quality: cleanness, cracks, segregation and shape. The most powerful approach to quality problems has been knowledge-based consisting of a combination of mathematical modelling, physical modelling, plant measurements and laboratory investigations.

Five fundamental aspects of continuous casting—fluid flow, heat flow, mechanical properties of steel at elevated temperature, stress generation and solidification—are discussed in this paper. Fluid flow, particularly in the tundish and mould, has a strong influence on the cleanness of the cast steel. The extraction of heat in the mould, sprays and radiation cooling zones is central to continuous casting and has been characterized quantitatively by mathematical models to predict the temperature field in the solidifying steel and in the mould wall. The ability of the models to simulate actual casting operations depends utterly on the accurate measurement of surface heat flux in the different cooling zones. The calculation of stresses generated thermally and/or mechanically in the steel as well as in the mould is necessary to comprehend the mechanisms by which cracks form or shape defects arise. Formulation of correct mechanisms also depends on an understanding of the high-temperature mechanical properties of steel. Finally knowledge of solidification phenomena under continuous-casting conditions leads to an understanding of segregation problems and the formulation of corrective measures. The knowledge-based approach is applied to two quality problems: transverse cracks in slabs and off-squareness/off-corner cracks in billets. But it is advocated that the methodology applies to the processing of any material.

INTRODUCTION

Continuous casting has emerged as one of the great technological developments of this century, replacing ingot casting and slabbing/blooming operations

180

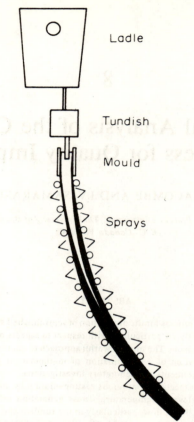

Ladle

Tundish

Mould

Sprays

Fig. 1. Schematic diagram of continuous-casting machine [1].

for the production of semi-finished shapes: slabs, blooms and billets. The process, as shown in Fig. 1, has been adopted worldwide by the steel industry over the last two decades owing to its inherent advantages of low cost, high yield, flexibility of operation, and ability to achieve a high quality cast product. Increasingly emphasis has focussed on the quality aspects of the process, to the extent that continuous-cast steel can be direct rolled or at least hot charged to a reheating furnace, without cooling for intermediate inspection. The attainment of quality rests on several factors: human (management, teamwork, commitment), the ability to monitor and control the operation effectively and, most importantly, on a fundamental understanding of the process.

This paper focuses on the fundamental aspects of continuous casting that underpin quality. The role of fundamental knowledge in analysing the process for the achievement of quality can be seen in Fig. 2. Central to this knowledge-based methodology is the mathematical and physical model that links fundamentals to the continuous-casting process. The models are invaluable

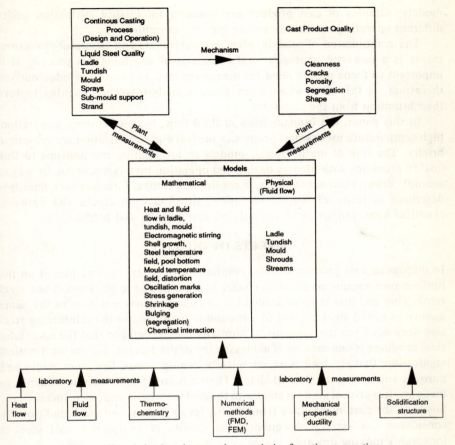

Fig. 2. Knowledge-based approach to analysis of continuous casting.

in the formulation of mechanisms of quality problems and thereby effectively link the process to product quality. By establishing the relationship between quality and process design/operation in this way, the minimization of quality problems can be achieved rationally and quantitatively.

Although at the heart of the process analysis, the development of mathematical and physical models is not sufficient, by itself, to link product quality to the casting operation. Measurements also are needed without which the modelling effort may result in misleading or incorrect results. In the laboratory, the characterization of the high-temperature mechanical properties of steel and regions of low ductility are necessary for stress models and an understanding of crack formation under continuous-casting conditions, for example. In the plant, measurements on an operating caster are needed to characterize heat extraction in the mould and sprays for the definition of boundary conditions in heat-flow models; samples of cast product are required for quality evaluation under different operating conditions, to cite but two examples.

The combination of models, plant measurements and laboratory measurements is a powerful approach to the analysis of any materials process. It is important to focus on the need for measurements, and the knowledge derived therefrom, at this time when many process analysts seem unwilling to turn their attention from the computer.

In this paper, the fundamentals of fluid flow, heat flow, stress generation, high-temperature mechanical properties of steel and solidification are presented briefly. The role of mathematical models in formulating mechanisms to link quality problems with process design and operation then is discussed, by way of example drawn from our research on continuous casting. First however, quality is described in terms of desired cleanness, freedom from cracks and porosity, chemical homogeneity (segregation) and shape of the cast product.

ASPECTS OF QUALITY

In discussing cast product quality, emphasis ultimately must be placed on the performance requirements of the steel in a particular application. Thus steel reinforcing bar that is to be imbedded in concrete does not require the same quality as rolled steel applied to automobile manufacture-the reinforcing steel also does not fetch the same price! Similarly it is recognized that the successful steel producer is one capable of achieving the desired quality for the most critical application, that is, clean steel, chemically homogeneous, free from cracks and porosity and having the desired shape. Then it is possible, at least in principle, to make any less critical grade of steel with the desired quality and price. Increasingly however, the customer is specifying quality levels that exceed those traditionally considered necessary; consequently the quality of continuous—cast steel is inexorably moving upward.

Cleanness

An example of a cleanness problem which could adversely impact on quality can be seen in Fig. 3. Shown is a sulphur print of a transverse section of a billet cast on a curved mould machine, Fig. 1. A band of black spots, each of which is a large oxide inclusion (~ 250 μm), appears adjacent to the inside radius face of the billet. The inclusions have segregated to this location because being less dense than steel, they have risen upward through the liquid pool until becoming trapped by the solidification front advancing from the inside radius face. The inclusion band is a zone of weakness in the billet and could cause crack problems, for example, during subsequent forging of bar rolled from the steel. Similar bands of aluminate inclusions can be found in aluminium-killed slabs cast on a curved mould machine [3]. The control of the overall inclusion content requires attention to the following details:

    i) clean steel, preferably via a ladle furnace delivered to the caster;

Fig. 3. Sulphur print of a transverse section of a 102 mm × 152 mm billet cast on a curved mould machine showing band of reoxidation inclusions (black spots) adjacent to the inside radius face [2].

ii) minimization of oxygen transfer from the air or refractories;

iii) prevention of exogenous inclusion pickup from refractories or ladle, tundish and mould powders;

iv) control of fluid flow in the tundish and mould to maximize inclusion float-out; and

v) adoption of optimum mould powder as well as start-up/shut-down procedures.

The analysis of fluid flow with physical and mathematical models will be addressed in the next section.

Cracks

Figure 4 shows schematically many of the different types of cracks that may be found on the surface or in the interior of continuously cast billets and slabs [4]. Surface cracks are a serious quality problem because the cracks oxidize and give rise to oxide-rich seams in the rolled product. Internal cracks also can be a problem particularly if during rolling, they do not close leaving voids in the steel product. The number of potential crack types is larger than would be found normally in a cast ingot because the continuously cast strand is forced to move through the machine and is subjected to withdrawal, bending/straightening and bulging stresses. Moreover as the strand moves from one cooling zone to the next, changes in heat extraction cause shifts in thermal gradients through the

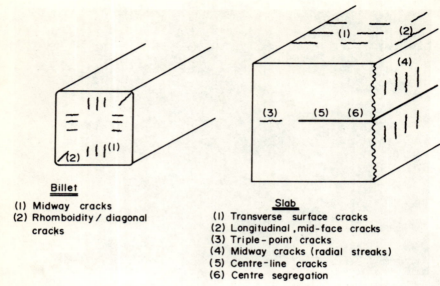

Billet

(1) Midway cracks
(2) Rhomboidity / diagonal cracks

Slab

(1) Transverse surface cracks
(2) Longitudinal , mid-face cracks
(3) Triple-point cracks
(4) Midway cracks (radial streaks)
(5) Centre-line cracks
(6) Centre segregation

Fig. 4. Schematic drawing of different types of cracks observed in continuously cast billets and slabs [4].

solidifying shell and stress generation resulting from differential expansion or contraction.

An understanding of the mechanism of formation of each crack type and the influence of casting parameters on cracking requires a fundamental analysis of stress generation and knowledge of the ductility of steel at elevated temperatures. For example, it has been found that high sulphur contents or excessive superheat adversely influences the formation of internal cracks [2]. This is readily understood when it is appreciated that high superheat favours the formation of a columnar structure having easy crack paths while a high sulphur content reduces the high-temperature low ductility of the steel in the temperature range where most cracks form. These points will be taken up again in later sections. The mechanism for the formation of the different cracks shown in Fig. 4 has been established in numerous studies, many of which were reviewed earlier by Brimacombe and Sorimachi [2]. In most instances, mathematical models have played a significant role in the elucidation of the mechanisms which may be quite complex, as well as in the implementation of remedial action. In a modern continuous-casting operation seeking to make high-quality steel, the goal must be zero cracks in the cast product.

Macrosegregation and porosity

Macrosegregation and attendant centreline porosity are other aspects of quality. Figure 5 shows concentration profiles for sulphur and carbon measured between

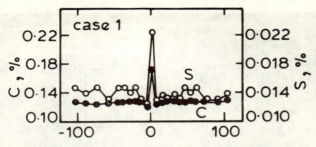

Fig. 5. Transverse concentration profiles of carbon and sulphur in an as-cast slab [5].

the inside and outside radius faces of a slab. At the centreline, the concentrations of these elements are significantly higher than in the remainder ·of the slab. Similar observations have been made on billets [6]. The presence of macrosegregation can cause problems in, for example, the controlled cooling of wire rod rolled from high-carbon billets where martensite may form preferentially to pearlite in regions of manganese segregation [7]. Similarly hydrogen-induced cracks and welding cracks may arise in plate rolled from slabs containing manganese segregation [7].

The origins of the macrosegregation lie in the distribution of elements like carbon, manganese, sulphur and phosphorus between solid and liquid during the freezing of steel, which is well known, as well as the solidification structure and convection near the bottom of the liquid pool. An understanding of the solidification process on the scale of individual dendrites derived from relatively simple solidification models [9], together with numerous plant trials, have shed light on the mechanism of the macrosegregation. The role of bulging in the case of slabs and of structure for both billets and slabs will be addressed later in the paper. The importance of steel superheat and fluid flow induced by electromagnetic stirrers will then be seen.

Shape

Shape defects have posed a greater quality problem in the continuous casting of billets than of slabs or blooms. In the latter, an off-corner longitudinal depression is sometimes seen running the length of the cast section but it normally can be controlled by adjustment of the mould taper. Off-squareness seen in billets, Fig. 6, is more serious because it usually is accompanied by off-corner, sub-surface cracks. Moreover, severely off-square billets may ride up on one another in pusher-type reheat furnaces and corners may be folded over during hot rolling to generate seams in the final product.

To understand the genesis of off-squareness requires knowledge of heat extraction in billet moulds, model-predicted thermal distortion of the moulds and the interaction between the mould and solidifying shell at the meniscus during the mould oscillation cycle [10-12]. The mechanism of off-squareness, which is eluci-

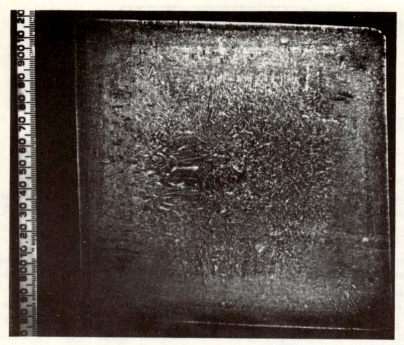

Fig. 6. Macroetched transverse section of 143 mm square billet exhibiting off-squareness.

dated in a later section, clarifies the effect of such variables as carbon content of the steel, cooling water velocity (and quality), mould taper and mould oscillation stroke/frequency so that an optimum mould system can be defined [13].

FLUID FLOW

The flow of molten steel through nozzles and in the ladle, tundish and mould is driven by gravity and governed by the equations of continuity and motion. Thus, in principle at least, it is possible to model the flow mathematically and to predict the velocity field and residence-time distribution in, for example, the tundish although the calculations are complicated by turbulence and the three-dimensional nature of the flow [14-16]. Thus mathematical models have been applied to the study of the effect of dams and weirs [15] and of inclined walls [16] on flow in tundishes. An example of the predictions by El-Kaddah and Szekely is shown in Fig. 7.

But the greatest use to date has been made of physical models for the establishment of the principles of tundish design, in particular, and flow at different stages from the ladle into the mould. It is important to note that a unique optimum tundish design suitable for any casting machine is not achievable since the heat size, cast section dimensions, number of strands, strand spacing and

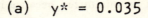

(a)   y* = 0.035

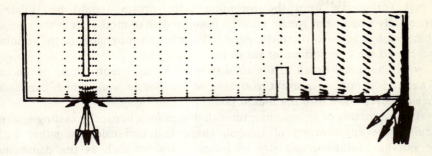

(b)   y* = 0.8

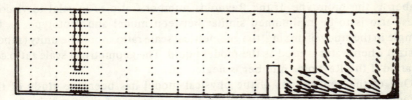

Fig. 7. Velocity vector plot in the longitudinal plane of a tundish in the presence of a dam and weir [15].
(a) near the inlet stream
(b) near the side wall

pouring rate vary from one operation to another. Nonetheless key features of a well-designed tundish can be identified.

i) Inclusion float out should be maximized. Thus the tundish volume should be large although other mitigating factors limit its size. For example, if heats having different composition are to be cast in sequence, without interruption of the casting operation, the volume of the "mixed" steel in the tundish must be minimized because it is off-grade and may represent a yield loss.

ii) The residence time of steel flowing to each of the strands must be the same to ensure uniformity of steel quality.

iii) Short-circuiting of steel through the tundish should be minimal, i.e. to the extent possible, plug flow should be achieved.

iv) The flow pattern in the tundish should permit much of the steel to move close to the surface where inclusions floating out can be absorbed by the tundish slag. This is necessary because the distance the inclusions travel by buoyancy alone is small in the time available in the tundish.

v) Dead volume should be minimized because it effectively reduces the residence time of the steel.

vi) Turbulence from the incoming ladle stream should be isolated, particularly when open stream pouring from tundish to mould (as in the case of billet casting) because the turbulence influences the roughness of the tundish stream and the entrainment of gas in the mould pool.

vii) The metal depth in the tundish must be sufficiently great, in excess of about 500 mm, to prevent vortexing at the nozzle wells which can draw the slag cover down into the mould pool.

Determination of an optimum tundish design for a particular casting system requires the specification of tundish shape and dimensions together with the number, location, and size of internal devices such as the dams and weirs mentioned earlier, for flow control.

The principles on which the physical models of tundishes and moulds are based, have been set out by Heaslip et al. [17]. The flows are all gravity driven, so that the Froude number ($= u^2/gL$) is an essential similarity parameter. Thus any physical model of a tundish and mould must have the same Froude number as the prototype. If the Reynolds number (ratio of inertial to viscous forces) also is to be maintained similar between model and prototype, as for example in the study of mixing, the geometrical scale ratio should be unity (model full size) when water is used as the model liquid. The geometrical scale ratio can be varied if the Reynolds number is neglected.

One of the applications of the physical model has been the study of the influence of ladle stream turbulence on the roughness of the tundish streams [17]. Figure 8 shows three streams from the tundish under conditions where the ladle stream was turned off (a) or on (b). With the ladle stream off, all three tundish streams are seen to be smooth; however when the ladle stream was on, and falling between Strands 1 and 2 (left to right), the streams issuing from these nozzles became very rough. This was important because stream roughness affects the quantity of air entrained by the tundish stream in the mould, in the absence of stream protection, and hence the extent of reoxidation of the steel. The reoxidation causes large oxide inclusions to form, giving rise to scum on the surface of the pool, which is swept over the side to become entrapped as a slag patch in the billet surface. Thus the water model helped to link a surface quality problem to fluid flow.

HEAT FLOW

The high productivity associated with the continuous-casting process is directly linked to the high rates of heat extraction in the mould and sprays. The ability to cast quality steel, however, critically depends on the control of heat transfer in each of these cooling zones; and hence, it is important to have a fundamental understanding of the governing phenomena. Determination of the heat flow quantitatively is also important because it allows the prediction of the shell

Fig. 8. Tundish streams 1, 2 and 3 (left to right) modelled by water (a) with ladle stream off (b) with ladle stream falling between Nozzles 1 and 2 [17].

profile, the pool depth and temperature distribution as a function of casting variables. In modern slab casters this predictive capability is built into process control computers that monitor the progress of casting; for example the sprays are automatically adjusted when the caster is slowed down for tundish nozzle or width changes. It will also be evident that a knowledge of the thermal history of the shell permits linkages to be established between a given casting defect and adverse thermal conditions in the caster.

Mould heat transfer

In the mould, heat is transferred from the steel to the cooling water via: an air gap separating the mould and the strand; the mould wall; and the mould/cooling water interface. Of these, the air gap constitutes the largest resistance to heat flow [18]. Because the gap size is small, particularly in the upper region of the mould, conduction through the gap is the dominant mode of heat transfer; and consequently the magnitude of the heat transferred is almost inversely proportional to gap width.

The strand/mould gap is a complex function of several variables so that its formation has not yet been accurately predicted mathematically. It is clear that

Table 1. Parameters influencing mould heat transfer

Parameter	Influence	Design and operating variables
Shrinkage of steel	This is the sum of the thermal and phase transformation shrinkage which is governed by the temperature distribution of the shell in turn affected by mould heat transfer.	Steel composition particularly carbon content [20, 21].
Mechanical properties of the steel	The evolution of the gap depends on the ability of the shrinking shell to resist ferrostatic pressure which tends to push it towards the mould wall [19]. Affected by shell temperature and mould heat extraction.	Steel composition particularly C, S and P contents [2].
Mould shape	The narrow face of a slab mould and the walls of a billet mould are tapered to compensate for shrinkage. In addition billet moulds distort because of differential thermal expansion coupled with the lack of constraint [22]. In slab moulds, distortion is less of a problem but could occur between the bolts which connect the copper plates which form the mould to the steel backing plates [23].	Taper. (Billet moulds) Copper properties [13] Type of constraint at top of mould tube. Mould cooling water velocity. Mould wall thickness. (Slab moulds) Copper properties. Bolt spacing and cooling channel geometry. Copper thickness. Cooling water velocity.
Depth and uniformity of oscillation marks	Oscillation marks locally enlarge the strand/mould gap width. In billet casting with oil lubrication, it has been postulated that these marks occur by mechanical interaction of the mould with the billet during the down-stroke of the oscillation cycle [11]. In slab casting with mould powders they form because of the pressure generated in the mould flux channel between mould and strand during the down stroke of the oscillation cycle [24].	Steel composition particularly carbon. Oscillation stroke and frequency. Properties of mould lubricant both oils and powder. (Billet moulds) Magnitude of taper at meniscus level [11]. All the variables that affect billet mould distortion listed for mould shape [13].

the gap width varies in both the longitudinal and transverse directions resulting in a non-uniform heat-extraction pattern [19].

Table 1 lists the parameters that influence mould heat transfer, together with a mechanistic description of their influence and the design and operating variables which affect each parameter. It is seen that all the parameters which affect gap width, namely steel shrinkage, mechanical properties of the steel, mould distortion and depth of oscillation marks, are in turn directly or indirectly affected by mould heat transfer which makes accurate computation of mould heat transfer from first principles very difficult.

Mould temperatures and mould heat-flux profiles

Owing to the difficulty of measuring or calculating gap widths, efforts have been

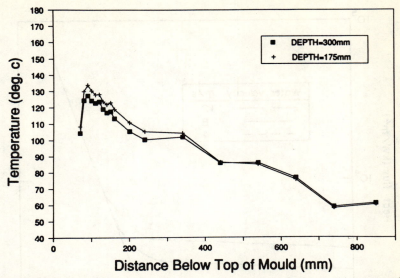

Fig. 9. Time-averaged axial temperature profiles in the narrow face of a slab mould, corresponding to different depths of nozzle submergence [27].

directed at calculating heat fluxes from the time-averaged response of thermocouples embedded in the mould wall a set distance away from the hot face [10]. Measurements have been made in billet moulds over a number of years [11, 25, 26] and more recently in slab moulds [27] and horizontal continuous casting moulds [28].

The time-averaged temperature profiles measured 6 mm from the inside of the narrow face of a slab mould for different depths of nozzle submergence are shown in Fig. 9 [27]. It is evident that the depth of submergence has an effect on mould temperatures over the top 400 mm of the mould, but little effect beneath this position. The mould temperature is highest in the meniscus region (~ 90 mm) and decreases with distance below the meniscus due to the increase in steel/mould gap width. These temperature profiles will be converted to hot-face heat-flux profiles with the aid of a three-dimensional mathematical heat-flow model of the mould which is currently being developed [27].

In the case of billet casting, two-dimensional, finite-difference models have been employed to back calculate the axial mould heat-extraction profiles from mould temperatures measured down the centreline of a given face. The procedure, which has been described elsewhere [10], relies on an accurate knowledge of the water velocity in the cooling channel. Measurements have been conducted with pitot tubes to determine the velocity at various locations on each face of the mould at individual plants as part of the test programme on measurement of mould temperatures. The heat-transfer coefficient at the cold face which critically influences mould temperature for a given steel-to-mould

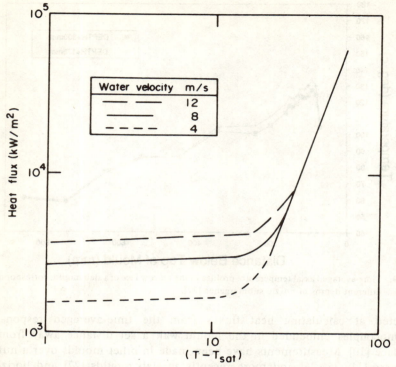

Fig. 10. Forced convection-boiling curves for subcooled water flowing at three different velocities [22].

heat flux may be determined from empirical correlations of forced convection and boiling, graphically shown in Fig. 10, for a given set of conditions. It has been shown that the mould cooling, particularly for billet casting, should be maintained in the forced convection regime in order to achieve a stable operation [22]. Boiling causes large fluctuations in mould wall and cooling water temperature, resulting in time-dependent variations in billet mould shape, which leads to non-uniform heat transfer around the billet periphery, off-squareness and off-corner cracking. These subjects will be taken up again later in the paper.

The heat-flux profile obtained for a medium-carbon billet cast through a single tapered mould, A, is compared with that pertaining to a double tapered mould, B, in Fig. 11. It is evident that there are significant differences in the pattern of heat extraction in the meniscus region. In the double-tapered mould the steep upper taper reduces the effects of thermal mould distortion (to be addressed later) and the resulting negative taper [22] on the air gap and heat extraction. However the effects of mould lubricant on heat transfer are enhanced as there is greater opportunity for intimate contact between the mould and strand if lubrication is inadequate. For instance the second peak in heat flux of

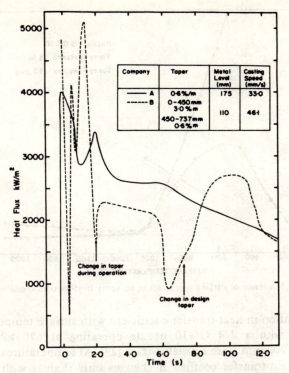

Company	Taper	Metal Level (mm)	Casting Speed (mm/s)
—— A	0·6%/m	175	33·0
---- B	0 - 450 mm 3·0 % m	110	46·1
	450 - 737mm 0·6 %·m		

Fig. 11. Axial heat-flux profiles measured in single-taper (A) and double-taper (B) billet moulds [25].

approximately 4000 kW/m² in the profile from Company B is thought to be due to sticking. The extremely high third peak of 5100 kW/m² is most likely due to improved contact between the mould and steel at the location where the mould tucks back in below the point of maximum outward bulging. This phenomenon occurs later in time for Company A because of the slower casting speed. The decrease in heat transfer from dwell times of 1.7 s to 8.5 s (188 mm to 502 mm) with Mould B is due to a loss of taper in this mould due to permanent distortion. This was detected from mould internal dimension measurements made between heats. The increasing heat flux below 6.3 s in Mould B is probably due to bulging of the shell against the mould stemming from the action of ferrostatic pressure on the strand, over a region where the taper is clearly insufficient. Thus it is seen that in billet moulds, heat transfer is extremely sensitive to mould shape because of the dominant role of the gap on heat transfer.

Spray heat transfer

Beneath the mould the strand is cooled by banks of pressure-atomized water sprays or air-mist sprays; the latter are employed mostly on slab casters. Figure 12

194

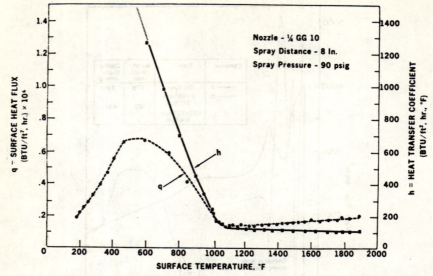

Fig. 12. Influence of surface temperature on spray heat-transfer coefficient [29].

shows the variation in heat-transfer coefficient with surface temperature during spray cooling with a 1/4 GG10 nozzle operating at 620 kPag (90 psig), approximately 200 mm from a billet surface [29]. At temperatures above 550°C, (1022°F) the heat-transfer coefficient changes only slightly with temperature, which is characteristic of film boiling [30]. At lower surface temperatures the heat-transfer coefficient increases sharply as the mechanism changes to transition boiling wherein the steam barrier breaks down. The critical temperature at which this occurs is termed the Liedenfrost temperature which increases with increasing water flux [31]. Spray cooling in the transition boiling regime leads to high heat-extraction from the billet surface and attendant overcooling, which is undesirable.

Determination of spray heat-transfer-coefficients

Laboratory experiments and in-plant studies have been undertaken to determine the relationship between the spray heat-transfer coefficient and spray water flux [29, 31–45].

$$h = \alpha \dot{W}^n \qquad (1)$$

The findings of one investigation by Jeschar et al. [32] are shown in Fig. 13 where n is seen to have a value of unity. In other studies, the exponent has varied from 0.5 to 1.0. The water flux, $\dot{W}$, is dependent on spray water pressure, nozzle-to-steel distance and nozzle type and has been determined experimentally for a variety of nozzles [29].

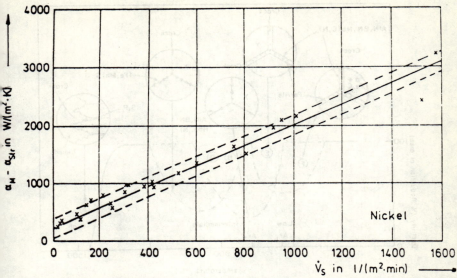

Fig. 13. Spray heat-transfer coefficient (radiation subtracted) as a function of water-flux density under conditions of film boiling [32].

Spray design principles

For a given caster, the spray system can be designed by combining an understanding of the genesis of particular defects with a mathematical model capable of simulating the thermal history of steel for a given spray configuration [4]. It involves the following steps:

    i) Determination of the thermal conditions, e.g. strand surface temperature distribution, or cooling rate in the sprays that minimize the formation of a given defect.

    ii) Calculation of the heat-transfer coefficient distribution, using the optimum thermal conditions from (i) as input to establish the heat-extraction requirements of the sprays. This is usually accomplished with the aid of a heat-flow computer model of the continuous-casting process.

    iii) Based on empirical relationships between heat-transfer coefficients and spray-water flux, Eqn. (1), calculation of the desired water flux through the secondary cooling zone.

    iv) Finally, determination of the nozzle type, water pressure, nozzle-to-nozzle and nozzle-to-strand distances from empirical correlations relating these variables to the spray-water flux.

    Examples of this approach applied to both billet and slab casters have been presented previously by Brimacombe et al. [4].

196

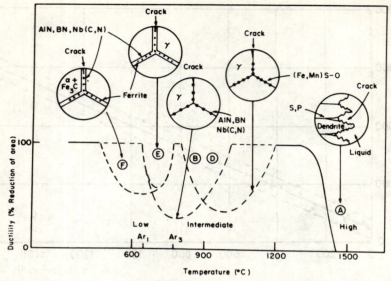

Fig. 14. Schematic representation of temperature zones of reduced hot ductility of steel related to embrittling mechanisms [46].

MECHANICAL PROPERTIES OF STEEL AT ELEVATED TEMPERATURE

Hot ductility of steel

It is well documented that steel has reduced ductility over specific temperature ranges which has important implications for crack formation. Figure 14 shows schematically the temperature zones of reduced hot ductility of steel and their corresponding embrittling mechanisms based on a review of the literature by Thomas et al. [46]. It is evident that there are three zones of reduced ductility; of these only the high- and low-temperature zones contribute importantly to crack formation in continuous casting, and will be discussed in greater detail [2]. The intermediate-temperature zone has rarely been shown to be responsible for cracking in continuous casting [2].

High-temperature zone

This zone of low ductility, which has been the subject of numerous investigations [47–58], is operative at temperatures within 30 to 70°C of the solidus temperature, and the associated strain-to-fracture of steel is less than 1 pct. [46]. The ductility loss is associated, as depicted in Fig. 14, with the microsegregation of sulphur and phosphorus at solidifying dendrite interfaces. This solute enrichment locally lowers the solidus temperature giving rise to a "zero" ductility temperature below the bulk solidus temperature (corresponding to the mean composition), as shown in Fig. 15. Tensile strains applied to the steel in this temperature zone cause the

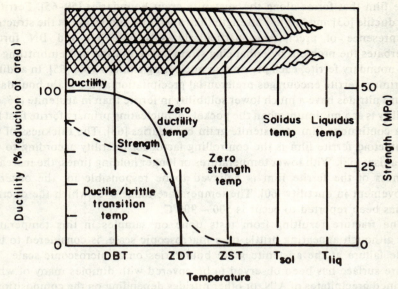

Fig. 15. Mechanical properties in the high-temperature zone of reduced ductility and corresponding schematic presentation of solid/liquid interface during casting [46].

dendrites to separate and the resulting fracture surface exhibits a smooth, rounded appearance characteristic of liquid film failure. The presence of manganese is beneficial, particularly at Mn/S ratios greater than 20 [53], since it preferentially combines with sulphur to form less harmful MnS precipitates, minimizing liquid film formation. Increasing contents of S [47], P, Sn [55] and Cu [56] all worsen the ductility which has also been shown to be relatively insensitive to strain rate and subsequent thermal treatment.

With the exception of transverse cracks in slabs, all cracks observed in continuously cast steel form in this zone of low ductility. Thus the depth of the cracks from the surface becomes very important since it indicates the shell thickness at the time of crack formation. Then if the shell profile is known from model predictions, the distance below the meniscus where a crack forms can be determined by matching crack depth with the profile. This can be a very useful quality tool.

Low-temperature zone

The low-temperature zone of low ductility in steel occurs in the two-phase austenite-ferrite region below the Ar_3 temperature. It corresponds to Zone E in Fig. 14 [46]. The mechanism of embrittlement, which has been reviewed by Thomas et al. [46], has been attributed to strain concentration in the primary ferrite film that forms along the austenite grain boundaries [59–65]. Ferrite is more ductile [63] and has less strength than austenite and weakens the structure. The presence of precipitates particularly AlN, Nb(CN) and BN further exacerbates the problem by enhancing strain concentration and embrittling the grain-boundary ferrite, each precipitate nucleating a micro-void [65]. In addition the primary ferrite encourages preferential precipitation at the grain boundaries because nitrides have a much lower solubility in ferrite than in austenite [66, 67]. Ductility is at a minimum when the pockets of nucleating primary ferrite first link into a continuous film at austenite grain boundaries [68]. The thickness of this proeutectoid ferrite film is the controlling factor for ductility according to this mechanism [68]. With lower temperatures or longer holding times, the increased thickness of the ferrite film is believed to be responsible for the observed improvement in ductility [60]. The temperature range over which the ductility loss has been reported to occur is $500 \sim 900°C$.

The fracture resulting from tests done on samples in this temperature zone, although appearing brittle on a macroscopic scale, is considered to be a ductile failure at the austenite grain boundaries on a microscopic scale. The fracture surface has been observed to be covered with dimples many of which contained precipitates of AlN, or other nitrides depending on the composition of the steel [46]; a few sulphides and oxides also have been found. In steels containing Nb or B, Nb(CN) and BN precipitates often have been found both on the fracture surface at the prior austenitic grain boundaries and within the matrix.

STRESS GENERATION

During its passage through the caster, the solidifying strand is subjected to varying thermal conditions and mechanical loading both of which contribute to the generation of stresses and strains. In order for cracking to occur at a given location, the two conditions that must be satisfied are:

    i) the stress-strain state must be tensile in nature; and

    ii) the fracture strength, or the strain-to-failure of the steel must be exceeded.

It was shown in the previous section that steel is particularly susceptible to cracking in certain temperature ranges owing to reduced ductility. In this section, factors that give rise to a tensile stress/strain state are identified.

Thermal stresses

A fundamental tenet governing thermal stress generation is that it requires free expansion or contraction to be constrained or the gradients in the material to be non-linear. In the continuous casting of steel, the conditions in the strand approach that of generalized plane strain which allows for some longitudinal expansion (casting direction) minimizing the generation of longitudinal stresses and strains. Furthermore the longitudinal thermal gradients are quite shallow as compared to those in the transverse plane.

Thus transverse cracking, which requires a longitudinal stress or strain component, can rarely be linked to adverse thermal conditions and is almost always mechanical in origin. This defect will be discussed later. In the transverse plane, however, there is greater restraint to free-expansion which, coupled with the steep and frequently non-linear temperature gradients, gives rise to high transverse thermal stresses. Moreover sudden changes in heat-extraction rates cause the thermal gradients to shift, particularly at the surface. Preferential expansion or contraction of these regions also generates stresses in the transverse plane. Owing to the high temperatures in the solidifying strand, rapid relaxation of these stresses can occur due to creep; nonetheless the resulting strain, if excessive relative to the strain-to-fracture and if tensile in nature, can lead to longitudinal cracks. Although many researchers have calculated the stress-strain state in a solidifying strand for simplified conditions, computations have not been made for all possible situations relating to quality. This is partly due to the complexity of the problem, coupled with the lack of constitutive equations to describe the mechanical behaviour of steel at very high temperatures, and the difficulty of quantitatively describing the mechanical boundary conditions at the mould/steel interface.

Thus only a qualitative description of the adverse thermal conditions and the resulting stresses and associated defects are presented in Table 2. Off-squareness due to uneven cooling in the mould or sprays [13, 22], gives rise to tensile stresses at the obtuse angle corners, or off-corner regions, of billets which could result in longitudinal corner cracks or off-corner cracks [13]. Reheating of the surface of billets below the mould or the sprays may also cause an expansion of the surface layers which imposes tensile stresses at the solidification front where the steel has the lowest ductility [4]. If there is binding at the corners of the billet within the mould then excessive cooling at some location of a face could cause localized tension and longitudinal depressions and cracking. Overcooling in the meniscus region or in the top spray zones generates surface tensile strains leading to longitudinal facial cracks in slabs.

Mechanical stresses

There are numerous sources of mechanical stresses in the solidifying strand and these are also listed in Table 2. Sticking in the mould due to improper lubrication, oscillation conditions, or excessive taper causes the withdrawal forces to act on the strand. This generates axial tensile stresses and strains that concentrate at locally thin regions of the shell such as at oscillation marks to cause transverse depressions or transverse cracks. In slab casting, bulging between the rolls followed by the rolling out of each bulge results in cyclic stresses in the strand [70]; the stresses at the solidification front where the steel has the lowest ductility is compressive at the position of maximum bulging and tensile beneath the roll at the solidification front as squeezing of the bulge occurs. Unbending on a liquid core has also been shown to give rise to tensile stresses and strains in the upper

Table 2. Stresses acting on the solidifying steel strand during continuous casting

Origin of stress	Cause of related factors	Nature of stress field	Type of quality problem
Thermal	Off-squareness due to uneven cooling in mould or sprays (billets) [13, 22]	Tensile stresses at obtuse angle corners and compression at acute angle corners.	Longitudinal off-corner cracks or diagonal cracks at obtuse angle corners.
Thermal	Reheating of surface in sub-mould region or beneath sprays (billets) [4].	Expansion of surface region imposes tensile stresses and strains at solidification front and compressive stresses on surface.	Longitudinal internal cracks midway through section; termed midway cracks.
Thermal	Excessive cooling over some region of the face coupled with binding of corners. Results in tensile stresses in region where localized cooling prevails.	Thermal stresses and strains where there is localized cooling.	Longitudinal depression with subsurface longitudinal facial crack.
Thermal	Excessive cooling near meniscus or in upper spray zones, particularly of 0.09–0.12% carbon slabs [69].	Tensile stresses generated on surface [69].	Longitudinal facial cracks in slabs [69].
Mechanical	Sticking in the mould due to improper lubrication, oscillation conditions or excessive taper.	Axial tensile stresses.	Transverse depressions and transverse cracks.
Mechanical	Bulging between rolls and the subsequent rolling out of each bulge in slab casting [70].	Axial tension near solidification front [70].	Internal cracks (Radial streaks in longitudinal sections).
Mechanical	Unbending on a liquid core in slabs.	Axial tension in the upper shell of the wide face [71].	Radial streaks.
Mechanical	Bulging in lower regions of the mould or in the sub-mould [12]. Improper lower taper. (Billets)	Transverse tensile stress adjacent to solidification front at off-corner location in billets [12].	Off-corner internal cracks [12].

shell and compression in the lower shell of the wide faces of a slab [71]. In billet casting, bulging of the shell in the lower region of the mould, if the taper is insufficient, causes a hinging action at off-corner sites and tensile strain at the solidification front [12].

SOLIDIFICATION

Two aspects of solidification that must be understood in continuous casting are the structure (columnar versus equiaxed) and the growth of the solid shell encasing the liquid pool. The cast structure is critically important to quality because it influences both internal crack formation and macrosegregation. The uniformity of growth of the solidifying skin, similarly, can exacerbate the formation of longitudinal surface cracks in slabs and the frequency of transverse depressions and breakouts in billet casting (Fig. 15).

Cast structure

A number of factors have a significant influence on the size of the central equiaxed zone relative to that of the surrounding columnar zone in a continuously cast section shown in Fig. 16:

   i) superheat of the steel
   ii) steel composition
  iii) fluid flow in the liquid pool
  iv) section size
   v) machine design (curved vs. straight machines) [72].

Of these, steel superheat (usually measured in the tundish) has a dominant effect. As can be seen in Fig. 17, the columnar zone is favoured at the expense of the equiaxed zone with increasing steel temperature (liquidus + superheat) [73].

Fig. 16. Macroetch of a transverse section of a billet showing columnar and equiaxed cast structures.

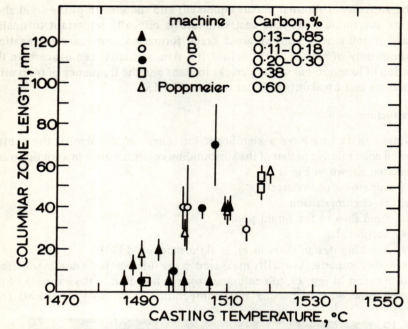

Fig. 17. Variation of length of columnar zone with casting temperature for several grades of steel [73].

From numerous studies, the largest effect is seen when the superheat is less than 30°C. Thus low superheat is desirable (to the extent possible without freezing off nozzles) to maximize an equiaxed structure which is more crack resistant and promotes reduced macrosegregation.

The effect of carbon content on the columnar zone length of continuously cast billets is shown in Fig. 18 [12]. Thus the equiaxed structure is favoured in the medium-carbon range, from about 0.17 to 0.38 pct C. Similar observations have been made in a recently completed study on electromagnetically stirred billets [26]. Increasing phosphorus content from 0.008 to 0.02 pct in 0.13–0.20 pct-carbon billets also causes the columnar zone to shrink [12].

Induced fluid flow, such as by electromagnetic stirrers, can markedly enhance the growth of the equiaxed zone, particularly if conducted in the mould at moderate superheats. Stirring below the mould also can interrupt the growth of columnar dendrites, although caution must be exercised to minimize the formation of white solidification bands (as observed in sulphur prints of as-cast sections).

A large section size also is favourable to the growth of an equiaxed zone. This is a major reason for the installation of casting machines for the production of large (e.g. 400 × 500 mm) blooms.

Interestingly the design of the casting machine, i.e. whether it is straight

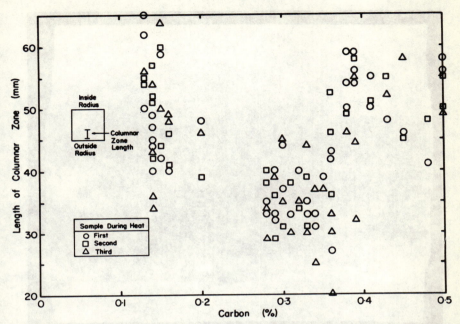

Fig. 18. Influence of carbon content of the steel on the columnar zone length, measured from the outside-radius face [12].

or curved, also influences the cast structure and this provides an important clue about the phenomena governing the columnar-to-equiaxed transition. Figure 19 shows a longitudinal section of a 133-mm square billet to which radioactive tracer (Au^{198}) was added during casting through a curved-mould machine, Fig. 1. The dark region in Fig. 19 is tracer rich and was liquid at the lime of tracer addition. From a careful examination of this autoradiograph, the length of the columnar zone adjacent to the inside-radius (top) face is seen to extend to near the centreline of the billet whereas the columnar zone adjacent to the outside radius (bottom) face is considerably shorter. This observation is characteristic of steel sections cast on a curved-mould machine relative to the axisymmetrical structure obtained with straight machines. Moreover, from Fig. 19, a band of white (tracer-free) crystals is seen lying in the bottom half of the billet section. These crystals must have descended, by convection and under the influence of gravity, from a tracer-free zone into the lower tracer-rich region of the liquid pool and settled preferentially against the solidification front advancing from the outside radius face. In this way, the growth of the columnar dendrites was stopped while opposite, adjacent to the inside radius face, the columnar dendrites grew unimpeded. Owing to the proximity of the white crystal band to the surface (~ 26 mm) the crystals must have been generated in the upper part of the continuous-casting machine (within 2 to 3 m from the meniscus) and most likely in the mould.

Fig. 19. Autoradiograph of longitudinal section from a 133 mm square billet cast in a curved-mould machine. Dark region is Au^{198}-rich and was liquid at time of tracer addition [73].

The influence of the factors cited earlier on the cast structure then can be explained in terms of the generation and survival of these unattached crystals. For example, superheat has such a strong effect because the crystals remelt in excessively hot liquid so that columnar growth can proceed unimpeded. Similarly medium-carbon steels exhibit a larger equiaxed structure because crystals that do enter a region of hot liquid, must remelt by the solid-state transformation of γ to δ phase which involves carbon diffusion and a small but important kinetic limitation [12]. These effects and the role of other mechanisms for the columnar-equiaxed transition in continuously cast billets subjected to in-mould EMS, have been the subject of a recent study [26].

Shell growth

In the mould region, the rate of shell growth is governed by the external heat extraction addressed in an earlier section. Thus all the variables that influence the mould heat flux distribution also directly impact on shell growth. Lower in the casting machine where the shell has thickened considerably, conduction through the solid steel itself becomes rate determining.

An important aspect of shell growth, particularly in the mould, is that

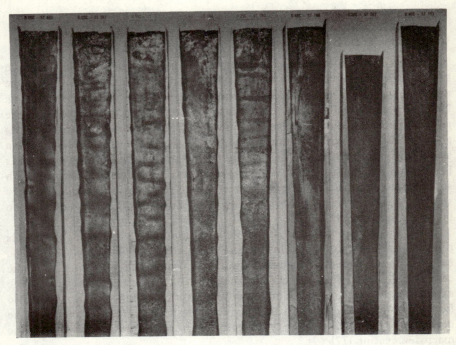

Fig. 20. Solidified metal shells obtained from controlled breakouts for steels containing 0.0 to 0.9 pct. carbon [74].

it may be non-uniform. The shell may be locally thin adjacent to deep oscillation marks where the steel/mould gap is large, as mentioned earlier. The carbon content of the steel also influences the uniformity of shell growth as can be seen in Fig. 20. Shown are breakout shells from experimentally cast billets having carbon contents ranging from 0 to 0.9 pct [74]. The shells have been sectioned longitudinally and set out in order of increasing carbon content. Thus it becomes obvious that the shell growth is most non-uniform in the 0.1-pct carbon breakout shell due to remarkable wrinkling of the surface. The wrinkles and associated gaps reduce the mould heat extraction as described earlier. The origin of this behaviour is believed to be shrinkage associated with the δ-γ transformation, which at 0.1 pct carbon proceeds in the solid state at the highest temperature [21].

Macrosegregation

In continuously cast billets, the severity of centreline macrosegregation and porosity is linked to the cast structure: the greater the columnar zone, in general, the worse are these quality problems. Thus all the factors, mentioned earlier, which favour an equiaxed structure are beneficial to quality [75]:

    i) low superheat

206

  ii) medium-carbon steel
 iii) EMS, particularly in-mould
 iv) large section size

The difficulty of controlling macrosegregation in billets then arises, in part, from the generation/survival of unattached crystals but also from the packing of the crystals in the lower part of the liquid pool where the solid fraction is increasing. Under adverse conditions, periodic bridging of crystals near the bottom of the pool causes isolation of pockets of liquid which freeze to form shrinkage cavities and regions of macrosegregation (C, Mn, S, P). For the casting of high-carbon steels which are particularly susceptible to this problem, another approach has been taken recently to reduce macrosegregation. The billets are cast with high superheat to eliminate the crystals (and equiaxed structure) while high intensity sprays are applied to maximize the rate of solidification thereby reducing the tendency for segregation.

Macrosegregation at the centreline of continuously cast slabs originates primarily from bulging very close to the completion of solidification [76]. The bulging draws enriched residual liquid downward where it freezes at the centreline; cracking may also occur at the same time if the bulging is severe. The extent of equiaxed structure is an important factor in the segregation, as it is also for "semi-macrosegregation" which appears as spots in the slab interior. The morphology and packing of the crystals as well as EMS influence the extent of the macrosegregation [77].

MECHANISMS LINKING QUALITY TO CASTING OPERATION

As stated earlier, the knowledge-based approach, involving mathematical models, plant tests and laboratory measurements, is a powerful tool in the achievement of quality in continuous casting. To illustrate this methodology, two quality problems are examined: transverse cracks in slabs and off-squareness/off-corner cracks in billets.

Transverse cracks in slabs

Transverse cracks appear at the base of oscillation marks in slabs. As shown in Fig. 21, the cracks may be visible [78] but frequently they can only be detected by an inspection scarf pass [2]. In general, the transverse cracks appear on the inside radius face of a curved mould casting machine which has a single unbending point between the curved region and straightening rolls. In a vertical-bending machine having an additional upper-bending point, the cracks may form on either the inside or outside faces. Steel composition has a strong influence on the formation of transverse cracks. Steel containing aluminum, niobium, vanadium and over 1 pct manganese is particularly susceptible to cracking [2].

The morphology of the crack surface has been examined with the SEM and reveals primarily intergranular fracture although a smooth appearance, indicative

Fig. 21. Narrow side of continuously cast slab exhibiting transverse corner cracks at the base of an oscillation mark [78].

of liquid-film separation (high-temperature low ductility) may be observed close to the top of the crack [78]. An example is shown in Fig. 22. This information together with other knowledge, derived in part from mathematical models, can be utilized to propose a fundamental mechanism for the formation of these cracks. The fact that at least some of the cracks exhibit a smooth interdendritic surface characteristic of the high-temperature zone of reduced ductility, Fig. 14, indicates that they may initiate in the mould close to the meniscus. However, it also is apparent that the cracks are propagated (and in other cases initiated) in the low-temperature zone of reduced ductility since the crack surface also exhibits intergranular facets, Fig. 22, and crack formation is exacerbated by additions such as aluminium, niobium and vanadium.

The origin of the longitudinal stress that causes the cracks is mechanical in nature as described earlier. The initiation of interdendritic hot tears, in the mould region, must arise from friction between the reciprocating mould and descending steel strand. Thus the lubricity of the mould flux could be an important factor. The propagation of existing hot tears or generation of new transverse cracks in the low-

208

Fig. 22. Surface of transverse crack revealing both smooth interdendritic and faceted intergranular regions [78].

temperature zone of reduced ductility obviously occurs during the mechanical bending or straightening of the strand when the surface/sub-surface is at about 600–900°C. During the straightening of slabs cast on a curved mould machine, the longitudinal mechanical stress is tensile on the inside radius face so that transverse cracks would preferentially appear on the upper side of the slab, as observed.

A mathematical model of heat flow in the solidifying strand based on the unsteady-state, two-dimensional heat-conduction equation [79]

$$\frac{\partial}{\partial x}\left(k\frac{\partial T}{\partial x} \right) + \frac{\partial}{\partial y}\left(k\frac{\partial T}{\partial y} \right) = \varrho C_p \frac{\partial T}{\partial t} \tag{2}$$

can be applied to design the spray cooling system (including the effect of support rolls) in order that the surface temperature of the strand is above 900°C during bending or straightening operations. The methodology involved, and spray heat-transfer data needed for the design, have been described earlier [4].

Because the transverse cracks normally appear in oscillation marks, it follows that factors affecting oscillation marks could also impact on surface quality. Logically, the depth of oscillation marks should be minimized since they locally increase the steel/mould gap and reduce heat extraction. Deep oscillation marks are the sites of locally high temperature and reduced shell growth; both effects enhance the propensity for transverse crack formation under the influence of mechanical tensile stresses in the mould. An example of the temperature distribution in the vicinity of a deep oscillation mark calculated with a mathematical model [78] is shown in Fig. 23.

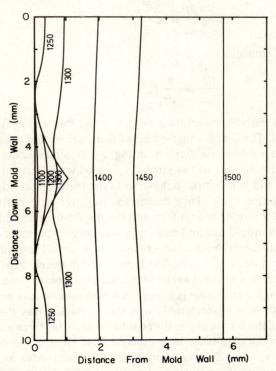

Fig. 23. Temperature distribution in mould flux and steel 10 s below meniscus [78].

Plant trials and a mathematical model [24] have shown that the oscillation-mark depth can be reduced by:

i) decreasing the viscosity of the mould flux
ii) reducing the negative-strip time in the oscillation cycle of the mould.

The negative-strip time is the period during which the mould is moving downward more rapidly than the descending strand and can be calculated from the following expression

$$t_N = \frac{1}{\pi f} \text{ arc cos} \left(\frac{v_s}{\pi f S} \right) \tag{3}$$

These effects have been explained with the mathematical model formulated by Tackeuchi and Brimacombe [24] who considered the flow of molten mould flux and the development of pressure in the gap between the steel meniscus and the reciprocating mould wall. The temperature of the mould flux was calculated with an equation similar to Eqn. (2) and mould heat fluxes derived from mould temperature measurements. Pressure development in the flux channel was computed from the simplified equation of motion

$$\frac{dP}{dz} = \mu_f \frac{\partial^2 u_z}{\partial y^2} + \varrho_f g \tag{4}$$

and equation of continuity

$$\frac{dQ_R}{dz} = \frac{d}{dz} \left(\int_0^{h(z)} u_z \, dy \right) = 0 \tag{5}$$

where Q_R is the relative consumption rate of mould flux and $h(z)$ is the width of the flux channel. The picture that emerges from the model predictions can be represented by the schematic diagram in Fig. 24. Depicted is the velocity of the mould, v_m, relative to that of the strand, v_s, (downward defined negative) as a function of time and below it the behaviour of the meniscus at different stages in the mould oscillation cycle. Thus during the negative-strip period $(v_m - v_s < 0)$ when the mould is forcing mould flux into the mould-meniscus channel (no slip assumed between mould wall and molten flux), a positive pressure develops in the molten flux which causes the meniscus to move away from the mould wall. During the positive-strip time $(v_m - v_s > 0)$, flux is pulled from the channel and a negative pressure is generated which draws the meniscus back toward the mould wall. The solidifying shell at the meniscus is either bent as shown, or may be overflowed by liquid steel, to form a characteristic oscillation mark. Thus the up-and-down vertical motion of the mould is translated into a transverse force acting negatively and positively on the meniscus via the mould flux. The total force acting on the meniscus when $v_m - v_s < 0$, was computed with the model as a function of negative-strip time and the results are shown in Fig. 25. If, as would be reasonable

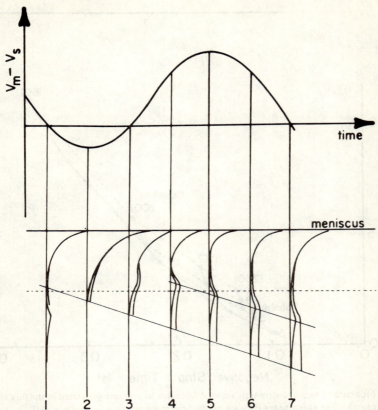

Fig. 24. Schematic representation of the formation of an oscillation mark with mould flux as lubricant [24].

to expect, oscillation-mark depth is proportional to the total force, the depth is predicted to decrease as the negative-strip time is reduced. Decreasing mould flux viscosity also is predicted to reduce the oscillation-mark depth as observed.

Translated into practice, the model predictions show that increasing the oscillation frequency and/or reducing the stroke length will reduce the depth of oscillation marks, Fig. 25, by reducing the negative-strip time. Increasing casting speed will have a similar effect. Thus to maintain control of oscillation-mark depth, the negative-strip time must be controlled even through inevitable speed changes. An adjunct to control of t_N is the prevention of fluctuation of the metal level at the meniscus which is equivalent to a sudden change in casting speed. Consequently, constant metal level must be maintained with minimal surface waves as could be generated by inert gas flushed through the submerged entry nozzle to prevent inclusion buildup. Otherwise, the depth of oscillation marks may vary with axial position on the slab.

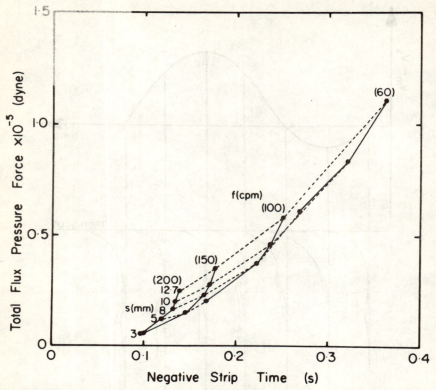

Fig. 25. Influence of negative-strip time on total force due to pressure generated in the flux channel (numbers in parentheses indicate oscillation frequencies), $v_s = 1$ m/min [24].

Use of a low-viscosity mould flux is beneficial but there are limits since these fluxes, containing more CaF_2 and alkali metal oxides, also transmit more heat to the mould. The enhanced cooling of the solidifying shell near the meniscus can generate excessive transverse tensile stresses and longitudinal facial cracks, particularly in 0.09 to 0.12 pct-carbon steels. The importance of mould flux viscosity also raises questions about alumina accumulation in the mould flux and its effect on the viscosity. For example, in Fig. 21, the disappearance of well-defined oscillation marks, and the appearance of transverse cracks may have resulted from high viscosity and loss of lubricity due to alumina buildup. As a result, all the factors that influence steel cleanness (tundish design, reoxidation protection, etc.) can bear directly on oscillation-mark depth and the appearance of transverse cracks. All these aspects of casting have been shown to be important for surface quality, in practice [80].

Off-squareness/off-corner internal cracks in billets

The off-squareness and attendant off-corner cracks seen in Fig. 6 are among the most common quality problems encountered in billet casting. To overcome them, control of oscillation marks and events at the meniscus again are seen to be important leading one to the inevitable conclusion that much of quality in continuous casting is made within centimetres of the meniscus.

Off-squareness can be linked to the mould in several ways although it is clear that severe off-squareness must involve the sprays as well. If one looks through the spray chamber to the bottom of the mould during casting, the two visible corners of the emerging billet may exhibit very different temperatures. One corner may be exceedingly hot (yellow-white in colour) while the other is cold (black). Following these billets to the run-out table, the hot corner always has an obtuse angle while the cold corner has an acute angle. The corner temperature at the mould exit and the concomitant off-squareness of the billet commonly changes during casting. Another link between the mould and off-squareness is the effect of cooling-water flow rate. Operators have found that, particularly with high-carbon billets, off-squareness can be reduced, at least temporarily, by lowering the cooling-water flow rate through the mould. In addition off-squareness can be reduced by the simple expedient of changing the mould.

Knowledge of the thermomechanical behaviour of the mould, gained from mathematical models and plant trials, is necessary to understand the origins of off-squareness and off-corner cracks. The rapid decline in the mould heat flux below the meniscus, as seen in Fig. 11, causes relatively steep temperature gradients to be established in the axial as well as through-thickness direction in the mould wall. Calculated axial temperature profiles at the hot and cold faces of a mould wall are shown in Fig. 26. Then, even though the maximum heat flux is at the meniscus, Fig. 11, the peak mould temperature is actually below this level because a significant quantity of heat is conducted up the mould wall. Typically the maximum mould temperature is found several centimetres below the meniscus.

The shape of the axial mould heat-flux profile not only determines the temperature distribution in the mould wall but also the resulting differential thermal expansion of the copper which is free to move. The hottest region of the mould wall expands most and thus, during casting, the mould tube assumes a slightly bulged shape as can be seen in Fig. 27. Shown in Fig. 27 is the predicted distortion at the mid-face and off-corner of a mould wall having a thickness of 6.35 mm. The maximum in the bulge, calculated relative to the original position of the wall before casting commenced, appears about 75 mm below the meniscus. Consequently above the peak bulge, the mould has a negative taper, the magnitude of which depends, in part, on the initial taper in the upper region of the mould, but which easily can reach 1–2%/m. Below the maximum bulge, the mould adopts an extra positive taper of about 0.3–0.4%/m.

The shape of the distorted mould, in particular the negative taper extending 75 to 100 mm below the meniscus, as well as the use of oil for lubricant, makes it possible for the mould and shell solidifying at the meniscus to interact

214

TEMPERATURE DISTRIBUTION

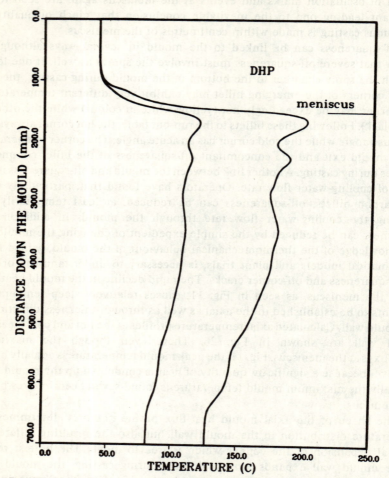

Fig. 26. Axial temperature profiles at the hot and cold faces of a mould wall [81].

mechanically during the mould oscillation cycle. In the period of negative strip, when the mould is moving downward more rapidly than the steel strand, the mould wall jams against the newly solidified shell at the meniscus, causing it to buckle as shown in Fig. 28. Presented in Fig. 28 are the oscillating mould velocity and the constant casting speed, v_c; negative strip occurs in the time interval *BC.* Following negative strip, molten steel can flow into the widening gap between the partially solidified meniscus and the mould wall to form an oscillation mark. Ideally the mould/shell interaction should occur uniformly around the periphery of the billet so that the shape of the oscillation marks and mould heat

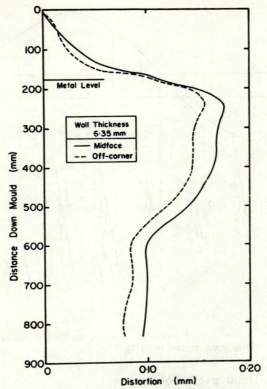

Fig. 27. Predicted distortion at the mid-face and off-corner of a mould tube having a wall thickness of 6.35 mm. Section size is 140 mm square [82].

extraction are also uniform but this usually is not the case. Frequently the oscillation marks are not uniform in depth across the billet face. Under these conditions, the rate of solidification and growth of the steel shell cannot be uniform. A similar interaction does not occur in slab casting because the copper is bolted to steel backing plates and mould flux is used as lubricant.

This is the genesis of off-squareness and off-corner internal cracks. If the shell growth is retarded in off-corner locations as described above, any bulging* of the shell in the lower part of the mould, or below it, will not extend to the corners but only to the thin, weak off-corner zone. The result is a hinging action, as depicted schematically in Fig. 29, at off-corner locations, and the generation of a tensile strain in the region of low-ductility adjacent to the solidification front. Thus cracks can form and continue growing inward following the solidification front, as it advances, as long as the strain is maintained. Based on this mechanism, off-corner cracks should appear at sites that have the deepest oscillation

*Evidence of bulging is the presence of white solidification bands in the macro-etches of many billets [12].

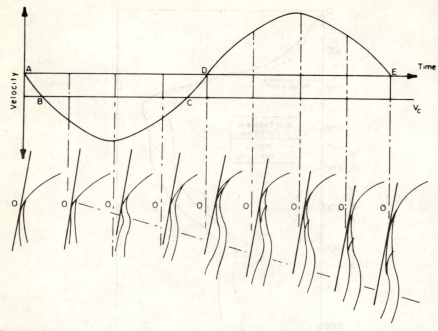

Fig. 28. Mechanism of oscillation-mark formation based on mechanical mould/shell interaction during negative strip (time interval BC) [11].

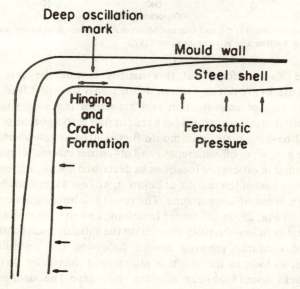

Fig. 29. Formation of off-corner internal crack due to bulging and hinging of the shell in the lower part of the mould [13].

marks; and this has been found to be the case [82]. Thus the achievement of uniform, shallow oscillation marks should reduce the incidence and severity of off-corner cracks. Also measures to minimize shell bulging in the lower region of the mould should decrease this cracking problem. In particular, the implementation of improved mould taper should have a beneficial influence. Recent calculations have revealed that, with the commonly employed single mould taper of 0.6–0.7%/m, the shell/mould gap can be about 0.6–0.7 mm in the lower part of the mould, depending on cooling water flow rate and carbon content. A gap of this magnitude is conducive to bulging of the shell which could be reduced by tapering the mould properly.

The non-uniform oscillation marks are at the root of off-squareness as well. In an earlier study [12] it was found that the obtuse angle corners of off-square billets usually had the deepest oscillation marks, and hence reduced local heat extraction in the mould, relative to other areas around the periphery of the billet. Under such conditions these corners, emerging from the mould, would be thin and hot, as observed. Cracks could form most easily adjacent to these corners owing to the locally weakened shell. By the same argument, corners with shallow oscillation marks experience higher heat extraction in the mould, more rapid solidification and are cold at the mould exit. Thus the solid shell leaving the mould may appear like the cross-hatched section in Fig. 30. At this point the off-squareness would be small since it cannot exceed that of the mould itself. But once the billet section reaches the sprays, the cooling of the shell again would be non-uniform due to the varying shell thickness. Thus the cold corners would experience a more rapid cooling rate than the hot corners, and the resulting differential thermal contraction forces the billet into an off-square shape.

From this mechanism, a number of key elements can be identified for optimum mould operation in order to minimize and control the mechanical interaction between the mould and the shell [13]:

i) increase wall thickness of the mould to at least 12.5 mm for small billets;

ii) apply a multiple taper to the mould to follow the shrinkage of the shell more closely;

iii) support the mould tube either on all four sides near the top, or uniformly top and bottom;

iv) ensure the cooling-water velocity is uniform around the mould periphery through tight tolerances on mould/water jacket dimensions and location of mould tube within water jacket;

v) employ copper grades like Cr-Zr that have a high softening (distortion) resistance;

vi) reduce negative-strip time to ~ 0.1 s;

vii) raise cooling-water velocity above 12 m/s to eliminate boiling and minimize mould distortion;

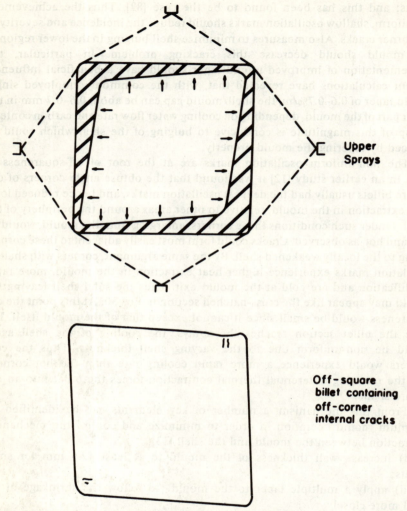

Upper
Sprays

Off-square
billet containing
off-corner
internal cracks

Fig. 30. Schematic diagram showing billet with non-uniform shell thickness being distorted into off-square shape by spray cooling [12].

viii) maintain careful control of water quality to prevent fouling of the copper (< 5 ppm hardness, addition of corrosion inhibitors, etc.);

ix) control metal level and hold at 100 mm below top of mould to maintain reasonable distance from mould tube supports; and

x) hold constant casting speed to make most effective use of taper.

LESSONS FOR THE DEVELOPMENT OF MATERIALS PROCESSES

Viewed more broadly, continuous casting is nothing more than a materials process, a unit operation that converts liquid steel into a semi-finished shape, hopefully with the desired quality. Thus the knowledge-based approach to the achievement of quality, advocated in this paper, applies to a wide range of materials processes whether they produce an intermediate or a finished product. Ultimately what is sought is the quantitative linkage of the properties of the product to the process design/operation as well as to compositional and microstructural variables of the material. This can be achieved rationally by the incorporation of basic knowledge into mathematical models as depicted in Fig. 2.

Obviously the analysis and design of materials processes following this methodology is strongly interdisciplinary. In the case of continuous casting, numerous fields of study bear on the process-fluid/heat flow, stress analysis, physical metallurgy, numerical analysis and mathematical modelling-and must be tapped for the solution of *real* problems. The same is true of other materials processes. Thus a team effort is required to draw upon the different disciplines and to integrate the relevant knowledge such that a new understanding of the process is achieved. As in any successful research effort, creativity also remains an essential ingredient to elucidate new mechanisms, to explain different phenomena in new ways, or to come up with an entirely new process.

If these statements seem obvious, they are made nonetheless because they need emphasis. Too few research efforts recognize the importance of mathematical modelling, the need to draw upon different disciplines in a concerted manner and the absolute necessity of conducting plant, as well as laboratory trials, to provide the knowledge necessary to place confidence in the models. The development of materials processes will be hampered unless these lessons are well learned and practiced.

ACKNOWLEDGEMENTS

Over the years, the authors have received the generous support of Stelco Inc., Canadian and American steel companies, the Natural Sciences and Engineering Research Council of Canada, the American Iron and Steel Institute and the B.C. Science Council. We also have benefited from the guidance and talents of colleagues in university and industry. We owe them all our gratitude.

REFERENCES

1. Brimacombe, J.K. In: *Encyclopedia of Materials Science and Engineering*, Pergamon Press, pp. 2312-2317 (1986).
2. Brimacombe, J.K. and K. Sorimachi. *Metall. Trans.* B, **8B**, 489-505 (1977).
3. Irving, W.R. and A. Perkins. In: *Continuous Casting of Steel*, Biarritz, The Metals Society/IRSID, pp. 107-115 (1976).
4. Brimacombe, J.K., P.K. Agarwal, L.A. Baptista, S. Hibbins and B. Prabhakar. Iron and Steel Society, Warrendale, Pa., Vol. 63, pp. 235-252 (1980).
5. Myoshi, S. In: *Continuous Casting of Steel*, Op. cit. pp. 286-291 (1976).
6. Moore, J.J. *Iron and Steelmaker*, 7 (2), 8-16 (1980).

220

7. Van Vuuren, C.J.C.J. *Steelmaking Proceedings,* Op. cit. **61,** 306–334 (1978).
8. Saeki, T., T. Komai, K. Miyamura, S. Mizoguchi and H. Kajioka. *Steelmaking Proceedings,* Op. cit. **68,** 229–235 (1985).
9. Flemings, M.C. *Solidification Processing,* McGraw-Hill, New York (1974).
10. Brimacombe, J.K., I.V. Samarasekera, N. Walker, I. Bakshi, R. Bommaraju, F. Weinberg and E.B. Hawbolt. *ISS Trans.,* **5,** 71–78 (1984).
11. Samarasekera, I.V., J.K. Brimacombe and R. Bommaraju. *ISS Trans.,* **5,** 79–94 (1984).
12. Bommaraju, R., J.K. Brimacombe and I.V. Samarasekera. *ISS Trans.,* **5,** 95–105 (1984).
13. Brimacombe, J.K., I.V. Samarasekera and R. Bommaraju. *Steelmaking Proceedings,* Op. cit. **69,** 409–423 (1986).
14. Tanaka, S., M. Lye, M. Salcudean and R.I.L. Guthrie. In: *Proc. International Symposium on the Continuous Casting of Steel Billets,* The Metallurgical Society of CIM, Montreal, pp. 142–161 (1985).
15. El-Kaddah, N., and J. Szekely. In: *Continuous Casting '85',* The Institute of Metals, London, pp. 49.1–49.6 (1985).
16. He, Y. and Y. Sahai. *Metall. Trans. B,* **18B,** 81–92 (1987).
17. Heaslip, L.J., A. McLean and I.D. Sommerville. *Continuous Casting:* Volume One. Op. cit. 67–84 (1983).
18. Watanabe, S., K. Harada, N. Fujita, Y. Tamura and K. Noro. *Tetsu-to-Hagane,* **58,** S393–S394 (1972).
19. Grill, A., K. Sorimachi and J.K. Brimacombe. *Metall. Trans.,* **7B,** (2), 177–189 (1976).
20. Singh, S.N. and K.E. Blazek. *Open Hearth Proc. AIME,* **59,** 264–283 (1976).
21. Grill, A. and J.K. Brimacombe. *Ironmaking and Steelmaking,* **2,** 76–79 (1976).
22. Samarasekera, I.V. and J.K. Brimacombe. *Ironmaking and Steelmaking,* **9,** 1–15 (1982).
23. Chizhikov, A.I., V.L. Iokhimovich, G.P. Rachuk, L.I. Morozenskii and A.I. Mikhailova. *Stal in English,* (12), 921–924 (1968).
24. Takeuchi, E. and J.K. Brimacombe. *Metall. Trans. B,* **15B,** 493–509 (1984).
25. Samarasekera, I.V. and J.K. Brimacombe. Mould Design to Prevent Breakouts in the Continuous Casting of Steel Billets, Report to Stelco Edmonton Steel Works, Unpublished work (1987).
26. Bommaraju, V.S.S.R. Mould Behaviour, Heat Extraction and the Quality of Continuously Cast Billets with In-Mold Electromagnetic Stirring, Ph.D. Thesis, University of British Columbia (1988).
27. Mahapatra, R.B., J.K. Brimacombe and I.V. Samarasekera. Unpublished Research (1987).
28. Lima, E., J.K. Brimacombe and I.V. Samarasekera. Unpublished Research (1987).
29. Mizikar, E. *Iron and Steel Engineer,* **47,** 53–60 (1970).
30. Kreith F. *Principles of Heat Transfer,* 3rd Ed., Intext Educational Publishers (1976).
31. Hoogendorn, C.J. and R. den Hond. 5th Int. Heat Transfer Conf. Paper B3.12, Tokyo (1974).
32. Jeschar, R., U. Reiners and R. Scholz. *Steelmaking Proceedings,* Op. cit. **69,** 511–521 (1986).
33. Mitsutsuka, M. *Tetsu-to-Hagane,* **54,** 1457–1471 (1968).
34. Junk, H. *Neue Heutte,* [HB Transl. No. 8740] **17,** 13–18 (1972).
35. Müller, H. and R. Jeschar. *Arch. Eisenhuttenwes,* **44,** 598–594 (1973).
36. Alberny, R., A. Leclerq and J. Basilis. *Circulaire d'Informatio-techniques,* 3 (315), 763–776 (1973).
37. Alberny, R. *Info. Symp. on Casting and Solidification of Steel,* Committee of European Communities, Luxembourg, IPC Science and Technology Press, Vol. 1, pp. 278–335 (1977).
38. Sugitani, Y., K. Takashima and S. Kawasaki. *Tetsu-to-Hagane,* 59, pp. S388–S389 (1973).
39. Ishiguro, M. et al. *Tetsu-to-Hagane,* [HB Transl. No. 8735], **60,** S464–S465 (1974).
40. Kawakazu, T., T. Kitagawa, E. Sakamoto and F. Miyashita, et al. *Tetsu-to-Hagane,* [HB Transl. No. 9319], **60,** S103–S104 (1974).
41. Nilles, P., P. Dauby, B. Mairy and A. Palmaers. Proc. Open Hearth Conf., Chicago, *ISS-AIME,* **61,** 399–410 (1978).
42. Bolle, E. and J.C. Moureau. Int. Conf. on Heat and Mass Transfer Metallurgical Processes, Dubrovnik, Yugoslavia (1979).

43. Bolle, L. and J.C. Moureau. *Proc. of Two Phase Flows and Heat Transfer,* NATO Advanced Study Institute, **III**, 1327–1346 (1976).
44. Sasaki, K., Y. Sugitani and M. Kawasaki. *Tetsu-to-Hagane,* **65**, 90–96 (1979).
45. Hibbins, S.G. and J.K. Brimacombe. *Trans. ISS,* **3**, 37–51 (1983).
46. Thomas, B.G., J.K. Brimacombe and I.V. Samarasekera. *ISS Transactions,* **7**, 7–18 (1986).
47. Kinoshita, K., G. Kasai and T. Emi. *Tetsu-to-Hagane,* (62), S#505, 43/1–26 (1976).
48. Weinberg, F. *Metall. Trans. B,* **10B**, 219–227 (1979).
49. Miyazaki, J. et al., *100th ISIJ Meeting,* #S806 (1980).
50. Bhattacharya, U.K., C.M. Adams and H.F. Taylor. *Trans. AFS,* **60**, 675–686 (1952).
51. Christopher, C.F. *Trans. AFS,* **64**, 293–310 (1956).
52. Bishop, H.F., C.G. Ackerlind and W.S. Pellini. *Trans. AFS,* **65**, 247–258 (1957).
53. Morzenskii, L.I., O.A. Mitenov and V.K. Krutikov. *Stal in English,* **4**, 272–276 (1965).
54. Adams, C.J. Proc. Nat. Open Hearth Basic Oxygen Steel Conference, **54**, 290–302 (1971).
55. Stephenson, E.T. *Journal of Metals,* 48–51 (1974).
56. Cooper, R.B. and T.H. Burns. *Metals Eng. Quarterly,* 41–44 (1974).
57. Fredriksson, H. and J. Stjerndahl. *Met. Trans.* A, **8A**, 1107–1114 (1977).
58. Suzuki, H.C., S. Nishimura and Y. Nakamura. *100th ISIJ Meeting,* #S805 (1980).
59. Ericson, L. *Scan. J. Metallurgy,* **6**, (3), 116–124 (1977).
60. Suzuki, H.G. *Tetsu-to-Hagane,* **56**, (8), 1–9 (1981).
61. Ouchi, C. and K. Matsumoto. *Trans. Iron Steel Inst.* Japan, **22**, 181–189 (1980).
62. Kinoshita, K., G. Kasai and T. Emi. *Tetsu-to-Hagane,* **62**, #S505, 43/1–26 (1976).
63. Robbins, J.L., O.C. Shepard and O.D. Sherby. *JISI,* **99**, 175–180 (1961).
64. Robbins, J.L., O.C. Shepard and O.D. Sherby. *Trans. ASM,* **60**, 205–216 (1967).
65. Yamanaka, K. et al. *ISIJ Trans.,* **20**, 810–816 (1980).
66. Wilcox, J.R. and R.W.K. Honeycombe. Hot Working and Forming Processes, Metals Society Publication No. 264, London, pp. 108–112 (1980).
67. Fudaba, K. and O. Akisue. *99th ISIJ Meeting,* #S362 (1980).
68. Suzuki, H.G., S. Nishimura and S. Yamaguchi. *ISIJ Trans.,* **22**, 48–56 (1982).
69. Brimacombe, J.K., F. Weinberg, E.B. Hawbolt. *Metall. Trans. B,* **10B**, 279–292 (1979).
70. Vaterlaus, A. and M. Wolf. On Strand Deformation and Internal Crack Formation, Slab Seminar II, in 7th Concast Technology Convention, (1984).
71. Uehara, M., I.V. Samarasekera and J.K. Brimacombe. *Ironmaking and Steelmaking,* **13**, (3), 138–153 (1986).
72. Lait, J.E. and J.K. Brimacombe. *ISS Trans.,* **1**, 1–13 (1982).
73. Van Drunen, G., J.K. Brimacombe and F. Weinberg. *Ironmaking and Steelmaking,* **2**, 125–133 (1975).
74. Singh, S.N. and K.E. Blazek. *Open Hearth Proceedings,* **57**, 16–36 (1974).
75. Mori, H., N. Tanaka, N. Sato and M. Hirai. *Trans. ISIJ,* **12**, 102–111 (1972).
76. Irving, W.R., A. Perkins and M.G. Brooks: *Ironmaking and Steelmaking,* **11**, 152–162 (1984).
77. Fujimura, T., E. Takeuchi and J.K. Brimacombe. "Segregation Phenomena in the Continuous Casting of Steel Slabs", Japan-Canada Seminar on Secondary Steelmaking. Refining, Casting, Physical Metallurgy and Properties, Tokyo, The Iron and Steel Institute of Japan/The Canadian Steel Industry Research Association, pp. C-5-1-C-5-15 (1985).
78. Takeuchi, E. and J.K. Brimacombe. *Metall. Trans. B.* **16B**, 605–625 (1985).
79. Brimacombe, J.K. *Can. Met. Quart.,* **15**, 163–175 (1976).
80. Irving, W.R., A. Perkins and R. Gray. *Ironmaking and Steelmaking,* **11**, 146–151 (1984).
81. Samarasekera, I.V. and J.K. Brimacombe. *Metall. Trans. B,* **13B**, 105–116 (1982).
82. Samarasekera, I.V., R. Bommaraju and J.K. Brimacombe. *Proc. Intl. Symposium on Continuous Casting of Billets,* Vancouver, Canada, CIMM, pp. 33–58 (1985).

222

NOMENCLATURE

C_p	Specific heat, kJ/kg °C
f	Mould oscillation frequency, Hz
g	Gravitational constant, 9.8 m/s²
h	Spray heat-transfer coefficient, kW/m² °C
$h(z)$	Width of mould flux channel, cm
k	Thermal conductivity, kW/m °C
n	Exponent in Eq. (1)
P	Pressure in mould flux, dyne/cm²
Q_R	Relative consumption rate of mould flux, cm³/s
S	Stroke of mould oscillation, cm
t	Time, s
t_N	Negative-strip time, s
T	Temperature, °C
v_m	Velocity of mould, cm/s
v_s	Velocity of strand, cm/s
W	Water flux, 1/m²s
x	Transverse direction
y	Transverse direction
z	Axial direction
α	Constant of proportionality in Eq. (1)
ϱ	Density, kg/m³
ϱ_f	Density of mould flux, g/cm³
μf	Viscosity of mould flux, poise

9

Some Unresolved Theoretical Problems in Crystal Growth Part II: Consequences of Properly Accounting for Surface Creation

WILLIAM A. TILLER

Department of Materials Science and Engineering, Stanford University,
Stanford, CA 94305-2205, U.S.A.

ABSTRACT

In present-day theoretical treatments of crystal surface morphology, two major problems exist with respect to surface creation: (1) the continuum model rather than the terrace/ledge kink *(TLK)* model is generally used and this tends to eliminate important crystallographic features in the process, and (2) only a curvature term is used to represent surface creation and this is shown to be inadequate for evolving crystal shapes. Equations describing the proper accounting for surface creation during crystallization are given.

Qualitative and quantitative application of the key concepts that allow microscopic orientational and macroscopic morphological features to be predicted are given. Consequences are given for: (1) rumpled interface formation, (2) the crystallographic aspects of interface stability, (3) sphere-like growth forms, (4) dendrite filament growth, and (5) layer source locations and external crystal shapes for *CZ* growth.

INTRODUCTION

In the topic area of interface dynamics and interface stability, the present analytical methodology has become quite sophisticated and is moving rapidly to include important non-linear fluid convection factors involving large computational efforts. It is thus timely to correct any flaws or weaknesses in the basic physics currently being utilized in this topic area so that the numerical analyses can be expected to have meaningful convergence with physical reality.

Over a decade ago, this author attempted to point out two important corrections needed in the mathematical description of modern crystal morphology problems to properly account for the creation of a new surface [1–3].

However, these suggestions did not take root in the consciousness of either the theoretical or the experimental crystal growth community, probably because this author was not sufficiently clear in his descriptions. At this point in time, it seems important to "reseed" the terrain by reintroducing these ideas via more concrete examples that reveal their importance to a full understanding of crystal growth phenomena. That is the major purpose of this paper.

Most of the modern theoretical schools of crystal morphology development treat the surface as a continuum and introduce surface excess energy effects via the Herring curvature expression [4]. Although this is fine for a static interface, it is not adequate for a moving interface where the shape contours are changing with time. In this paper, the proper replacement expression will be given.

The second interfacial aspect in need of correction relates to the use of the continuum interface model rather than the terrace/ledge/kink *(TLK)* model to describe interfacial energy effects. Although the former model is more mathematically convenient, it is usually physically incorrect in some important ways and produces important consequences for crystal morphology predictions. Although practitioners have long discussed the concept of an atomically rough interface and the uniform attachment mechanism for the crystallization process, except perhaps for $^4He,$ one is hard put to find clearcut *experimental* evidence to support the existence of such an interfacial condition during crystallization from any media. On the other hand, there is overwhelming experimental evidence for the operation of the *TLK* model.

One should not reject the possibility that metallic systems and transparent organic systems that freeze with smooth surface contours at the optical microscopy level are, in fact, *TLK* systems with atomic scale ledge heights and layers that have little tendency to bunch. As can be seen in Appendix A, a *TLK* system can yield an isotropic value of γ, the interfacial energy, provided the ledge energy, γ_l, and the variation of γ_l with ledge spacing, λ_l, vary with orientation relative to the face energy, γ_f, in the proper fashion. Although one may never be able to experimentally obtain an atomic scale resolution of a crystal/melt interface for metallic systems, there is certainly pedogogical merit in considering all crystals to grow via the *TLK* mechanism with metallic systems occupying that domain of $(\gamma_l, \gamma_f, \partial\gamma_l/\partial\lambda_l)$ space leading to almost spherical γ-plots. At least with such a picture, some crystallographic features would be restored to the crystallization process for all systems.

In the main body of the paper, these two cases will be addressed via the analysis of four different cases: (1) surface creation in the development of a rumpled interface, (2) influence on interface breakdown conditions, for non-faceted and faceted interfaces, (3) influence on dendritic growth, and (4) influence on Czochralski-grown crystal shape.

ANALYSIS

From a thermodynamic point of view, the overall free energy driving force for the crystallization process ΔG_{∞} can be given as a sum of the driving force consumed in the change of state variables (P, T, C, ϕ) from the far-field region to the interface region, ΔG_{sv}, plus the driving force available at the interface, ΔG_i,

$$\Delta G_{\infty} = \Delta G_{sv} + \Delta G_i \qquad (1a)$$

ΔG_i, in turn, can be given as the sum of the driving force needed for molecular transition to the crystal phase at the interface, ΔG_K, plus that needed for the storage of excess energy in the form of defects or surface, ΔG_E, so that

$$\Delta G_i = \Delta G_K + \Delta G_E \qquad (1b)$$

In this paper, we neglect any defect formation aspects in ΔG_E and focus only on the surface storage aspects. As such, no consideration is given to heat flow or matter transfer aspects which enter the problem area via the ΔG_{sv} term. The only solute aspects that interest us here are those associated with surface adsorption processes which alter the ledge and face aspects of the interfacial energy and thus influence ΔG_E.

To understand how variation in ΔG_E influence the dynamic interface temperature, T_i, perhaps it is best to begin with a familiar example. The development of the Gibbs-Thomson equation for a circular cylinder growing from a radius R to $R + dR$ shows us that, at equilibrium, the ratio of the total free energy change to the volume change is zero leading to

$$\Delta G_E = \gamma / R \qquad (2a)$$

and

$$T_E = T_E^o - \left(\frac{\gamma}{\Delta S_F} \right) K \qquad (2b)$$

where γ is the surface tension, $k = R^{-1}$ is the curvature of the cylinder and T_E^o is the equilibrium temperature for a planar surface. Herring [4] generalized the expression for any type of cylinder to the form

$$\Delta G_E = (\gamma + \gamma'') / R_1 \qquad (2c)$$

where γ'' is the torque term in the direction of the principal radius of curvature for the cylinder R_1.

For dynamic shape changes during crystal evolution with non-constant curvature, it is necessary to account for local changes in surface energy associated with correlated local changes in volume transformed. It is this transformed volume that must be undercooled sufficiently to provide the volume free energy change needed to compensate for the change in surface energy storage [3]. For a moving interface, on a point to point basis, the key quantity of interest is the ratio of rates $\Delta \dot{G}_E / \Delta \dot{v}$ where $\Delta \dot{v}$ is the local rate of volume transformation for unit area of interface. The local rate of surface free energy change per unit area of interface, $\Delta \dot{G}_E$, involves (i) creation of new layer edge, (ii) interaction of

226

layer edges, and (iii) annihilation of ledges which can be expressed in the following way

$$\Delta \dot{G}_E = \delta \dot{G}_C + \delta \dot{G}_I + \delta \dot{G}_A \tag{3a}$$

For a general surface, some regions are totally dominated by ledge creation, others by ledge interaction and still others by ledge annihilation. Defining s as a surface coordinate on a general cylindrical surface relative to some origin and θ as the angle of layer flow relative to the facet plane, the local interface temperature, T_i, for such a dynamic surface is given by

$$T_i(s, \theta) = T_E(s, \theta) - \Delta T_k(s, \theta) - \frac{\Delta \dot{G}_E(s, \theta)}{\Delta S_F \Delta \dot{v}(s, \theta)} \tag{3b}$$

where ΔS_F is the entropy of fusion and T_E is the equilibrium temperature for that surface curvature.

As a simple illustration for calculation purposes, let us consider the two-dimensional example of a small amplitude harmonic undulation from flatness for an interface between a crystal and its melt for the three cases illustrated in Fig. 1,

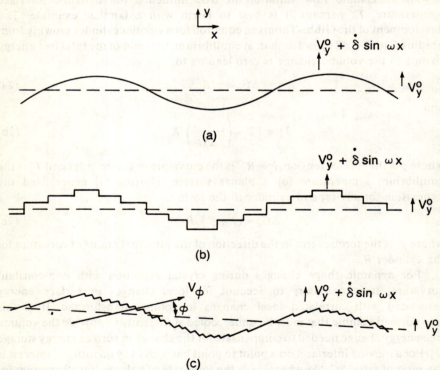

Fig. 1. Interface undulation development relative to a planar interface moving at velocity, V_y^o, based on several models: (a) continuum model, (b) terrace/ledge/kink model with terrace plane parallel to the planar interface and (c) terrace/ledge/kink model with terrace plane making an angle ϕ with the planar interface.

(a) use of the continuum surface model, (b) use of the *TLK* model where the planar surface is of the facet plane orientation ($\theta = 0$), and (c) use of the *TLK* model where the planar surface is a vicinal surface making a specific angle $\theta = \phi$ with the facet plane. These three cases are illustrated in Fig. 1.

Case (a): one sees periodic regions of positive and negative curvature for the interface of shape $y(t) = y_0(t) + \delta(t) \sin \omega x$; thus, the average curvature of the interface is zero and, if the Herring expression of eqn. 2c is used to account for the formation of surface, the average interface undercooling will be zero regardless of the amplitude δ of the wave. This fact reveals the inadequacy of using eqn. 2c to account for surface area changes during crystallization.

To properly account for surface creation in the Fig. 1a case, we have for a time increment dt and the surface element dx at x,

$$\frac{\Delta \dot{G}_E}{\Delta \dot{v}} = (\gamma + \gamma'') \frac{\Delta s}{\Delta v} \tag{4a}$$

$$\approx (\gamma + \gamma'') \frac{\{[1 + (y'(t + dt))^2]^{1/2} - [1 + (y'(t)^2]^{1/2}\} dx}{y(t + dt) - y(t)} \tag{4b}$$

where

$$y(t) = y_0(t) + \delta(t) \sin wx, \tag{4c}$$

$$y'(t) = w\delta \cos wx \tag{4d}$$

and

$$y'(t + dt) = w(\delta + \dot{\delta} dt) \cos wx. \tag{4e}$$

Inserting eqns. 4c-4e into eqn. 4b and expanding leads to

$$\frac{\Delta \dot{G}_E}{\Delta \dot{v}} = (\gamma + \gamma'') \frac{\{w^2 \delta \dot{\delta} dt \cos^2 wx + ...\} dx}{\{V_y^\circ + \dot{\delta} \sin wx\} dx dt} \tag{4f}$$

$$= (\gamma + \gamma'') \frac{w^2 \delta \dot{\delta}}{V_y^\circ} \cos^2 wx \left\{ 1 - \frac{\dot{\delta}}{V_y^\circ} \sin wx + ... \right\} \tag{4g}$$

Inserting eqn. 4g into eqn. 3b yields

$$T_i = T_E^\circ - \Delta T_K(x, t) - \frac{(\gamma + \gamma'')}{\Delta S_F} w^2 \delta \left\{ \sin wx + \frac{\dot{\delta}}{V_y^\circ} \cos^2 wx ... \right. \tag{5}$$

$$\left. \left[1 - \frac{\dot{\delta}}{V_y^\circ} \sin wx \right] + ... \right\}$$

where $V_y^\circ = dy_0/dt$ is the velocity of the original flat interface. From eqn. 5, the average surface creation contribution is given by

$$\left(\frac{\Delta \dot{G}_E}{\Delta S_F \Delta \dot{v}}\right)_{AVG} = \left(\frac{\gamma + \gamma''}{2}\right) w^2 \delta(\dot{\delta}/V_y^\circ) + \dots \qquad (6)$$

If the $\dot{\delta}/V_y^\circ$ term in eqn. 5 is neglected, surface creation remains unaccounted for and one is left with the Herring result [4] for a static interface wherein the average curvature is zero. However, including the $\dot{\delta}/V_y^\circ$ term provides a positive surface creation contribution which is proportional to the undulation amplitude, δ, as seen in eqn. 6. From eqn. 6, one notes that rapid local shape changes on a slowly drifting main interface ($\dot{\delta}/V_y^\circ \gg 1$) have much larger energy consequences than the same shape changes on a rapidly drifting interface ($\dot{\delta}/V_y^\circ \ll 1$).

Case (b): in this case it is necessary to consider two-dimensional nucleation at the peak of each undulation. From nucleation theory one finds that

$$V = \alpha_0 A I \qquad (7a)$$

with

$$I = I_0 \exp[-\pi \alpha_0 \gamma_i^2 v_m / RT\Delta G_i]. \qquad (7b)$$

Here, A is the area available for nucleation, a_0 is the height of the two-dimensional pill box shaped nucleus, $I_0 \sim 10^{28}$, v_m is the molar volume, γ_i is the ledge energy and ΔG_i is the available driving force per cm^3 for the nucleation process. Combining eqn. 7b and 7a, ΔG_i can be obtained in terms of V. Calling $A \sim 1\ cm^2$ for the flat facet plane moving at V_y° and $A = 2X_C$, where $X_C \sim (a_0/\omega^2\delta)^{1/2}$, for the tip of the undulations moving at $V_y^\circ + \dot{\delta}$, the excess ΔG_i needed to drive the tip nucleation process over that for the flat plane process may be readily obtained and one finds for the undulated interface that

$$T_i = T_E^\circ - \Delta T_K^P - \frac{\pi \alpha_0 v_m \gamma_i^2}{\Delta S_F RT_i} \left\{ \frac{\ln\left[(1 + \dot{\delta}/V_y^\circ)/2X_C\right]}{\ln[V_y^\circ/\alpha_o I_o]\ln[(V_y^\circ/\alpha_o I_o)(\dot{\delta}/V_y^\circ)/2X_C]} \right\} \qquad (8)$$

ΔT_K^P is the value for the flat interface. The denominator of eqn. 8 can be approximated with reasonable accuracy since $I_0 \sim 10^{28}$ and the numerator expands as a simple ascending power series in $\dot{\delta}/V_y^\circ$.

For a simple temperature variation along the interface, the interface shape is expected to quickly adjust away from the simple harmonic form. In the troughs, $\Delta \dot{G}_E$ will be negative due to ledge annihilation so the ledge velocity will accelerate in this region even for negligible supercooling. One can readily show that T_i from eqn. 8 is considerably below that for eqn. 5 unless γ_i is extremely small. Not surprisingly, the TLK model result is very different from the continuum model result.

Case (c): for the case where the median plane of the undulations makes an angle ϕ with the layer plane, the layer edges are close enough to interact strongly. During undulation formation, the number of ledges per unit distance perpendicular to the layer flow direction, n_i does not change as δ increases from zero to a reasonably large value. Thus, one finds that, on a point to point basis,

$$T_i = T_E^\circ - \Delta T_K^P - \left(\frac{\partial \gamma_l}{\partial \theta}\right)_\varnothing \frac{w\dot{\delta}\cos wx}{\Delta S_F V_\varnothing} \tag{9a}$$

where $V_y^\circ + \dot{\delta} \sin wx = V_\varnothing \sin \varnothing$. So long as $\varnothing$ is not near a cusp in the γ-plot, eqn. 9a should hold everywhere and

$$\Delta \dot{G}_E = n_i\gamma_l \approx \left(\frac{\partial \gamma_l}{\partial|\theta|}\right)_\varnothing \frac{w\dot{\delta}\cos wx}{V_\varnothing} \tag{9b}$$

In this case, the undulation drifts in the $\varnothing$ direction as the amplitude grows or shrinks. The sign of the effect depends on the portion of the wave under consideration while the magnitude of the effect depends on the velocity factor, $\dot{\delta}/V_\varnothing$, and the anisotropy factor $(\partial\gamma_l/\partial\theta)_\varnothing$, which may be close to zero for metals. One thus expects that, for metal crystals grown from the melt, negligible excess driving force in needed to rumple such an interface provided $\varnothing$ is not near the facet plane orientation.

Equations 8 and 9a represent the total range of crystallographic variation possible for undulation formation via the *TLK* model and both extremes are quite different quantitatively from the continuum model result of eqn. 5.

The crystallographic aspects of interface instability

Since the ΔG_E term in eqn. 5 and 9a are proportional to $\dot{\delta}$ and, since the marginal stability condition for planar interface breakdown occurs where $\dot{\delta}/\delta = 0$, the surface creation effect does not alter the onset condition for cell formation for off-facet orientations. However, in the domain where cell formation occurs, the surface creation term definitely alters the *rate* of cell development and is an important contribution to cell morphology considerations, especially during the transient stages of cell development.

For a faceted interface, even at $\dot{\delta}/\delta = 0$, eqn. 8 shows that the onset condition for cell formation may be significantly affected, depending strongly on the magnitude of γ_l and only weakly on the magnitude of X_C. In the ΔT_K^P term of eqn. 8, $A = 1$ cm^2 was used and this is likely to be an overestimate. A value of $A = A_P$ should be used in general and this can be included in eqn. 8 via replacing ln $[V_y^\circ/a_o I_o]$ by ln $[V_y^\circ/a_o I_o A_P]$. The value of A_P can be readily found since it is just the area of a growing pillbox wherein the probability of a new nucleation event is unity. Using eqn. 7 and lateral pillbox growth at a rate $V_l = \beta \Delta T_K^P$, one finds that

$$A_p = \pi^{1/3}\left(\frac{\beta\alpha_o}{V_y^\circ}\right)^{2/3} \exp\left[2\pi\alpha_o v_m \gamma_l^2 / 3\Delta S_F RT_i \Delta T_K^P\right] \tag{10}$$

An alternate approach for determining the condition of marginal stability for a faceted surface may be made by recognizing that $2\pi/\sqrt{A_P} = \omega_C = (2\sqrt{\pi}/a_o\beta\Delta T_K^P)V_y^\circ$ is a critical frequency for the perturbed system. A typical situation is illustrated in Fig. 2 where curve a represents the marginal stability condition for an interface where ϕ is not at the facet plane and $\omega = \omega^*$. Curves b

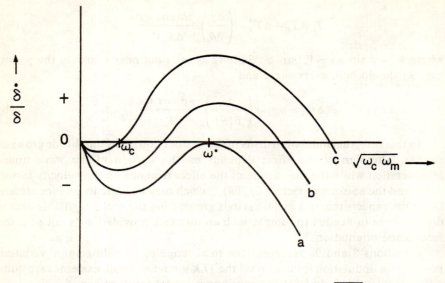

Fig. 2. Perturbation development, $\dot{\delta}/\delta$, as a function of average frequency $\sqrt{\omega_C\omega_m}$ for three cases: (a) marginal stability of an interface of type c in Fig. 1, (b) unstable interface for type c in Fig. 1 but stable for type b, (c) marginal stability for interface of type b in Fig. 1.

and c both indicate instability for this non-faceted interface but curve c indicates the condition of marginal stability for that faceted interface with critical frequency $\omega_C = 2\pi/\sqrt{A_P}$. If the system is driven a little harder, this faceted interface will begin to develop undulations. With this physical insight, one can immediately use conventional perturbation theory to write an instability condition for faceted interfaces; i.e.,

$$\frac{G_L}{V_y^o} + \frac{\Delta H_F}{2K_L} < \frac{-m_L(1-k_o)C_i}{D_L}\,\hat{K}S_C \tag{11a}$$

where

$$S_C = \frac{1}{2}\left[\frac{\alpha_C-1}{\alpha_C-(1-k_o)}\right] + \frac{\gamma(D_L/V_y^o)\omega_C^2}{2\Delta S_F m_L C_i(1-k_o)} \tag{11b}$$

and

$$\alpha_C = \frac{1}{2}\left\{1 + \left[\left(\frac{2D_L}{V_y^o}\right)^2(\omega_i^2+\omega_m^2)_C\right]^{1/2}\right\} \tag{11c}$$

Here, $\omega_i = \omega_m = \omega_C$ and all the other symbols have their usual meaning. Since ω_C may be much less than ω^*, the second term in eqn. 11b can be readily shown to be small in comparison with the first term which is $(D_L\omega_C/V_y^o)^2/2[k_o+(D_L\omega_C/V_y^o)^2]$.

This first term will, itself, be very small where $k_o \gg (D_L \omega_C / V_y^o)^2$. Thus, one can see that, under the proper conditions, $S_C \ll S^*$ and G_L / V_y^o must be reduced by a large factor before instability sets in on the faceted interface.

Even in metallic systems, where γ_l is small and $2D$ nucleation is relatively easy, $\omega_C < \omega^*$. Thus, by considering the orientation dependence of interface instability, careful experiments should reveal important information concerning 2D nucleation kinetics of the facet plane. Certainly, from this author's early experiments on interface instability for lead crystals grown from the melt [5], the $\{111\}$ facet plane was qualitatively observed to resist breakdown at a lower value of G_L than all other orientations, for fixed V and C_∞.

Sphere-like growth forms

Neglecting undulation formation, if one adopts the continuum model for interfaces, a crystal sphere with isotropic γ will continue to grow as a sphere because the crystal surface is an isotherm. For the TLK model with a cubic system (FCC, DC, SC, etc.), in the small driving force regime with small γ_l, the growing sphere becomes sphere-like with: (i) small flats at the 8 $\{111\}$ poles for the generation of new layers, (ii) faint ridges along [110]-type lines where layer edges from two sources annihilate, and (iii) faint nipples at the 6 $\{100\}$-type poles where layer edges from four different sources annihilate. As γ_l increases, the terrace and ledge roughening decreases so that the sphere-like form changes to cube-like with large $\{111\}$ flats, strong [110]-type ridges and well developed corner nipples because, now, a large driving force is needed to generate new layers at the $\{111\}$ poles by a two-dimensional nucleation mechanism. This pattern of morphological change is illustrated in Fig. 3.

If one starts with the perfectly spherical crystal of Fig. 3a then, for large γ_l but very low supersaturation, σ, insufficient driving force exists for layer creation on the $\{111\}$ poles so only the existing ledges and kinks can lead to atom attachment. Flats begin to develop on the sphere at the $\{111\}$ poles by movement of the existing ledges and this eventually leads to the bipyramid form of Fig. 3c bounding the original sphere. For increasing σ, a critical value, $\sigma = \sigma_C$, will be reached at which the nucleation frequency becomes non-zero and the crystal becomes of slowly growing bipyramid form. For large σ, the face nucleation frequency can become so great that the body can develop sphere-like characteristics. It can also develop a dendritic character from the (100) corners.

One of the interesting features here, even for small γ_l, is that the growing sphere has pre-made distortions pointing in the dendrite directions, as a natural consequence of the TLK model, with the nipple amplitude growing as the overall sphere growth rate increases. It is through such symmetrical layer-fed shape distortions that the unique crystallographic features of dendrite growth are manifest. The breakaway condition of unique dendrite development from the sphere awaits only the critical driving force needed for the local layer generation rate at the nipple to exceed that which feeds the main sphere.

232

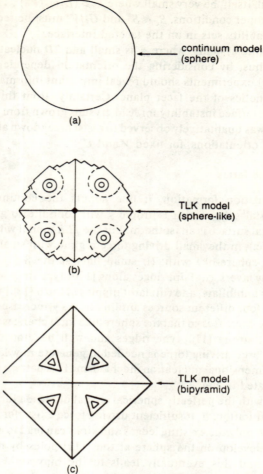

Fig. 3. Sphere-like growth forms: (a) true sphere based on the continuum model, (b) development of flats, ridges, and nipples on a sphere based on the *TLK* model with small γ_l and (c) bipyramid growth form based on the *TLK* model with large γ_l.

Dendrite filament growth

In Fig. 4, the growth velocity for three FCC filaments of the same circular paraboloidal shape are compared. Filament a uses the continuum model with $\gamma = 0$ and $\beta = \infty$ so it is an isothermal filament. Filament b uses the continuum model but with $\gamma > 0$ and $\beta < \infty$ so it is a non-isothermal filament with the tip being the coolest spot on the body. Because of this pattern of temperature variation on the filament, local excess heat transport to the tip occurs and it slows down the rate of tip motion relative to the isothermal filament as indicated in Fig.

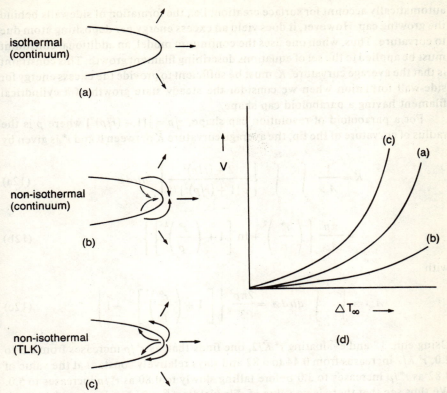

Fig. 4. Comparison of equal shape paraboloid of revolution filament growth forms: (a) isothermal filament based on the continuum model, (b) non-isothermal filament based on the continuum model, (c) non-isothermal filament based on the *TLK* model for the *FCC* system and (d) dendrite growth velocity, *V*, for these three filament forms as a function of bath supercooling, ΔT_∞.

4d. Filament *c* uses the *TLK* model with a $\langle 100 \rangle$ axis and $4 - \{111\}$ poles. It has $\gamma_l > 0$ and $\beta < \infty$. In this case, the tip is the warmest spot on the filament body because maximum ledge annihilation occurs at that location while the $\{111\}$ poles are the coolest spots on the body. Thus, the local excess heat transport at the tip is in such a direction as to allow the filament to speed up relative to the isothermal filament. All other things being equal, filament *c* will be the one preferred in nature because it grows faster than the other two for the same driving force and the same radius of curvature at the tip. It also gives a well defined dendrite direction.

In the quantitative analysis of dendrite growth, the *TLK* model of the tip has not been utilized because of its mathematical complexity and theorists have used the mathematically simpler but physically incorrect, continuum model for the surface. The dilemma with this approach is that it does not automatically account for surface creation; i.e., the formation of sidewalls behind the growing cap. However, it does yield an excess energy per depositing atom due to curvature. Thus, when one uses the continuum model, an additional constraint must be applied to the set of equations describing filament growth. This constraint is that the average curvature, K, must be sufficient to provide the excess energy for side-wall formation when we consider the steady state growth of a cylindrical filament having a paraboloid cap shape.

For a paraboloid of revolution cap shape, $z/p = \frac{1}{2}[1 - (r/p)^2]$ where p is the radius of curvature of the tip, the average curvature $\bar{K}$ between 0 and r^* is given by

$$\bar{K} = \frac{1}{A_{or^*}} \int_a^{r^*} \int_0^{2\pi} \left\{ \frac{2 + (r/p)^2}{p[1 + (r/p)^2]^{3/2}} \right\} dlrd\phi \qquad (12a)$$

$$= \frac{\pi p}{A_{or^*}} \left\{ \left(\frac{r^*}{p} \right)^2 + \ln \left[1 + \left(\frac{r^*}{p} \right)^2 \right] \right\} \qquad (12b)$$

with

$$A_{or^*} = \int_0^{r^*} \int_0^{2\pi} dlrd\phi = \frac{2\pi p^2}{3} \left\{ \left[1 + \left(\frac{r^*}{p} \right)^2 \right]^{3/2} - 1 \right\} \qquad (12c)$$

Using eqn. 12 and evaluating $r^* \bar{K}/2$, one finds that, as r^*/p increases from 0.5 to 2.0, $r^* K/2$ increases from 0.44 to 0.82 and stays relatively constant at the value of 0.82 as r^*/p increases to 3.0 before falling slowly to 0.80 as r^*/p increases to 5.0. We thus see that there is no value of r^*/p yielding $\bar{K} = 2/r^*$ as it must if side-wall creation is to be fully accounted for. The maximum value of $\bar{K}$ occurs at $r^*/p \sim 3.0$. It is interesting to note from the data presented by Glicksman [6], that the dendrite tip closely approaches a paraboloid of revolution and that the onset of sidebranching occurs at $r/p \sim 3$.

This surface creation constraint, in the continuum model, would seem to place an additional constraint on the filament growth so that a maximum V with respect to p is not possible. Only a maximum V with respect to orientational variation is possible so that the dendrite growth direction is selected. Since the onset of side-branching seems to occur at about the same location on the paraboloidal cap as the maximum $\bar{K}$, the marginal stability condition of Langer and Müller-Krumbhaar [7] appears to be a manifestation of this surface creation constraint for the continuum surface model in that it selects r^* for sidebranching and thus constrains ϱ via $r^*/p \sim 3$.

Layer source locations and external crystal shape for CZ growth

The contour of a static interface ($V = 0$) for growth in the Czochralski (CZ) pulling mode is schematically illustrated in Fig. 5a, i.e., a large flat segment of interface is closely coincident with the freezing point isotherm and a rim, $\sim 10 - 100$ μm

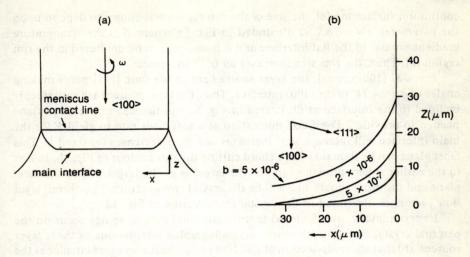

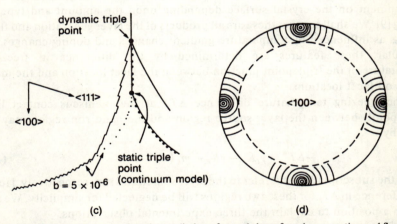

Fig. 5. The *TLK* model approach to the prediction of crystal shape from layer creation and flow in the meniscus rim region: (a) cross section of interface shape based on the continuum model (exaggerated rim region), (b) quantitative rim shape, $Z(X)$, for three values of the parameter b, (c) side view of layer creation and layer flow in the rim region based on the *TLK* model and (d) interface cross section revealing the location of the $4 - \{111\}$ layer sources on the $\{100\}$ interface.

wide, connects the main interface with the solid gas interface at the triple junction. In Fig. 5a, the size of the rim region has been expanded for communication purposes. Although the rim region is small in dimension, it is generally the location of the layer sources that feed the main interface and provide the surface creation for the sidewalls of the crystal. It is these layer sources that generate the external shape of the crystal (square, triangular, hexagonal, etc.). Using the continuum surface model, the size of the rim region was shown to depend upon the parameter $b = \gamma/G\Delta S_f$ as illustrated in Fig. 5b where G is the temperature gradient normal to the flat interface and it is assumed to be unaltered in the rim region [8]. Thus, the rim size increases as $G^{-1/2}$ increases.

For a Si (100) crystal, the layer sources are at the four {111} poles making angles of $\phi = 54.74°$ to the (100) interface. Thus, the layer sources are very steeply inclined to the interface as illustrated in Fig. 5c and the size of the facet plane needed to provide 2 D-pillbox nucleation at a sufficient rate to yield V for the main interface will increase as γ_l increases and as G decreases for fixed V. This facet plane can be thought of as a chord cutting the rim contour in Fig. 5b. Just as in the earlier sphere growth example, the larger is γ_l, the larger will be the facet plane and the more square-like will be the crystal cross-section. The overall layer flow picture is illustrated in the crystal cross-section of Fig. 5d.

Experimentally, one finds (a) that longitudinal flats or spines occur on the external crystal surface at the same crystallographic orientations as these layer sources, (b) that the cross-section of a Si(100) crystal becomes more circular as the crystal rotation rate, ω, is increased, and (c) that the cross-section of large diameter Si(111) crystals is triangle-like but with the presence of large flats or oval development on the crystal surface depending upon the amount and type of doping [9]. We shall see that these are all products of the layer generation and flow process as influenced by temperature gradient changes and doping changes. In particular, these features are determined by the difference in freezing temperature of the triplepoint junction between the facet location and the main rim at off-facet locations.

The freezing temperature difference ΔT^* at the meniscus contact line (triplepoint) between the layer source region and the main rim region may be given by

$$-\Delta T^* = \gamma_l K_{ls} - \gamma \bar{K}_R - m_L (C_{ls} - C_R) \tag{13}$$

where the subscripts ls and R refer to the layer source and rim, respectively. Here, the difference in ΔT_K for these two regions will be neglected for simplicity. We are now in a position to explain the three experimental observations.

a) For pure dislocation-free S_i, the presence of a ridge versus a flat depends primarily on the size of the facet required to produce the needed nucleation frequency of 2D-pillboxes relative to the radius of curvature of the ledges for the average rim material at the triple-point. For a steep temperature gradient in the meniscus region, it is very likely that $K_{ls} >> \bar{K}_R$ and $\Delta T^* < 0$. Thus, the inclined ledge flow from the source will lift the meniscus above that for the main rim material before freezing and a raised ridge or spine of Si will be formed at these locations. It is interesting to note that, if dislocation generation occurs in the Si interface region by some mechanism, a more efficient layer source is present so the 2D nucleation mechanism ceases to operate at some of the {111} poles. This causes the disappearance of a local spine or spines on the crystal surface and has been taken to indicate the loss of the dislocation-free state.

For a shallow temperature gradient in the meniscus region, the facet plane will become greatly enlarged in order to have sufficient undercooling for the needed nucleation so it is likely that $\bar{K}_R > K_{ls}$ and $\Delta T^* > 0$. Thus, the inclined ledge flow will intersect the meniscus and freeze at a position below that of the main rim. This leads to the development of a flat at the layer source location. The larger is γ_l, the larger will be the size of the flat.

When the crystal diameter is being decreased rapidly, these layer source planes become exposed as inward-inclined facets. When the crystal diameter is being increased rapidly (as in seeding, etc.), the unused facet planes on the other half of the crystallographic sphere can come into play and become exposed as outward-directed facets. For $Si(100)$, these inward-and outward-directed flats lie along the same riblines; however, for $Si(111)$ the inward- and outward-directed flats lie along symmetrical sets of riblines rotated from each other by 60° and yield six riblines in the seeding region but only three riblines on the crystal.

b) When the crystal rotation rate, ω, is increased, the convected heat flux from the liquid to the interface increases and this causes the main interface to recede to a position higher in the meniscus. At this location, convection is diminished but a larger G is needed to satisfy the heat balance. This effect causes the layer source areas to decrease in size for the same V so the crystal becomes more round. The larger is γ_l for a material, the greater must ω be to restore roundness to the crystal cross section.

c) During the growth of large diameter Si crystals, Lin and Hill [9] observed the crystal cross-section to change with doping as indicated in Fig. 6. For lightly doped or heavily boron doped crystals, the periphery of the crystal was observed to be somewhat triangular with three large flats at the $\langle 211 \rangle$ type directions. However, for heavily antimony-doped crystals, the outer crystal morphology was significantly changed and large peaks had developed to replace the large flats.

To grow large diameter crystals, one generally reduces ω for mechanical stability reasons and reduces G for thermal stress reasons. Thus, the layer source facet areas should be large and produce relatively large flats on the external

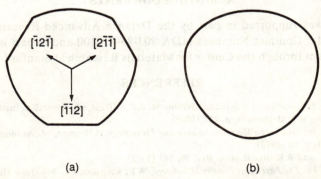

(a)   (b)

Fig. 6. Observed cross sections of large diameter $Si(111)$ crystals: (a) intrinsic and heavily boron-doped and (b) heavily antimony-doped.

238

surface of the crystal in the manner already indicated for pure materials. This will also be the expectation for lightly doped materials but not necessarily for heavily doped materials because solute partitioning in the rim region will be more restricted than for the main interface so a higher interface liquid concentration will be found in the rim region (for $k_o < 1$). For the same reasons, an even higher liquid concentration will be found at the layer source location where the layers flow upwards to meet the meniscus.

Since boron has a value of $k_0 \sim 0.8$ in Si and a small value of m_L, it should only affect γ_l or γ in a small way and a negligible solute redistribution effect should occur for boron so the heavily doped Si should behave in essentially the same way as lightly-doped Si. For Sb-doping, the picture is quite different since $k_0 \sim 0.02$ and m_L is fairly large. Considerable solute enrichment will occur in the layer source region compared to the rim region and ΔT^* becomes dominated by the solute term. Thus, for heavy Sb-doping, $|\Delta T^*|$ is so large that the outward inclined ledges keep growing and lifting the meniscus far beyond that for the average rim material until it finally freezes to generate the cross-sectional shape shown in Fig. 6b.

In closing this section, if these considerations are extended to a material like $GaAs$, which is thought to have a much larger value of γ_l than Si, one would expect all the shape anisotropies, flat formation, peak formation, etc. to be much more pronounced than in Si.

CONCLUSIONS

The TLK model of interfaces rather than the continuum model should be utilized in all morphological considerations of crystal growth in that it more readily allows surface creation to be properly accounted for and it is the model more consistent with physical experience. When this is done, important crystallographic and crystal shape features become theoretically manifest and many hitherto unexplained features of crystal morphology find a consistent rationale.

ACKNOWLEDGEMENTS

This work was supported in part by the Defense Advanced Research Projects Agency under Contract Numbers MDA 903-85-K-0100 and in part by the NSF-MRL Program through the Center for Materials Research at Stanford University.

REFERENCES

1. Tiller, W.A. In: *Interfaces* Conference, Melbourne, R.C. Gifkins, ed., Australian Institute of Metals and Butterworths, Melbourne, p. 257 (1969).
2. Tiller, W.A. In: *Treatise on Materials Science and Technology,* H.Herman, ed., Academic Press, New York, Vol. 1, p. 66 (1972).
3. Tiller, W.A. and B.K. Jindal. *Acta Met.* **20,** 543 (1972).
4. Herring, C. In: *The Physics of Powder Metallurgy,* W.E. Kingston, ed., Mc-Graw Hill, New York, (1951).

5. Tiller, W.A. The Effect of Growth Conditions upon the Solidification of a Binary Alloy, Ph.D. Thesis, The University of Toronto, April (1955).
6. Glicksman, M.E. *Mat. Sci. and Eng.* **65**, 45 (1984).
7. Langer, J.S. and H. Müller-Krumbhaar. *Acta Metall.*, **26**, 1681, 1689, 1697 (1978).
8. Bolling, G.F. and W.A. Tiller. *J. Appl. Phys.* 31, 1345 (1960).
9. Lin, W. and D.W. Hill. *Silicon Processing,* ASTM STP804, D.C. Gupta, ed., American Society for Testing and Materials, Philadelphia, Pa. (1983).

APPENDIX A

TLK model requirements for an isotropic γ-plot in 2-D

If we consider a vicinal face of a crystal making an angle θ with the facet plane for a cubic system, we have a picture of terraces and ledges with ledge height h and ledge spacing λ_l such that

$$\frac{h}{\lambda_l} = \tan \theta \tag{A.1}$$

It is common practice to describe the interfacial energy, $\gamma(\theta)$, in terms of the terrace or face energy, γ_f, and ledge energy, $\gamma_l(\theta, h)$ by the equation

$$\gamma(\theta) = \gamma_f \cos \theta + \gamma_l(\theta, h) \sin \theta \tag{A.2}$$

where all of the ledge-ledge interaction is included in γ_l and γ_f is the value for the perfect face of infinite extent.

For a circular γ-plot, we require that

$$\frac{\partial \gamma}{\partial \theta} = 0 = -\gamma_f \sin \theta + \frac{\partial \gamma_l}{\partial \theta} \sin \theta + \gamma_l \cos \theta \tag{A.3a}$$

which leads to

$$\tan \theta = \frac{\gamma_l}{\gamma_f - \dfrac{\partial \gamma_l}{\partial \theta}} \tag{A.3b}$$

Considering the relationship between $\tan \theta$ and λ_l, we find that

$$\frac{\partial \lambda_l}{\partial \theta} = -\frac{h}{\sin^2 \theta} = -\frac{h}{1 + \left(\dfrac{\lambda_l}{h}\right)^2}. \tag{A.4}$$

so that eqn. A.3b becomes of the form

$$\frac{\partial(\gamma_l/\gamma_f)}{\partial(\lambda_l/h)} = \left\{ 1 + \left(\frac{\lambda_l}{h}\right)^2 \right\} \left\{ \left(\frac{\lambda_l}{h}\right)(\gamma_l/\gamma_f) - 1 \right\}. \tag{A.5}$$

Defining $x = \lambda_l/h$ and $y = -\gamma_l/\gamma_f$, eqn. A.5 is of the form

$$y' = g(x)y - f(x) \tag{A.6a}$$

with

$$g(x) = x(1 + x^2) \text{ and } f(x) = (1 + x^2) \tag{A.6b}$$

so the solution is of the form

$$y = A_e \int g(x)dx + e^{\int g(x)dx} \int e^{-\int g(x)dx} f(x)dx \tag{A.7}$$

So long as $y = \gamma_i/\gamma_f$ varies with $x = \lambda_i/h$ according to eqn. A.7, the y-plot will be completely circular.

10

Modeling the Formation of Eutectic Microstructures

ALAIN KARMA

California Institute of Technology, Pasadena, CA 91125, U.S.A.

ABSTRACT

We present a numerical algorithm which can simulate the kinetics of eutectic growth. The algorithm uses simple rules which govern the attachment and escape of random walkers to and from the solid-liquid interface. These walkers mimic molecular diffusion between neighboring solid phases while the governing rules contain the important physics of the crystal growth and surface tensions between different phases. This algorithm is used to investigate the instabilities and growth mechanisms responsible for a variety of eutectic morphologies which have been observed in thin film experiments.

INTRODUCTION

The main goal of theoretical modeling of eutectic growth is to be able to predict accurately the microstructure which a given material will form under prescribed solidification conditions. In particular, one wishes to predict the geometrical nature of this microstructure, for example, whether it is lamellar, rod-like, or more complex, and its scale. Simple lamellar and fibrous structures can be characterized essentially by only one parameter, namely the lamellar or fiber spacing, while more irregular microstructures demand a more complex characterization, for example, in terms of a statistical distribution of fiber spacings and orientations.

The pioneering analysis of Jackson and Hunt [1] has provided the theoretical basis for our understanding of both lamellar and rod-like eutectic microstructures. Their analysis applies to spatially periodic eutectic structures forming under steady-state growth condition. During steady-state growth, the shape of the solid-liquid interface is preserved and it is essentially this time independence of the interface shape which renders steady-state growth amenable to analytical treatment.

However, in many important situations, the solid-liquid interface constantly changes its shape during the growth process. Non-stationary interface shapes are characteristic of systems where facets are present on the solid-liquid interface of one solid phase ($f-nf$ systems), but also occur frequently in systems where the solid-liquid interfaces of both solid phases are non-faceted (we shall only consider in what follows this second class of $nf-nf$ systems). For example, in three dimensions, rearrangements of lamellar structures are made possible by the motion of lamellar faults. These faults permit an interplay between lamellar and rod morphologies and are likely to play a role in selecting the operating point of the system (i.e., in selecting one of the two morphologies and its scale). Understanding their motion is of primary importance and requires a time-dependent description of eutectic growth. Also, in practice, steady-state growth is limited to a small range of melt compositions in the vicinity of the eutectic composition. Outside this range the solid-liquid interface becomes time-dependent and a wide variety of eutectic morphologies can form. Our knowledge of these morphologies has remained very limited. On the experimental side, the best documented non-steady-state behavior is one in which dendrites of one solid phase and lamellar eutectics both coexist [2]. This situation is observed at off-eutectic compositions when the growth of one solid phase becomes sufficiently enhanced to cause dendrites of this phase to emerge ahead of the eutectic interface. There is also experimental evidence for dynamical modes of growth in which dendrites are absent [3, 4]. One of them is an oscillatory mode on twice the lamellar spacing which causes lamellae from the minor phase (the phase of smaller volume fraction) to follow sinusoidal solidification paths, as opposed to the vertical paths followed during steady-state growth. Other modes, which seem to involve more convoluted solidification paths of these lamellae, generate complex eutectic patterns ranging from orderly to chaotic.

An important tool which is missing at this point to model this large class of time-dependent growth problems and understand in more detail these questions of "microstructure selection" is a numerical algorithm capable of simulating the motion of a eutectic solidification front (i.e., a tool to model non-steady-state growth). To construct such an algorithm, even for the simplest two-dimensional situation characteristic of thin films, represents a formidable task since the difficulty of incorporating in the growth problem the constraint of mechanical equilibrium at solid-solid-solid triple points (where three phases meet in space) adds to the already existing difficulty of tracking the motion of an interface.

In this paper I shall discuss a numerical algorithm which simultaneously overcomes both difficulties and can thereby simulate eutectic growth of $nf-nf$ systems [5]. An example of a simulated lamellar structure is shown in Fig. 1. This algorithm is based on a random walk model originally developed by Kadanoff [6], Tang [7], and Liang [8] to study a hydrodynamics problem (viscous flow in a Hele-Shaw cell). Here, in the context of eutectic growth, random-walkers on a lattice mimic molecular diffusion between neighboring solid phases. The rules governing the escape of walkers from the interface contain the physics of the

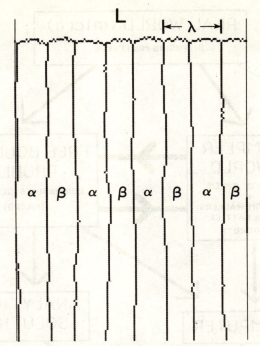

Fig. 1. Simulated lamellar eutectic morphology.

crystal growth and are constructed in such a way to preserve the balance of surface tensions at triple points. The main virtue of this algorithm is its ability to simulate almost arbitrarily complex growth processes and to help visualize the interfacial deformations underlying the formation of eutectic microstructures.

WHY A LATTICE-RANDOM-WALK ALGORITHM?

Although it is clear that the main purpose of the lattice random-walk algorithm is to simulate interfacial dynamics it is not obvious why one should in the first place develop such an algorithm instead of using a more traditional numerical method. The reason for this choice can be made clearer by examining the relationships between various levels of description (from "micro" to "macro") of the eutectic growth problem. For this purpose we have displayed in Fig. 2 a chart illustrating the links between these various levels of description.

At the microscopic level of description, which we shall refer to as the real world, we have a large ensemble of interacting solute and solvent molecules. Given the interaction forces between these molecules one could in principle simulate on a computer the growth process by molecular dynamics (MD). This is however practically impossible since even large-scale MD simulations can only track a relatively small number of particles and thus cannot describe interfacial

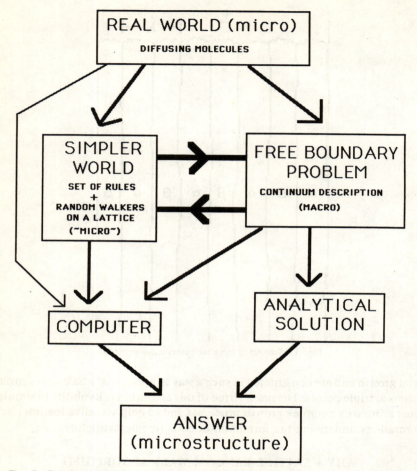

Fig. 2. Relationship of the lattice-random-walk algorithm to other levels of description.

deformations occurring on a scale of several microns. Of course, this has never been considered a serious problem since, for *nf* interfaces, one usually does not attempt to describe the growth problem at a microscopic level but at a macroscopic level where coarse grained variables such as the solute concentration can be defined.

At this hydrodynamic level of description the free boundary problem (FBP) associated with interface motion consists of the equations of heat and solute diffusion coupled to a set of boundary conditions on the interface. Two types of parameters enter in the FBP: control parameters such as the temperature gradient or the sample composition, and material parameters such as the solute diffusivities, the surface tensions between various phases, the liquidus and solidus slopes. Since material parameters are measurable experimentally and a set of control parameters can be assigned to a given experiment, the continuum description should in principle provide an adequate description of the motion of the solidification front (in the same sense that the Navier-Stokes equations provide an adequate description of hydrodynamics).

Is the continuum description really adequate?

It should be noted that, even if it provides an adequate description of the problem, it still remains extremely difficult to describe by a conventional front tracking algorithm a number of complex processes which often occur during the formation of eutectic microstructures. These processes are depicted schematically in Fig. 3. Figure 3a describes a time sequence where two solid phases approach each other and subsequently bind by surface tension. Figure 3b shows a time sequence of the termination of one solid phase and finally Fig. 3c depicts the

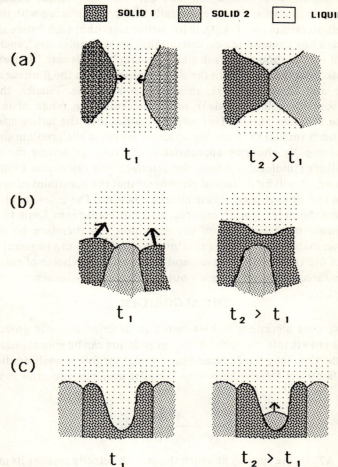

Fig. 3. Complex processes characteristic of eutectic growth.

nucleation of one solid phase inside a groove of the other solid phase (nucleation events are not described by the lattice-random-walk-algorithm in its present form but could be easily incorporated).

What renders simulations of eutectic growth difficult here is the presence, both spatially and temporally, of discontinuities in the interface shape (for example: the slope of the interface is discontinuous in the region of a triple point and the interface shape changes abruptly at the time of contact of two solid phases). The most natural way to proceed in order to avoid the difficulties caused by these discontinuities is to describe the growth problem at a more microscopic level. At the molecular level one can, at best, only describe a very small region around the triple point given finite computer time and memory. An alternate route is to simulate a "simpler world" which contains all the essential physics of the crystal growth and surface tensions between the various phases but is computationally tractable. This is essentially the approach taken with the lattice-random-walk algorithm (see Fig. 4). In the lattice algorithm each lattice site can be thought of as a box containing a certain number of (M) "molecules", each random walk in turn can be though of as displacing a single "molecule." For example a random-walk from the α-phase to the β-phase displaces to the β-phase an excess B-molecule and to the α-phase an excess A-molecule. Finally, the lattice algorithm contains a lengthscale r which measures the range of interaction between the separate phases. This length is larger than the lattice spacing but should be much smaller than any macroscopic length in the problem such as the lamellar spacing. By choosing appropriately the rules governing the escape of random walkers ("molecules") from the interface one can insure (within some restrictions which will be discussed elsewhere) that the constraint of mechanical equilibrium remains satisfied during interface motion. The rules are also chosen such that the "simpler world" reduces in some continuum limit to the free boundary problem. Parameters of the algorithm can therefore be related to materials parameters and a given experimental realization can be simulated. The dynamics of the triple point may depend on a particular choice of rules (i.e., of "interaction forces between phases") but is automatically incorporated!

THE ALGORITHM

The random-walk algorithm is best suited to describe eutectic growth in the limit of slow growth rate where the diffusion equation can be approximated by the Laplacian. In this limit the Jackson-Hunt model for free growth predicts a one parameter family of steady-states with lamellar spacings and growth velocities related by

$$v = v_m \left\{ \frac{2}{\wedge} - \frac{1}{\wedge^2} \right\}, \quad \wedge = \frac{\lambda}{\lambda_m} \tag{1}$$

where $\lambda_m \sim \Delta T^{-1}$ is the spacing at which the growth velocity reaches its maximum

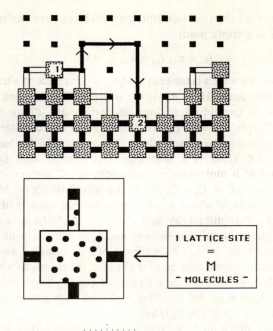

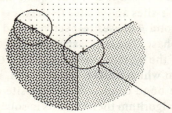

Fig. 4. Basic building blocks of the lattice-random-walk algorithm.

value $v_m \sim \Delta T^2$ (expressions for λ_m and v_m are given in Ref. 1). The motion of the eutectic interface is governed by,

$$\nabla^2 u = 0 \tag{2}$$

conservation of mass at both solid-liquid interfaces,

$$V_n = -\hat{n} \cdot \vec{\nabla} u \qquad \alpha\text{-phase} \tag{3a}$$

$$Q V_n = \hat{n} \cdot \vec{\nabla} u \qquad \beta\text{-phase} \tag{3b}$$

local thermodynamic equilibrium at the interface (Gibbs-Thomson relations),

$$u(\alpha) = \Delta T\, m_\alpha^{-1} - d_\alpha k \qquad \alpha\text{-phase} \tag{4a}$$

$$u(\beta) = -\Delta T\, m_\beta^{-1} + d_\beta k \qquad \beta\text{-phase} \tag{4b}$$

and the constraint of mechanical equilibrium which requires the sum of surface tensions to vanish at a triple point

$$\sigma_{La}\hat{t}_{La} + \sigma_{L\beta}\hat{t}_{L\beta} + \sigma_{\alpha\beta}\hat{t}_{\alpha\beta} = 0 \tag{5}$$

where $\hat{t}_{\gamma v}$ is a unit vector which is tangent to the $\gamma - v$ interface at a triple point and points away from this point. Here $\sigma_{\alpha\beta}$ is the $\alpha - \beta$ surface tension, σ_{LS} the $L-S$ surface tension, m_s is the liquidus slope of each phase dT/du defined to be positive, $d_S = \sigma_{LS} T_E/(m_S L_S)$ and L_S are respectively the capillary length and latent heat per unit volume of each phase ($s = \alpha, \beta$), k is the local interfacial curvature, $u \equiv (C - C_E)/\Delta C$ is a dimensionless composition field where C denotes the number of B molecules per unit volume, C_s is the concentration of each solid phase, and $\Delta C \equiv C_\beta - C_\alpha > 0$ is the miscibility gap. In addition we have defined $V_n = (-u_a)v_n/D$ where v_n is the local normal velocity of the interface, D the coefficient of solute diffusivity, and $Q = u_\beta/(-u_a)$ with $u_S \equiv (C_S - C_E)/\Delta C$ ($u_\alpha < 0$, $u_\beta > 0$). Finally, the exponentially decaying part of u in the upward z direction is translated, in the Laplacian limit of the diffusion equation, into a linear gradient boundary condition on u far ahead of the interface where, in steady state, $(\partial_z u)_\infty = (v/D)u_\infty$ with $C_\infty = C_\alpha \eta + C_\beta(1-\eta)$; $\eta \equiv \lambda_a/\lambda$ is the volume fraction of the α phase, λ_S ($s = \alpha, \beta$) the lamellar width of each phase, $\lambda = \lambda_a + \lambda_\beta$ the lamellar spacing and $u_\infty \equiv [C_\infty - C_E]/\Delta C$.

Simulations take place on a two dimensional square lattice where sites are divided into three categories: α-sites and β-sites (s-sites; $s = \alpha, \beta$) represent sites occupied by the solid α and solid β phases respectively and empty sites (e-sites) those occupied by the liquid phase. The lattice spacing is set equal to unity and W measures the lateral width of the system where periodic BC are imposed at the endpoints. The interface, from which walkers are released, is composed of all s-sites which have at least one bond connected to an e-site (i.e. the solid-liquid interface). The essence of the algorithm then consists in solving Laplace equation (eqn. 2) by releasing random walkers from the interface and moving the interface according to specific rules which are consistent with the mass conservation relations (eqns. 3a-3b). The Gibbs-Thomson relations (eqns. 4a-4b) are automatically satisfied by choosing the escape probability of a random walker from site s as

$$P(s) = |u(s)|/\max\{|u|\} \tag{6}$$

and by assigning to this walker a flux $f(s)$ equal to the sign of $u(s)$ ($f(s) = \pm 1$), where $\max\{|u|\}$ is the maximum value of $|u|$ on the inter-face. The intrinsic noise in the algorithm which comes from the random walks is diminished by moving the interface only when a bond connecting and s-site with an e-site has been visited a certain number of times (M or QM times, depending whether the s-site is of α or β type). Finally, the constraint of mechanical equilibrium (eqn. 5) is incorporated into the algorithm by using in eqn. 6 a form of the Gibbs-Thomson relation which takes into account the interaction energy between the three phases at triple points [5]. Away from these points this form reduces to eqns. 4a-b. In their vicinity, that is within a distance r of triple points where r measures the range of the interaction energy between different phases, this form causes the composition field $u(s)$ to become large when slight deviations from the constraint of mechanical equilibrium occur. This increase in composition in turn induces large normal composition gradients which cause interface motion to smooth out these deviations in a time much shorter than the time it takes for triple points to move a distance r.

Three types of walks can be distinguished and each type given clear physical interpretation. Walks that start from an s-site and end on an s-site of the same phase (α-α and β-β walks) do not contribute to net interface motion, since they conserve flux and only displace material. They simulate the effect of capillary forces at the solid-liquid interface. On the contrary α-β and β-α walks correspond to diffusion between neighboring lamellae of rejected A and B molecules and contribute to the net growth of both solid phases. Finally walkers are released from infinity at off-eutectic compositions to simulate the gradient BC on u. These walks enhance the growth of one solid phase preferentially. It should be noted that this way of incorporating in the algorithm the effect of the boundary-layer can only describe situations where the average vertical velocity of the eutectic front remains constant. This does not require, however, the front to grow in steady-state. The solid-liquid interface shape can vary in time, the average vertical velocity remaining constant. This limitation is not present in simulations performed exactly at eutectic composition ($u_\infty = 0$).

Also, the random-walk model simulates the deterministic continuum eqns. 2-5 only when both limits $M \gg 1$ (deterministic) and $1 \ll r \ll \lambda$ (continuum) one satisfied simultaneously. In simulations values of M in a range between 10 and 20 are typically sufficient for the interface evolution to be independent of the sequence of random numbers used to generate the walks, apart from noise induced global symmetry breaking (i.e. left-right symmetry of the "tilting mode", displayed in Fig. 5a), and therefore deterministic for all practical purposes. Because of limited computation time most simulations were performed with values of r and λ ($r \simeq 5$, $\lambda \simeq 60$) which introduce corrections to the continuum limit. Our hope is that these corrections only have quantitative effects and do not change the qualitative dynamical behavior of the model. This was found to be true in cases where a simulation was repeated with larger values of λ and r.

INSTABILITIES

In this study we have restricted our attention to a symmetrical binary eutectic phase diagram with $m_\alpha = m_\beta \equiv m$, $Q = 1$, and have chosen equal surface tensions $a_{\alpha L} = a_{\beta L} = a_{\alpha\beta}$. Three different instabilities were observed: the classic long-wavelength instability leading to termination of lamellae, the oscillatory instability (Fig. 5b), at twice λ predicted successfully by the discrete stability analysis of Dayte and Langer [9] and a new tilting instability on the lamellar spacing which forces lamellae from the phase of smaller volume fraction (β-phase

252

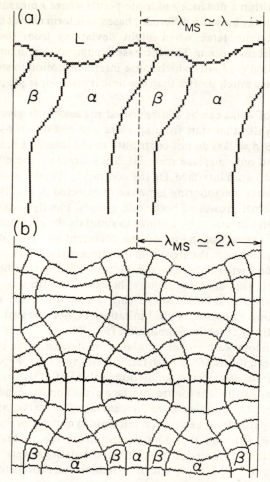

Fig. 5. Interfacial deformations associated with (a) tilting instability, and (b) oscillatory instability. Lamellae bend in response to a bulge in the interface to preserve balance of surface tensions at triple points. In (a) $W = \lambda = 61$ and $\wedge = 1.8$, (b) $W = 2\lambda = 121$ and $\wedge = 1.1$, and in both (a) and (b) $M = 15$, $r = 5$, and $u_\infty = -1/6$. (b) shows only the late stages of a simulation that started from slightly perturbed steady-state growth. Tilting can either be right or left of the vertical growth axis.

here) to bend coherently on one side of the vertical growth axis (Fig. 5a). At fixed $\wedge$ the tilting instability first occurs at a more off-eutectic composition (larger value of $|u_\infty|$) than the oscillatory one, and at fixed composition it occurs at a larger value of $\wedge$.

The appearance of both the oscillatory and tilting instabilities can be explained physically by observing that, at off-eutectic melt compositions, the composition gradient ahead of the interface (i.e., the solute boundary-layer)

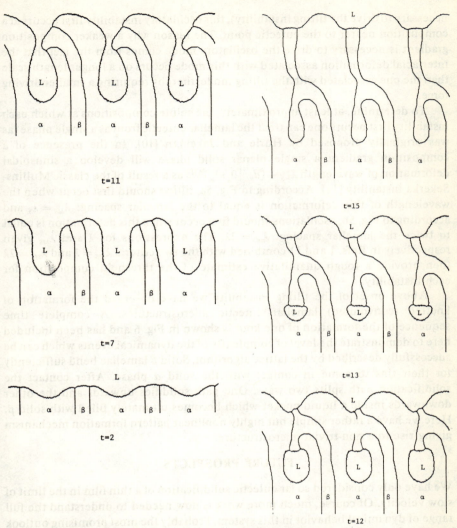

Fig. 6. Sequence of interfacial deformations occurring during the formation of a knotted microstructure. Each time unit corresponds to 8×10^5 random walks. $W = \lambda = 121$, $\wedge = 2$, $u_\infty = -1/4$, $M = 10$, and $r = 5$.

enhances the growth of one phase preferentially (α-phase here for $u_\infty < 0$). When this gradient becomes sufficienly strong, the α-phase bulges forward pushing away the β-phase. The bending of the solid-solid (α-β) interface arises as a response to the bulge in the α-phase to preserve the proper balance of surface tensions at triple points. Furthermore, since a weaker composition gradient is necessary to drive the oscillatory instability (as compared with the gradient necessary to drive the tilting instability), this oscillatory instability first occurs at a composition nearer to the eutectic point. The reason why a weaker composition gradient is necessary to drive the oscillatory mode comes from the fact that the interfacial deformation associated with this mode occurs on a longer length scale than the one associated with the tilting mode, thereby requiring a smaller driving force.

To determine, at least approximately, the solute compositions at which each instability first occurs one can treat the lamellar eutectic front as a single phase (as was originally proposed by Hurle and Jakeman [10]. In the presence of a composition gradient a single planar solid phase will develop a sinusoidal deformation of wavelength $\lambda_{MS} \sim (d_a/|\partial_z u)_\infty|)^{1/2}$ as a result of the classic Mullins-Sekerka instability [11]. According to Fig. 5a, tilting should first occur when the wavelength of this deformation is equal to the lamellar spacing: $\lambda_{MS} = \lambda$, and according to Fig. 5b, oscillations should first occur when this deformation is equal to twice the lamellar spacing: $\lambda_{MS} = 2\lambda$. The expressions for λ and λ_{MS} given respectively in Refs. 1 and 11 combined with the two criteria $\lambda_{MS} = \lambda$ and $\lambda_{MS} = 2\lambda$ then provide a rough quantitative estimate of the threshold composition for each instability.

Above onset of the tilting instability we have observed the formation of knotted (convoluted) lamellar eutectic microstructures. A complete time sequence of the formation of one knot is shown in Fig. 6 and has been included here to demonstrate the level of complexity of the dynamical events which can be successfully described by the lattice algorithm. Solid β lamellae bend sufficiently for their tips to come in contact with the solid α phase. After contact the solidification path splits two ways. One path solidifies upwards and the other downwards inside a liquid pocket which becomes ultimately filled with solid β. Here we have a rather simple but highly nonlinear pattern formation mechanism giving rise to a non-trivial microstructure.

FUTURE PROSPECTS

We have only considered so far eutectic solidification of a thin film in the limit of slow velocity. Of course, much more work is now needed to understand the full range of dynamical behavior in this system. Probably the most promising outlook for the near future is to extend lattice simulations to three dimensions. There one could hope to develop a more systematic understanding of pattern selection mechanisms for lamellar and fibrous structures. It is also conceivable that the model discussed in this paper could be modified to investigate the even vaster class of eutectic microstructures which form when the interface of either one or both solid phases is faceted [12]. Finally, it is essential to understand the dynamics of triple points and, in particular, its dependence on microscopic details. Only then will the limitation of a continuum description become clearer.

ACKNOWLEDGEMENTS

I wish to thank M.C. Cross for many helpful discussions. I am also grateful to Prof. J.S. Langer for communicating to me the experimental works of J. Van Suchtelen and W.F. Kaukler, and to W.F. Kaukler for providing additional helpful information. This research was supported by the California Institute of Technology through a Weingart Fellowship and through the Program in Advanced Technologies which is funded by GM, GTE, TRW and Aerojet.

REFERENCES

1. Jackson, K.A. and J.D. Hunt. *Trans. Metall. Soc. AIME* **236**, 1129 (1966).
2. Woodruff, D.P. *The Solid-Liquid Interface,* Cambridge University Press, (1973).
3. Kaukler, W.F. (private communication).
4. Van Suchtelen, J. (unpublished).
5. Karma, A. *Phys. Rev. Lett.* **59**, 71 (1987) (See also *Proceedings of the 1987 TMS Fall Meeting*, the Metallurgical Society AIME in: Solidification Processing of Eutectic Alloys.)
6. Kadanoff, L.P. *J. Stat. Phys.* **39**, 267 (1985).
7. Tang, C. *Phys. Rev.* **A31**, 1977 (1985).
8. Liang, S. *Phys. Rev.* **A33**, 2663 (1986).
9. Dayte, V. and J.S. Langer. *Phys. Rev.* **B24**, 4155 (1981).
10. Hurle, D.T.J. and E. Jakeman. *J. Cryst. Growth* **3, 4**, 574 (1968).
11. Mullins, W.W. and R.F. Sekerka. *J. Appl. Phys.* **35**, 444 (1964).
12. Fisher, D.J. and W. Kurz. *Proc. Quality Control of Engineering Alloys and the Role of Metals Science,* H. Nieswaag and J.W. Schut (Eds.), Dept. of Met. Sci. & Tech., University of Delft p. 59 (1977).

11

Thermodynamics and Kinetics of Glass Formation

A.L. GREER AND P.V. EVANS

University of Cambridge
Department of Materials Science and Metallurgy
Pembroke Street, Cambridge, CB2 3QZ, U.K.

ABSTRACT

The thermodynamic properties of undercooled alloy liquids are of great importance in determining their solidification behaviour, but are mostly unknown and not accessible to direct experiment. The implications of the few existing measurements are assessed, and models for estimation are suggested: the hole theory for congruently freezing systems and the regular associated solution otherwise. In the former case, it is shown that inverse melting is a possibility. The influence of the liquid thermodynamics on glass-forming ability, i.e. on crystallization kinetics, is considered.

1. INTRODUCTION

An important aim of solidification studies must be the prediction of the kinetics of the transformation from liquid to crystalline solid. For alloys there is particular current interest in rapid solidification and in glass formation i.e., the avoidance of crystallization. It would be desirable to be able to predict the conditions for glass formation in any given alloy system: composition range, critical cooling rate and other production variables. This goal is still some way off. Nevertheless, basic expressions are established for the the rates of the underlying nucleation and growth stages in the transformation; the overriding problem is in finding parameter values. In this paper, only the thermodynamic parameters of the undercooled liquid will be considered. These are only rarely accessible by direct experiment. The thermodynamic properties of the undercooled liquid are important in all types of rapid solidification processing where there may be a relatively high dynamic undercooling, but are especially important in glass formation. In glass-forming systems not only can high undercoolings be achieved, but the thermodynamic properties of the undercooled liquid are particularly unusual. The ability to form a glass is, of course, determined not only by thermodynamics. In this paper some attention will be given also to kinetics, both as influenced directly by the thermodynamics and as potentially linked to the thermodynamics through the structure of the liquid.

2. ORDERING IN ALLOY LIQUIDS

The entropy of fusion of metals and alloys can readily be measured at the equilibrium melting temperature, T_m; for pure metals typically $\Delta S^m = 8$ to 12 J K^{-1} mol^{-1}. It has been determined in a few deep undercooling experiments using droplet emulsions that the specific heats of the crystal and undercooled liquid are very similar [1]; the difference is often precisely zero at a temperature close to T_m. That being so, the entropy excess of the liquid pure metal over the crystal $\Delta S(T)$ is maintained substantially constant as the liquid is undercooled. On the other hand, it has been known for some time that in molecular liquids $\Delta S(T)$ can decrease strongly on undercooling [2]; indeed in some cases this effect is so marked as to lead to the apparent ("Kauzmann") paradox that the liquid would have a lower entropy than the crystal at an attainable undercooling. The effect is most evident in glass-forming systems, and it is the onset of the glass transition, breaking the equilibrium evolution of the liquid thermodynamic properties, that prevents the "paradox" from being realized. Recently it has become clear that alloy liquids can show this effect. The decrease in $\Delta S(T)$ is associated with an excess specific heat, $\Delta C_p(T)$, in the undercooled liquid. In a few cases $\Delta C_p(T)$ has been measured directly (next section), but even when this is not possible there is clear evidence for the excess in glass-forming alloys. The heat of crystallization of glassy alloys is generally found to be significantly less than (0.5 to 0.7 times) the magnitude of the heat of fusion at T_m, ΔH^m. This reduction in the excess enthalpy of the liquid $\Delta H(T)$ on undercooling is also related to $\Delta C_p(T)$ [3, 4], which is determined by this means to have an effective average value of (0.4 to 1.6) ΔS^m [3, 4].

The decrease in $\Delta S(T)$, which is particularly marked as glass-forming liquids are undercooled, reflects an increased order, and indeed there is substantial evidence from structural studies that highly undercooled liquid alloys and glassy alloys can be ordered. The glass-forming liquids of composition (in at.%) $Fe_{83}B_{17}$, $Ni_{80}B_{20}$ and $Mn_{76}Si_{24}$ show structure in the radial distribution function out to unexpectedly large distances [5], and this is found also for many glasses, e.g., $Ni_{81}B_{19}$ [6]. Ordering of the glassy structure, in some form distinct from incipient crystallization, has been associated with the property changes on annealing glassy alloys [7]. This structural relaxation has been divided into topological and compositional short range ordering (TSRO and CSRO) [8]. TSRO is associated most readily with changes occurring near the glass transition temperature, T_g. CSRO is more rapid and occurs reversibly at lower temperature.

The reversible ordering is probably due to exchange of different atomic species of similar chemical type on one type of site. This is distinct from the type of chemical order that is directly associated with the glassy structure; for example, in most metal-metalloid glasses there is good evidence that the metalloids are coordinated only by metal atoms [9]. This type of chemical order can exist not only up to T_g, but beyond in the undercooled liquid and equilibrium liquid (as evidenced by the liquid studies cited above). It can be found in metal-metal systems as well, for example, in Cu-Ti [10]. There is certainly evidence for ordering transformations in some rather special liquids. For example in $Te_{80}Ge_{20}$ there is a substantial peak in specific heat *above* T_m, which is well above T_g [11]. In that case the effect is associated with a metal-to-semiconductor transition, but substantial ordering, albeit less dramatic, may generally be expected in alloy liquids.

The excess specific heat of undercooled alloy liquids, reflecting the ordering and its temperature dependence, is in most cases not measured, if measurable at all. This paper is therefore concerned largely with the assessment of models which may be used to estimate the thermodynamic properties of undercooled liquids in the absence of direct measurement. The simple property of most interest is the amount, $\Delta G(T)$, by which the Gibbs free energy of the liquid exceeds that of the solid. When the crystal and liquid have the same specific heat, as is true to a good approximation for pure metals, $\Delta G(T)$ rises linearly as the liquid is undercooled. (This has been termed the Turnbull approximation.) For alloys where $\Delta C_p(T)$ is significant $\Delta G(T)$ rises less rapidly.

Phase diagram fitting seems an obvious way to determine the relative free energies of the phases in an alloy system as a function of both composition and temperature. It is important to recognize, however, that virtually all phase diagram fitting in effect takes all phases to have the same specific heat. Thus any prediction of relevance to undercooled alloy liquids is immediately ruled out. Even if this were not so, the fitting is interpolation and could only with great caution be used for extrapolation into non-equilibrium states. The excess specific heat of alloy liquids must, of course, be consistent with the equilibrium phase diagram. In cases where $\Delta C_p(T)$ is known, its incorporation into the free energy expression for the liquid does improve the fit to the diagram, for example, for Au-Si [12], but independently derived data are required—the fitting cannot be used to determine $\Delta C_p(T)$ and $\Delta G(T)$.

For congruently freezing systems (which we take here to include eutectics), the simplest approach is to assume that at any particular composition there is a constant ΔC_p. The success of this approach is assessed briefly in Section 3. The few measured examples, however, show that ΔC_p is far from constant. One model which can accommodate that is based on the hole theory of the liquid, and is assessed also in Section 3. When freezing is not congruent it is necessary to know $\Delta G(T)$ as a function of composition. The associated solution model may be useful here and is considered in Section 6.

3. HOLE THEORY OF THE LIQUID

In this theory, developed by Hirai and Eyring [13, 14] it is assumed that the free volume in a liquid is distributed as discrete holes of constant volume and formation energy. Recently Dubey and Ramachandrarao [15] have presented a rigorous derivation of $\Delta C_p(T)$ based on a minimization of the free energy. The final result, after some approximations is

$$\Delta G(T) = \frac{\Delta H^m \Delta T}{T_m} - \frac{\Delta C_p^m \Delta T^2}{2T}\left(1 + \frac{\Delta T}{6T}\right) \tag{1}$$

where ΔT is the undercooling $(T - T_m)$. In Fig. 1 this expression is tested against data on $Pd_{40}Ni_{40}P_{20}$ and is found to be in good agreement [16]. $Pd_{40}Ni_{40}P_{20}$ is particularly resistant to crystallization and its liquid specific heat can be measured by differential scanning calorimetry (DSC), both on undercooling the liquid by ≤ 40 K and by heating the glassy alloy beyond the glass transition for ~ 50 K. Interpolation of the specific heat is possible with some confidence is the 200 K interval in which crystallization is too rapid to permit measurement. Dubey and Ramachandrarao found good agreement with eqn. (1) also for $Au_{81.4}Si_{18.6}$ and $Au_{77}Ge_{13.6}Si_{9.4}$ [15].

The other curves in Fig. 1 show the predictions of various models based

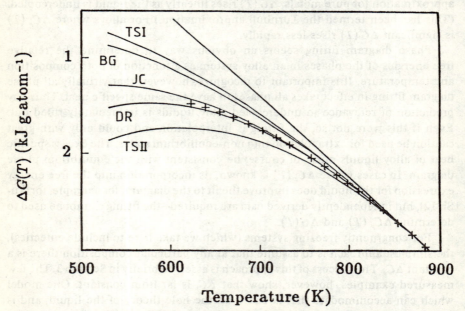

Fig. 1. The excess of the Gibbs free energy of liquid $Pd_{40}Ni_{40}P_{20}$ over that of the solid as a function of temperature, calculated from measured specific heat (crosses). The straight line shows the Turnbull approximation, and the other curves the expressions due to: DR, Dubey and Ramachandrarao [15]; TSI and TSII Thompson and Spaepen [16, 17]; BG, Battezzati and Garrone [20]; JC, Jones and Chadwick with $\Delta C_p = \Delta C_p^m$ [17].

on a constant ΔC_p. Jones and Chadwick [17] derived an expression for $\Delta G(T)$ based on a constant ΔC_p; here ΔC_p is taken as ΔC_p^m and this is seen to work quite well. The other treatments avoid input of a measured ΔC_p, but adopt various measures for estimating it: Hoffman [18] assumes $\Delta H = 0$ at $T \approx T_g$; Thompson and Spaepen [19] assume $\Delta S = 0$ at $T \approx T_g$; and Battezzati and Garrone [20] take $\Delta C_p = 0.8 \, \Delta S^m$. These various approaches are discussed in more detail elsewhere [16]; none is particularly successful in matching the $Pd_{40}Ni_{40}P_{20}$ data.

The success of the Dubey and Ramachandrarao model is emphasized by considering the ideal glass transition temperature, T_o, at which $\Delta S(T) = 0$. The excess entropy $\Delta S(T)$ of liquid $Pd_{40}Ni_{40}P_{20}$ derived from the measured $\Delta C_p(T)$ is illustrated in Fig. 2; some extrapolation shows that it goes to zero very close to the T_o value predicted directly by setting $\Delta S = 0$ in the hole theory [21] and to the measured kinetic T_g. It appears, notwithstanding any doubts about the physical picture of the liquid in the hole theory, that the theory as presented by Dubey and Ramachandrarao is capable of predicting the thermodynamic properties of undercooled liquids with some accuracy. The imput parameters in the model, T_m, ΔH^m and ΔC_p^m, permit varying degrees of deviation of the $\Delta G(T)$ curve from the linear Turnbull prediction; the model can be used to describe liquids with a wide range of glass-forming ability. Its main disadvantage is that it relates the properties of the liquid to those of the solid (be it single phase or a eutectic mixture of phases) at the same composition. A different approach is required if predictions are to be made for systems which do not freeze congruently.

4. INVERSE MELTING

The Dubey and Ramachandrarao model will now be applied to another transformation: inverse melting. This is a postulated equilibrium reversible transformation in which a crystal transforms to a liquid on cooling [22]. Such a transformation has been reported by Blatter and von Allmen [23] in $Ti_{70}Cr_{30}$ in which the crystal phase is a metastable supersaturated bcc solid solution (β-phase). Annealing a thin film of amorphous phase at 1073 K caused transformation to β-phase and subsequent annealing at 873 K caused reversion to the amorphous phase. At these comparatively low temperatures the stable intermetallic compounds cannot nucleate and spinodal decomposition of the β-phase is suppressed by strain effects. Thus inverse melting (to a liquid, or at low temperatures, an amorphous phase) occurs in this case without partitioning and the Dubey and Ramachandrarao approach can be used. The β-phase forms on rapid cooling of the equilibrium high temperature liquid. If the liquid reappears at low temperature, the free energy curves $G(T)$ for the liquid/amorphous phase and for the β-phase must intersect twice. At the lower temperature intersection (inverse melting point) the liquid/amorphous phase must have a lower entropy than the crystal. The data presented in Figs. 1 and 2 show, even for a very good glass former such as $Pd_{40}Ni_{40}P_{20}$, that $\Delta S(T)$ does not

262

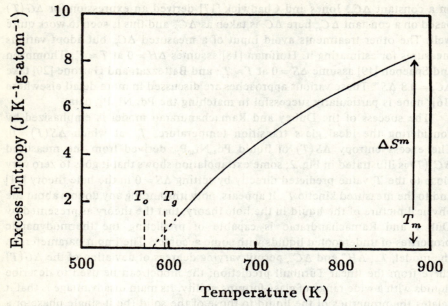

Fig. 2. The excess entropy of liquid $Pd_{40}Ni_{40}P_{20}$ calculated from measured specific heat. T_0 is the ideal glass transition temperature from the Dubey and Ramachandrarao model [15, 21], T_g is the dynamic glass transition temperature.

reach zero or become negative on cooling and that $\Delta G(T)$ does not even approach zero. The $\Delta G(T)$ for $Ti_{70}Cr_{30}$ calculated using eqn. (1) and taking $\Delta C_p^m = 0.8 \, \Delta S^m$ is shown in Fig. 3. In Ti-Cr some effect additional to the topological ordering fitted by eqn. (1) must be operating. It has been suggested that this is a chemical ordering in the liquid [22]. The entropy of chemical mixing in the β-phase can be taken to be ideal (there is a tendency for phase separation rather than ordering), while in the liquid/amorphous phase this can be taken as zero. The ideal mixing entropy for $Ti_{70}Cr_{30}$ would be 5.08 $JK^{-1} \, mol^{-1}$. This could easily outweigh the entropy of (partitionless) fusion reduced on undercooling from its value at T_m of 6.98 $JK^{-1} \, mol^{-1}$ by effects such as in Fig. 2. If so, the chemically ordered undercooled liquid could have an entropy substantially below that of the chemically disordered crystal. The effects of ordering in the liquid (similar to CSRO in glasses but at higher temperature), for simplicity of calculation assumed to set in at a critical temperature, on $\Delta G(T)$ are shown in Fig. 3. If the ordering occurs at a high enough temperature $\Delta G(T)$ can become zero between 873 K and 1073 K, the annealing temperatures used by Blatter and von Allmen [23]. Although this calculation is very crude, it does at least illustrate that inverse melting is thermodynamically possible, even without my vibrational and electronic contributions to $\Delta S(T)$. These extra contributions will be present, but are not expected to be large for systems in which the liquid and crystal phases are

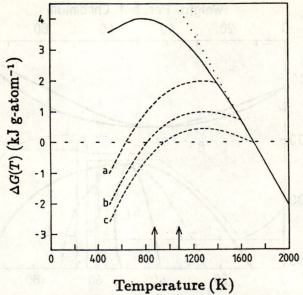

Fig. 3. The excess of the Gibbs free energy of liquid $Ti_{70}Cr_{30}$ over that of the crystalline solid solution according to the model of Dubey and Ramachandrarao [15] (solid line). The dashed lines show the effect of complete chemical ordering in the liquid (and none in the crystal) at (a) 1400 K, (b) 1600 K, (c) 1708 K. The arrows indicate the annealing temperatures used in [23] to demonstrate reversible amorphisation.

both metallic. The corresponding metastable phase diagram for Ti-Cr is shown in Fig. 4.

Inverse melting provides an interesting example of glass formation by a route other than rapid quenching. Indeed in $Ti_{70}Cr_{30}$ rapid quenching (except at the very highest rates obtainable using pulsed lasers) yields the β-phase because this simple crystal structure nucleates with ease at low undercooling. However, the amorphous phase can be obtained by annealing of the β-phase at a temperature sufficiently low (e.g. 873 K) for the amorphous phase to be more stable (Fig. 3). Since the β-phase can be obtained in bulk on comparatively slow cooling, this amorphization may be practically significant in providing a route to bulk glassy alloy formation without the necessity for a this dimension [24]. In addition, inverse melting may not be confined to Ti-Cr, as low temperature amorphization has been found also in Co-Nb, Cu-Ti, Fe-Ti, Mn-Ti and Nb-Ti [25].

5. CRYSTAL GROWTH

It is of importance to know how the kinetics of solidification are affected by the thermodynamics of the undercooled liquid, and in particular by the deviation of the behaviour from that expected for pure metals, as shown for example in Fig. 1. Consideration here will be limited to congruently freezing

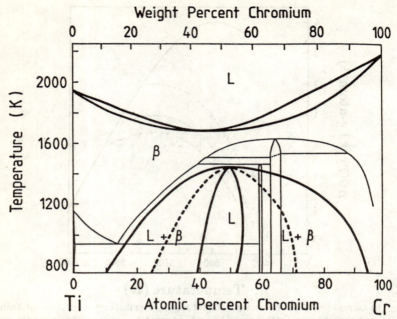

Fig. 4. A metastable phase diagram for Ti-Cr (bold lines) showing a reappearing liquid, superposed
on the equilibrium diagram. The T_0 lines for partitionless inverse melting are shown dashed.

systems with liquid thermodynamics described by eqn. (1). Special ordering
effects, such as those leading to inverse melting, will not be taken into account.
The liquid thermodynamics will affect the kinetics of both nucleation and growth,
but only growth will be considered here. The nucleation kinetics may not always
be of overriding importance in determining the glass-forming ability, and may not
be calculable in any case for heterogeneous nucleation. In melt-spinning, for
example, the wheel surface is likely to act as a substrate for heterogeneous
nucleation. Crystal growth into the ribbon will follow, but a largely glassy ribbon
may still be formed it the growth can be stifled by the rapidly falling
temperature. Whether or not this occurs will depend on the crystal growth
kinetics and on the temperature of the liquid at the start of solidification.

The crystal growth rate U in a congruently freezing system is given
by

$$U = U_o[1 - \exp(-\Delta G/RT)], \qquad (2)$$

where the kinetic prefactor U_o may be either diffusion-limited or collision-limited
[26]. The collision-limited case appears to apply to pure metals and possibly also to
disordered solid solutions; U_o is the speed of sound (typically a few thousand
m s^{-1}) and this large, temperature-independent value means that U is limited only
by the rate of heat extraction from the crystal-liquid interface. If crystal growth

were always collision-limited, glass formation by liquid quenching would be impossible. However, in many cases of alloy solidification, particularly when ordered solutions or compounds are formed, U_o is diffusion-limited. (As we are taking the case of no partitioning, the diffusion is not long-range, but short-range, to effect a change in local coordination.) U_o in this case is D/λ (where D is the liquid diffusivity at the interface, and λ is a diffusive jump distance) and is strongly temperature-dependent.

We have developed a one-dimensional numerical model for crystal growth in an undercooled liquid. As details have been given previously [27], only an outline is presented here. The model takes account of the release of latent heat at the crystal-liquid interface and solves the heat flow equation using an explicit finite difference scheme. The spatial element typically has dimension 0.1 to 1 μm. The crystal growth rate is calculated using eqn. (2), with the temperature taken to be that at the interface. Either the collision-limited or diffusion-limited case can be taken but in the present discussion of glass formation only the latter will be considered. The model is applied here to solidification of a splat from which heat is extracted on one side at a rate given by

$$Q = h(T_{sp} - T_{sub}), \tag{3}$$

where T_{sp} and T_{sub} are the temperatures of the splat and substrate at their interface, and h is the heat transfer coefficient. The initial, uniform temperature of the splat before contact with the substrate can be selected. If initially the liquid is undercooled, the calculation may apply to solidification of droplets hitting a substrate at the base of a drop-tube [28]; if superheated ($T > T_m$), to melt-spinning.

For the purposes of illustration we model a hypothetical alloy with parameters similar to those for $Fe_{80}B_{20}$ [29], but it is taken to freeze congruently. The parameters are: $T_m = 1448$ K; $T_g = 700$ K; $\Delta H^m = 13.76$ kJ mol^{-1}; thermal diffusivity, 1.2×10^{-5} m^2 S^{-1}; $D = D_o \exp [-B/(T - T_o)]$;

$B = 1300$ K; $T_o = 581$ K; $D_0 = 1.4 \times 10^{-8}$ m^{2}s^{-1}; $\lambda = 2.5 \times 10^{-10}$m;

$T_{sub} = 300$ K; $h = 10^7$ W m^{-2} K^{-1}; and $\Delta C_p^m = 0$ or 12 JK^{-1} mol^{-1}. The excess free energy of the liquid is given by eqn. (1). Nucleation is assumed to occur on contact with the substrate, provided at that point that $T_{sp} < T_m$.

Figures 5 to 7 show how the interface temperature evolves as solidification proceeds on a planar front through the splat. On the distance scale 0 μm corresponds to the top of the splat and 5 μm to the surface in contact with the substrate, i.e., solidification proceeds from right to left. There is a competition in determining the interface temperature between the release of latent heat and the external heat extraction. Three cases are considered. In the first, illustrated in Fig. 5, ΔC_p^m is taken to be zero, i.e. the Turnbull approximation. The behaviour is shown for a selection of initial melt temperatures. For initial melt temperatures ≤ 1140 K the interface temperature drops below T_g before solidification is complete, i.e., some glass will be formed. This initial temperature

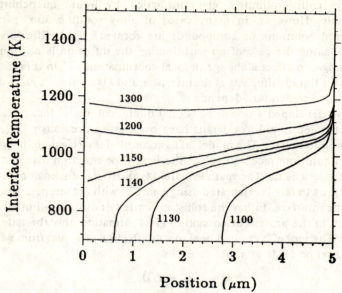

Fig. 5. The evolution of crystal-liquid interface temperature as solidification proceeds from right to left through a splat at the initial temperatures marked. The undercooled liquid is assumed to have no excess specific heat.

represents an undercooling of ~ 310 K, however. Next the effect of a reasonable ΔC_p^m is considered; a value of 12 JK^{-1} mol^{-1} is taken to represent a good glass former. In Fig. 6 the effect of this ΔC_p^m in lowering $\Delta G(T)$ and therefore the crystal growth rate is shown. Now some glass can be formed if the initial melt temperature is $\leq$ 1210 K (still an undercooling of ~ 240 K). But a $\Delta C_p(T)$ also has the effect of lowering $\Delta H(T)$ as the liquid is undercooled. To simplify the calculation a linear $\Delta C_p(T)$ was taken varying from ΔC_p^m at T_m to $2\Delta C_p^m$ at T_g. This is a good approximation to the $\Delta C_p(T)$ was taken varying from ΔC_p^m at T_m. This is a good approximation to the $\Delta C_p(T)$ predicted by the Dubey and Ramachandrarao model, and it was verified that this did not have a significant effect on the calculations, yielding for example virtually identical curves in Fig. 6. When the temperature-dependent latent heat $\Delta H(T)$ is taken into account the curves in Fig. 7 are obtained. The effect is dramatic: now a glass can be obtained even for an initial melt superheat of ~ 150 K. This is more realistic for $Fe_{80}B_{20}$, but still the predicted depth of crystallization is too great. This could be because U is in reality slower than expected from eqn. (2) because of the need for eutectic partitioning, or because the nucleation occurs only at a significant undercooling. The effect of nucleation undercooling has been investigated; surprisingly it has very little effect on the behaviour until its value exceeds ~ 100 K. It is seen that modifying the thermodynamics of the undercooled liquid does have a significant effect on the prediction of glass-forming ability. More realistic modelling of undercooled

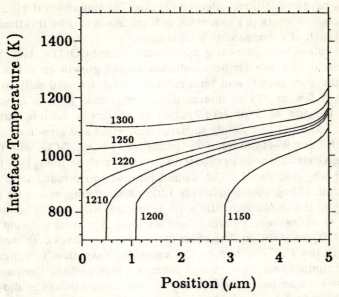

Fig. 6. As for Fig. 5, but the effect of an excess specific heat of the liquid ($\Delta C_p^m = 12$ J K^{-1} mol^{-1}) on the driving force for solidification is included.

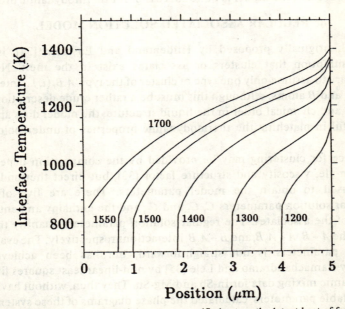

Fig. 7. As for Fig. 6, but the effect of the excess specific heat on the latent heat of freezing is also included.

liquids may have an additional effect on the glass-forming ability if it is considered that the atomic mobility in a somewhat ordered liquid may be less than that in a pure metal melt at a comparable temperature.

Recent pulsed laser quenching experiments on the CO-Ti system have given microstructural evidence for both collision-limited growth (at a few hundred m s^{-2}) of the pure metals and terminal solid solutions, and diffusion-limited growth (at a few m s^{-1}) of intermetallic compounds [30]. It is interesting that even the use of 5 ns laser pulses to achieve a cooling rate of 10^{10} to 10^{12} K s^{-1} does not completely suppress nucleation and growth of the Ti$_2$CO phase which has a 96-atom unit cell. On the other hand, TiCO$_3$ with the much simpler Cu$_3$Au structure can be suppressed even by melt-spinning [31]. Indeed, it is not clear why disordered TiCO$_3$ cannot form as an extended solid solution; phase diagram fitting shows that very little undercooling would be required. Complexity of the crystal structure with its possible effect on U_0 cannot be the only effect operating here. Possibly there is a local maximum in liquid viscosity near the TiCO$_3$ composition. The effect of viscosity is perhaps even more clearly illustrated in the Cu-Ti system by the especially easy glass formation at the equiatomic composition, even though there is a high melting compound there [32]. If there are maxima in viscosity at particular compositions in alloy liquids, then this could be taken as evidence for the existence of particular associates in the liquid [33]. The regular associated solution model has not yet been applied quantitatively to the prediction of viscosity, but this should be explored as it is already evident that the model is quite successful for thermodynamic properties.

6. REGULAR ASSOCIATED SOLUTION MODEL

The model, originally proposed by Hildebrand and Eastman [34], is based on the assumption that clusters or associates exist in the melt. Normally it is necessary to assume only one type of cluster of the type A_iB_j $(i,j...$ integers) in a melt of A and B atoms. Although this must be a rather crude description of the topological and chemical order in the liquid structure, the model does appear to be successful in matching the thermodynamic properties of undercooled alloy liquids.

Evidence for clustering may be provided by the composition dependence of, for example, viscosity and structure factor [33], but direct thermodynamic data are used to obtain the model parameters. There are five of these: three regular solution parameters C, C_1 and C_2, and the enthalpy and entropy of formation of the associates. The regular solution parameters quantify the non-ideality of the $A - B$, $A - A_iB_j$ and $B - A_iB_j$ interactions respectively. The evaluation of all five parameters is not straightforward, but has been achieved, for example, by Ramachandrarao and Lele [35] by non-linear least squares fitting to thermodynamic mixing data for In-Sb and Mg-Sn. They then, without having any more adjustable parameters, calculated the phase diagrams of these systems. It is remarkable, given that the equilibria represented in the diagram are at ~ 600 K

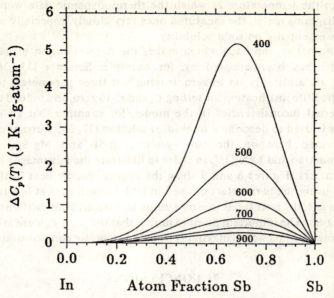

Fig. 8. Excess liquid specific heats in the In-Sb system, calculated using the regular associated solution model, using parameters in [35].

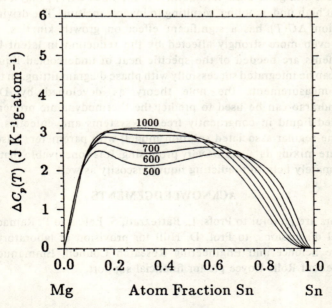

Fig. 9. Excess liquid specific heats in the Mg-Sn system, calculated using the regular associated solution model, using parameters in [35].

lower than the temperature at which the thermodynamic data were obtained, that the diagrams match the measured ones very closely, especially so for In-Sb, which has essentially no solid solubility.

An alternative approach which makes the determination of parameters a little easier has been adopted by, for example, Sommer [33, 36]. In this, C_1 and C_2 are arbitrarily set to zero, leaving just three parameters to fit. There seems to be little justification for setting C_1 and C_2 to zero, and indeed this leads to some internal inconsistencies in the model, for example that an ideal mass-action law is used to describe a semi-ideal solution [37, 38]. Further calculations are performed here on the two systems, In-Sb and Mg-Sn, studied by Ramachandrarao and Lele [35], in order to illustrate the advantages of fitting all five parameters. Figures 8 and 9 show the excess specific heat in these alloys, predicted using the parameters derived in [35]. In each case at low temperature the excess specific heat tends to a maximum at eutectic compositions, two in the case of Mg-Sn. The important point to note is that if C_1 and C_2 were arbitrarily set to zero these maxima would inevitably be at the associate composition (In_1Sb_1 and Mg_2Sn_1).

7. CONCLUSIONS

The amount $\Delta G(T)$ by which the Gibbs free energy of some undercooled alloy liquids exceeds that of the corresponding crystal is much less than expected. The lowering of the free energy difference is associated with an excess specific heat and good glass formability. Sometimes $\Delta G(T)$ may even become negative at high undercooling, resulting in inverse melting. The driving force for solidification $\Delta G(T)$ has a significant effect on growth kinetics, which are, however, even more strongly affected by the reduction in latent heat. More measurements are needed of the specific heat of undercooled liquids, and if available can be integrated successfully with phase diagram fitting. In the absence of such measurements the hole theory as developed by Dubey and Ramachandrarao can be used to predict the thermodynamic properties of the undercooled liquid in congruently freezing systems and eutectics. For other systems the regular associated solution model, with parameters based on high temperature mixing data, seems very promising. It is conceivable that the model could ultimately help in predicting liquid viscosity as well.

ACKNOWLEDGEMENTS

The authors are grateful to Profs. L. Battezzati, S. Lele and P. Ramachandrarao for fruitful discussions, to Prof. D. Hull for provision of laboratory facilities, and to the Science and Engineering Research Council, Emmanuel College, Cambridge and Rolls Royce plc for financial support.

REFERENCES

1. Perepezko, J.H. and J.S. Paik. *J. Non-Cryst. Solids* **61** and **62**, 113 (1984).
2. Kauzmann, W. *Chem. Rev.* **43**, 219 (1948).
3. Garrone, E., and L. Battezzati. *Phil. Mag.* **B. 52**, 1033 (1985).
4. Battezzati, L. and A.L. Greer. *Int. J. Rapid Solid.* **3**, 23 (1987).
5. Nassif, E., P. Lamparter, B. Sedelmeyer and S. Steeb. *Z. Naturforsch.* **38a**, 1093 (1983).
6. Lamparter, P., W. Sperl, S. Steel and J. Bletry. *Z. Naturforsch.* **37a**, 1223 (1982).
7. Greer, A.L. *J. Non-Cryst. Solids* **61** and **62**, 737 (1984).
8. Egami, T. *Mater. Res. Bull.* **13**, 557 (1978).
9. Gaskell, P.H. In: *Amorphous Materials; Modeling of Structure and Properties,* V. Vitek (ed.), Met. Soc. AIME, Warrendale, PA, p. 47.
10. Sakata, M., N. Cowlam and H.A. Davies. *J. Phys. F. Metal Physics* **11**, L 157 (1981).
11. Battezzati, L. and A.L. Greer. *J. Mater. Res.,* in press.
12. Evans, P.V. Ph.D. Thesis, Univ. of Cambridge (1988).
13. Hirai, N. and H. Eyring. *J. Appl. Phys.* **29**, 810 (1958).
14. Hirai, N. and H. Eyring, *J. Poly. Sci.* **37**, 51 (1959).
15. Dubey, K.S. and P. Ramachandrarao. *Acta Metall.* **32**, 91 (1984).
16. Evans, P.V., A. Garcia-Escorial, P.E. Donovan and A.L. Greer. *MRS Symp. Proc.* **57**, 239 (1987).
17. Jones, D.R.H. and G.A. Chadwick. *Phil. Mag.* **24**, 995 (1971).
18. Hoffmann, J.D. *J. Chem. Phys.* **29**, 1192 (1958).
19. Thompson, C.V. and F. Spaepen. *Acta Metall.* **27**, 1855 (1979).
20. Battezzati, L. and E. Garrone. *Z. Metallk.* **75**, 305 (1984).
21. Dubey, K.S. and P. Ramachandrarao. *Int. J. Rapid Solid.* **1**, 1 (1984).
22. Greer, A.L. *J. Less-Common Metals,* in press.
23. Blatter, A. and M. von Allman. *Phys. Rev. Lett.* **54**, 2103 (1985).
24. Blatter, A. and M. von Allmen. *J. Mater. Res.,* in press.
25. von Allmen, M. and A. Blatter. *Appl. Phys. Lett.* **50**, 1873 (1987).
26. Coriell, S.R. and D. Turnbull. *Acta Metall.* **30**, 2135 (1982).
27. Evans, P.V. and A.L. Greer. *Mater. Sci. Eng.* **89**, in press.
28. Cochrane, R.F., P.V. Evans and A.L. Greer. *Mater. Sci. Eng.* **89**, in press.
29. Spaepen, F. and C.J. Lin. In: *Amorphous Metals and Non-Equilibrium Processing,* M. von Allman, (Ed.), Les Editions de Physique, Les Ulis, pp. 65–72 (1984).
30. Vitta, S. Ph.D. Thesis, Univ. of Cambridge (1987).
31. Inoue, A., K. Kobayashi, C. Suryanarayana and T. Masumoto. *Scripta Metall.* **14**, 119 (1980).
32. Reeve, J., G.P. Gregan and H.A. Davies. In: *Rapidly Quenched Metals,* S. Steeb and H. Warlimont, (Eds.), North Holland, Amsterdam, pp. 203–206 (1985).
33. Sommer, F. *Z. Metallk.* **73**, 77 (1982).
34. Hildebrand, J.H. and E.D. Eastman. *J. Amer. Chem. Soc.* **37**, 2452 (1915).
35. Ramachandrarao, P. and S. Lele. *Mater. Sci. Forum* **3**, 449 (1985).
36. Sommer, F. *Z. Metallk.* **73**, 72 (1982).
37. Lele, S., G.V.S. Sastry and P. Ramachandrarao. *Z. Metallk.* **75**, 446 (1984).
38. Lele, S., G.V.S. Sastry and P. Ramachandrarao. *Z. Metallk.* **75**, 897 (1984).

12

Terminal Solid Solubility Extension under Rapid Solidification

K.C. RUSSELL[1,2] F.H. FROES[2] AND Y-W. KIM[3]

[1]Department of Materials Science & Engineering and Department of
Nuclear Engineering, M.I.T., Cambridge, Massachusetts 02139, U.S.A.
[2]AFWAL/MLLS, Wright Patterson AFB, Ohio, 45433, U.S.A.
[3]Metcut, P.O. Box 33511, Wright Patterson AFB, Ohio, 45433, U.S.A.

ABSTRACT

Kinetic, thermodynamic, and crystal chemistry criteria proposed to predict the occurrence of terminal solid solubility extension (TSSE) under rapid solidification are analysed. Solidification velocities great enough to give absolute interface stability have been found to give TSSE in several alloy systems. At sufficiently high interface velocities solute trapping may also give TSSE, even though the equilibrium microstructure is two phase. Criteria based on solidification taking place below T_o or T_s have been proposed but not tested thoroughly. Criteria based on the Darken-Gurry alloying coordinates or on atomic radius and enthalpy of mixing are found to have some predictive power for Al-based alloys. A new criterion based on Miedema's macroscopic atom model for solution thermodynamics is found to predict TSSE very well for Al-based alloys. The criterion is presently being extended to other binary alloys where there exists an adequate data base.

I. INTRODUCTION

Dispersed phases precipitated from the molten alloy are seldom very useful in strengthening structural materials, as such particles tend to be coarse and heterogeneously distributed. The fine, uniform particle dispersions which may be formed by precipitation from solid solutions are almost always much more beneficial [1]. In addition many desirable effects may be obtained from solute which is left in solid solution, including increases in strength and modulus, decreased alloy density, and avoidance of undesirable second phases [2-5]. Desirable changes may also be effected in magnetic and electrical properties. The amount of solute which can be retained in solid solution is thus a basic limiting factor in determining a number of important properties of alloys.

There is no unique definition of terminal solid solubility extension (TSSE). A simple, operational definition would define solubility extension as the amount of solute left in solution after the alloy is cooled to room temperature. This solubility is easily determined by lattice parameter measurement. However, such a definition would not include material in solution after solidification is complete, but which precipitated from solid solution during subsequent cooling. A second, conceptually simple but non-operational definition of the extended solubility is the concentration of solute in solid solution the instant solidification is complete. This solubility depends only on events which take place during melting and solidification. Unfortunately, experimental results are reported in terms of the first definition whereas theoretical analyses use the second. A calculation is in progress which will relate the two definitions by calculating the amount of material which precipitated out during solid state cooling. They key quantities are activation energy for diffusion, cooling rate, and temperature at which precipitation began.

The importance of TSSE is such that inevitably there have been a number of criteria proposed to predict the conditions of cooling and alloy composition which will give solubility extension. These criteria may be divided into three broad groups: kinetic, thermodynamic, and crystal chemistry. We will consider these proposed criteria one by one.

II. THERMODYNAMICS OF RS

All three criteria involve thermodynamics. It is therefore useful to introduce some of the thermodynamic concepts relevant to TSSE.

Figure 1 shows a simple binary eutectic phase diagram with limited terminal solid solubility and no intermediate phases. Under equilibrium cooling conditions C_α^e is the maximum solute concentration which may be retained in the α phase. One goal of highly non-equilibrium rapid solidification is to defeat this equilibrium limit and retain a greater solute content is solid solution.

The following temperatures are relevant in the solidification of a melt of composition $C > C_\alpha^e$:

- T_m^A, the melting point of pure A
- T_l, the liquidus temperature, where the first solid α forms
- T_e, the temperature of the eutectic reaction
- T_o, the temperature at which the free energies of solid and liquid of composition C' are equal. Below T_o it is thermodynamically possible to have massive solidification, where the solid is uniformly of composition C'.
- T_β^{nuc}, the temperature where a vertical line at C' intersects the extrapolated β liquidus. Below this temperature β phase may form from the liquid.

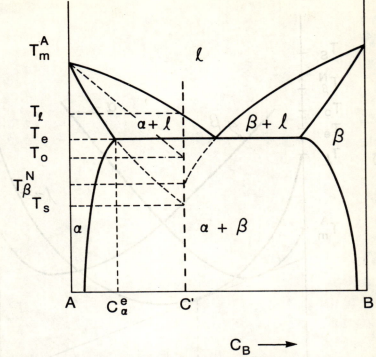

Fig. 1. Simple binary eutectic phase diagram showing temperatures relevant to rapid solidification of an alloy of composition C'. See text for definitions of temperatures.

- T_s, the extrapolated α solidus. Although massive solidification of α phase is thermodynamically possible at any temperature below T_o, between T_o and T_s formation of a mixture of B-poor α and B-rich liquid gives the lowest free energy. Below T_s, α of composition C' has a lower free energy than any combination of α and liquid.

Figure 2 shows schematic free energies corresponding to a simple eutectic phase diagram of the type shown in Fig. 1. Figure 2 corresponds to T_l, where the first solid forms from a melt of composition C_l. The effect of decreasing temperature may be represented in an approximate way by translating G_l upward with respect to G_α and G_β. The appropriate vertical translation of G_l to give alignment of the intersection of G_l with temperatures shown on the ordinate will give the configurations corresponding to T_l, T_e, T_o, T_β^N, and T_s.

At T_l, the tangent to G_l at C' is common to G_α. At T_e, equilibrium between the solid α, liquid, and solid B requires that G_α, G_l and G_β have a common tangent. At T_o, liquid and α have the same free energy so G_l and G_α intersect at C'. At T_β^{nuc}, it is first possible to nucleate β from liquid, so the tangent to G_l at C' is common to G_β. At T_s the tangent to G_l at C' is common to G_l, and equilibrium solidification gives α of the same composition as the liquid.

276

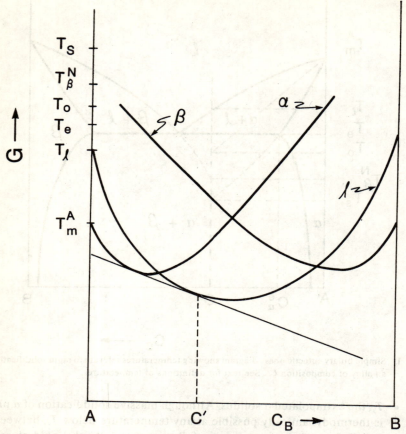

Fig. 2. Schematic sketch of free energy of liquid and solid α and β phases which would produce the phase diagram in Fig. 1. Shifting G_l upward relative to G_α and G_β corresponds to a lowering of the temperature.

A. Miedema macroscopic atom model

Determination of T_o, T_β^N and T_s requires knowledge of free energies of stable and metastable phases and of phase boundaries involving both stable and metastable phases. Phase boundaries involving only stable phases have in many cases been measured. Determination of metastable phase boundaries, such as extrapolations of the liquidus or solidus curves, must be done on the basis of the free energies of stable and metastable phases. Solid and liquid phase-free energy measurements are much less likely to be available than are the associated phase diagrams. Free energies of metastable phases are usually not amenable to measurement; accordingly, determination of such quantities as T_o, T_β^N, and T_s usually involves modelling of liquid and solid solution thermodynamics.

The ideal thermodynamic model would allow one to calculate accurately the free energies of any compound or solid or liquid solution from only the properties of the individual atoms. Most solution models fall far short of this ideal. Calculations based on atomic properties alone tend to be highly approximate, and the more accurate models tend to require information on solution energetics which is frequently not available.

This lack of either thermodynamic measurements or an accurate model has for years impeded theoretical studies of such kinetic processes as nucleation, growth, spinodal decomposition, and coarsening, as well as rapid solidification. Fortunately the "macroscopic atom" model developed over the past decade by Miedema and his co-workers [6-10] comes close to satisfying the stated ideal of an accurate model based on atomic properties only. Miedema's technique gives only the enthalpy; the entropy of the solution or compound may be estimated by one of several techniques to allow calculation of the free energy by the relation

$$\Delta G = \Delta H - T\Delta S \tag{1}$$

Miedema's approach gives mixing enthalpies of binary liquids or enthalpies of formation of binary compounds to an accuracy comparable to that of calorimetry. Mixing enthalpies of solid solutions are obtained to a lower accuracy. However, knowledge of the phase diagram and of the free energy of a coexisting liquid or compound phase allows calculation of the free energy of the solid solution.

Figure 3 shows the essentials of Miedema's "macroscopic atom" approach. Elemental A and B are taken to be separated into Wigner-Seitz atomic cells, and then reassembled into the liquid or solid solution or intermetallic compound.

In the case of liquid solutions or intermetallic compounds the enthalpy of mixing is given by

$$\Delta H\alpha - P(\Delta\phi^*)^2 + Q(\Delta n_{ws}1/3)^2 - R + \Delta H^{trans} \tag{2}$$

where

$\Delta\phi^*$ = difference in electron potentials of the elements
Δn_{ws} = difference between electron densities of the elements at the Wigner-Seitz cell boundary.

The first two terms contribute to the enthalpy in all cases. Unequal electron densities at the Wigner-Seitz boundary (n_{ws}) must be evened out, which gives a positive contribution to the enthalpy, and unequal electron potentials (ϕ^*) cause charge transfer and a negative enthalpy contribution.

The constant R is due to the electron hybridization which occurs when d-valence electrons of transition metals hybridize with s or p valence electrons of polyvalent non-transition elements. R is the order of 1 eV/atom (100 kJ/mol) for such alloys, depending somewhat on which non-transition element is involved, and is zero in all other cases.

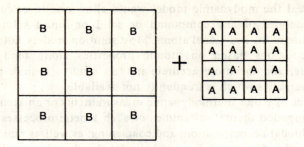

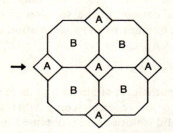

Fig. 3. Schematic illustration of Miedema's "macroscopic atom" thermodynamic model. Atoms of elemental A and B are separated at the Wigner-Seitz cell boundary and reassembled into a solid or liquid solution or a compound. The terms in the equation for the enthalpy of mixing are explained in the text.

The substantial negative contribution of R to the enthalpy of alloy or compound formation explains why transition metals have a strong tendency to form compounds with such polyvalent non-transition elements as boron, carbon, and nitrogen, while many other metals do not.

The calculation applies only for metallic elements; ΔH^{trans} is the enthalpy required to transform a non-metallic component (e.g. silicon) into a hypothetical metallic state.

The values of ϕ^* may be adjusted slightly to fit the measured heats of mixing, but n_{ws} is a basic, measurable quantum mechanical quantity. The constants P, Q, and R depend on which groups in the periodic table the alloying elements are from; the ratio P/Q is the same for all solutions and compounds.

Miedema has tabulated enthalpies of mixing for virtually all binary liquids involving metals and for a range of compounds of metals and metals, semi-metals, and non-metals [6, 7].

Equation (2) may be multiplied by an appropriate function of composition to give an equation for enthalpy of mixing versus composition.

Miedema has in fact revolutionized the thermodynamic modeling of solutions and compounds involving metals. His results are invaluable in predicting TSSE by RS, as they are in a variety of other fields.

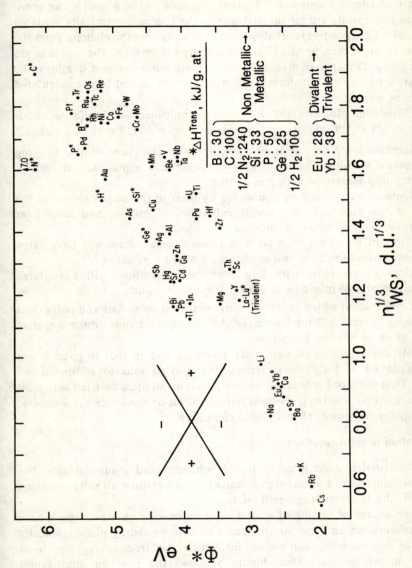

Fig. 4. Miedema plot of electron potential versus electron density at the Wigner-Seitz cell boundary. The figure completely characterizes the thermodynamic interaction of binary liquid alloys except for those which involve non-metals or combinations of transition metals and non-transition metals. Pairs of metals connected by lines of the slope shown will have a near-zero enthalpy of mixing and will form near-ideal liquid solutions and will not form compounds [6].

Figure 4 plots ϕ^* and $n_{ws}^{1/3}$ for most metals and a number of semi-metals and non-metals. Miedema's approach is limited to metals, so the ϕ^* and n_{ws} for semi-metals and non-metals are for the element in the thermodynamically unstable metallic state. The (positive) enthalpy needed to transform the element from the metallic state must then be added to the enthalpy of mixing. These values are shown in Fig. 4. Divalent elemental Eu and Yb are often trivalent in alloys; the appropriate enthalpy of transformation must then be added to the calculated enthalpies of mixing.

The Miedema plot has the following significance for liquid solutions and compounds which do not involve d-electron hybridization or non-metals:

1) Elements lying very near one another will have very similar alloying and compound-forming characteristics and near-zero enthalpies of mixing or for compound formation with one another.

2) Elements which may be connected by lines of the slopes shown at the left side of the plot will have small enthalpies of mixing and form near ideal solutions with complete liquid-phase miscibility.

3) Alloys of elements with large horizontal separations will have large positive enthalpies of mixing and tend toward phase separation.

4) Alloys of elements with large vertical separations will have large negative enthalpies of mixing and tend to form compounds.

The hybridization which occurs in alloys of transition metals and polyvalent non-transition metals will tend to make the mixing enthalpies more negative and favor more compound formation.

The Miedema plot is an extremely powerful tool in that it gives a two parameter plot (ϕ^* and $n_{ws}^{1/3}$) characterizing the alloying behavior of the various elements. That the plot becomes more complicated and must be used with a bit more care in alloys involving electron hybridization or non-metals is a modest price to pay for the power of this global technique.

B. Application to solid solutions

Miedema calculated enthalpies of liquid solutions and compounds to the accuracy of calorimetry but was significantly less successful with solid solutions due to difficulty in estimating strain energy.

The free energy of mixing of a solid solution may be obtained from phase diagram information and the thermodynamics of a coexisting phase, typically a liquid or intermetallic compound for which the free energy has been measured or can be calculated. Figure 5 shows the case for equilibrium between a line compound of free energy ΔG_c, and a solid solution α, of free energy ΔG_α. The co-existing compositions are X_c and X_e, respectively. Assuming the regular solution model [11] for simplicity:

$$\Delta G_\alpha = \Omega_a X(1-X) + RT[X \ln X + (1-X) \ln (1-X)] \tag{3}$$

where the regular solution coefficient, Ω_a is also the partial molar enthalpy of mixing of B in A. The standard states are pure, solid A and B.

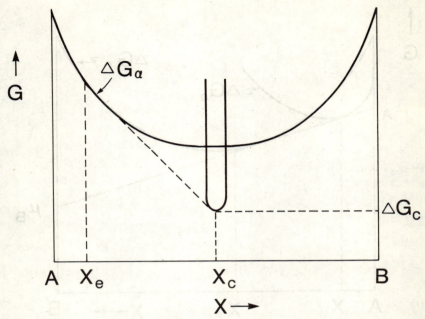

Fig. 5. Schematic figure showing equilibrium between a solid solution, α, and a compound, C. The regular solution coefficient, Ω_α, may be obtained from G_c and the co-existing compositions X_e and X_c.

The regular solution model has known shortcomings, but is usually adequate ·for relatively dilute α, say $X < 0.1$. ΔG_c may be obtained from Miedema's calculated enthalpies by adding a term for entropy of compound formation.

Equilibrium requires a common tangent to ΔG_c at X_c, and ΔG_α at X_e. Then:

$$\Omega_\alpha = \frac{\Delta G_c - RT[X_c \ln X_e + (1 - X_c) \ln (1 - X_e)]}{X_e^2 - 2X_c X_e + X_c} \tag{4}$$

or for the case of dilute α.

$$\Omega_\alpha = \frac{\Delta G_c}{X_c} - RT \ln X_e \tag{5}$$

The thermodynamic properties of solid α may be calculated in terms of quantities which may easily be measured or calculated.

The second case, that of α of composition X_e in equilibrium with a liquid of composition X_l is shown in Fig. 6. ΔG_l is assumed to be known, at least in the vicinity of X_e, either from measurement or from calculation.

We again assume a regular solution and take pure solid A and B as the standard states. Then we may write for liquid and solid:

$$\mu_B^l = \Omega_l (1 - X_l)^2 + RT \ln X_l + \Delta \mu_B^0 \tag{6a}$$

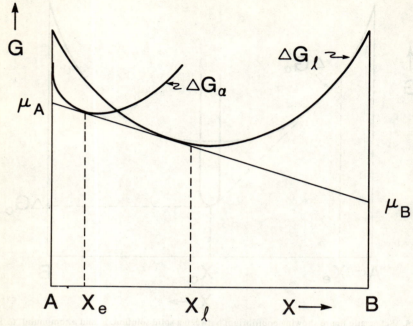

Fig. 6. Schematic figure showing equilibrium between a liquid and solid. The regular solution constant of α, Ω_a, may be obtained from G_l and the co-existing compositions X_a and X_l.

$$\mu_B^a = \Omega_a (1 - X_a)^2 + RT \ln X_a \tag{6b}$$

where

$$\Delta \mu_B^0 = L_B (T_m^B - T)/T_m^B$$

and L_B (> 0) is the heat of fusion of B and T_m^B is the melting point of B. Equating the chemical potential of B in co-existing liquid and α phases and assuming dilute solutions gives:

$$\Omega_a = \Omega_1 + \Delta \mu_B^o + RT \ln \frac{X_l}{X_a} \tag{7}$$

Now Ω_a may be calculated easily from Ω_l (obtained from Miedema's formulation), the partitioning coefficient, $k = \dfrac{C_a}{C_l}$, and the melting temperature and enthalpy of fusion of pure β. Eqn. (7) leads to the surprising conclusion that for dilute, regular solutions, for Ω_a to be temperature independent, k must be a function of temperature.

The technique just described, combined with phase diagram information which is usually available and Miedema's calculation of enthalpies of liquid and compounds can provide a fairly complete picture of the thermodynamics of any phase of interest. We now have the ability to do the thermodynamic calculations needed in discussions of criteria for TSSE.

III. CRITERIA FOR TSSE UNDER RS

A number of criteria have been proposed to predict whether or not TSSE will occur for a given combination of alloy and cooling history. In all cases the purpose is to predict whether nonequilibrium solute trapping will occur at the solid:liquid interface. Then the solute is "forced" into the solid even though local equilibrium would give segregation. We now consider these criteria in turn.

A. Thermodynamic criteria

Earlier we showed that above T_o, the temperature at which solid and liquid of the same composition have the same free energy, massive solidification cannot occur. Formation of solid the same composition as the liquid would give an increase in free energy and is therefore thermodynamically impossible. Below T_s, the extrapolated solidus, massive solidification leads to the minimum free energy state of the alloy. Solidification to form a two-phase mixture gives a free energy between that of the supersaturated liquid and that of the massively solidified phase. In addition to having a higher free energy than the massively solidified phase, the two-phase mixture requires long range diffusion in the liquid to form. Thus, although not forbidden, the formation of a two-phase mixture below T_s is at both a thermodynamic and kinetic disadvantage and seems unlikely.

Between T_o and T_s, massive solidification is thermodynamically possible, but results in a higher free energy than does the formation of the equilibrium two-phase mixture. However, massive solidification occurs without long range diffusion, required for equilibrium solidification. Thus, in the temperature region $T_s < T < T_o$, massive solidification has a kinetic advantage and thermodynamic disadvantage vis-a-vis two-phase equilibrium solidification.

One strong argument against massive solidification above T_s is that it violates local equilibrium, one of the most valued principles of the materials scientist. Local equilibrium requires each component to have the same chemical potential in the liquid as in the solid. Between T_o and T_s, equilibrium requires that solid and liquid have different compositions, as given by the phase diagram.

However, Baker and Cahn [12] showed that local equilibrium at the interface does not always apply during solidification. Zinc-rich Zn-Cd alloys have a retrograde solidus, i.e. the concentration of Cd in solid in equilibrium with Zn-rich alloys reaches a maximum above the eutectic temperature and then *decreases* with decreasing temperature. This behavior is in sharp contrast to that shown in Fig. 1, where the solidus concentration increases with decreasing temperature.

Baker and Cahn rapidly quenched Zn-Cd alloys of Cd contents greater than the maximum in the solidus and obtained a single-phase solid solution. Because of the retrograde solidus, massive solidification must have taken place in the two-phase region where local equilibrium could not be attained. The Cd atoms underwent an increase in chemical potential during solidification as they were dragged into the solid phase by the Zn.

Local equilibrium has been observed to apply in many solidification experiments. It is thus not clear whether the Baker-Cahn result is an anomaly, or whether rapid solidification will frequently violate local equilibrium.

It is useful to note the controversy of a few years ago over local equilibrium in solid-state massive transformation. Ultimately, the solid state massive transformation was found to occur usually, but not always under conditions of local equilibrium at the interface. It remains to be seen whether or not the same is true for massive solidification.

Anantharaman et al. [13] presented a review of rapidly solidified Al-based alloys in which they proposed that solutes with favorable Hume-Rothery size factors (within 15% of the solvent) should be amenable to TSSE. They noted, however, that six of the 17 such solutes (Au, Sn, Ti, V, Zn, and Zr) had TSSE to 4 a/o or less. They sought a basis for this anomaly in thermodynamic considerations, as reflected in the phase diagram.

They noted for simple eutectic systems showing small deviations from ideal solution behavior (such as Fig. 1), it should be easy to supercool the liquid below T_o and obtain massive solidification. For solid solutions with large positive deviations from ideality, however, they noted that even at absolute zero the free energy of the solid might be greater than that of the liquid, rendering massive solidification and TSSE thermodynamically impossible. Anantharaman et al. also argued that in some cases TSSE will be limited by nucleation of an adjacent stable or metastable solid phase. They proceeded to develop their ideas for several cases depending on whether the terminal solid solution and adjacent phase are formed eutectically or peritectically.

Anantharaman et al. thus implicitly assumed that only cooling to T_o was required for massive solidification, and that violation of local equilibrium was not a consideration.

B. Kinetic criteria

Various authors have attempted to prescribe conditions for segregation-free solidification in terms of various kinetic parameters, particularly liquid and interfacial diffusion coefficients and atomic attachment rates at the solid-liquid interface.

The criteria are of two basic types, appealing to either: (1) Absolute interface stability, or (2) Solute trapping.

Segregation-free solidification may occur even with equilibrium solute partitioning between liquid and solid if the interface is stable. Such solidification is illustrated in Fig. 7. The liquid concentration at the interface rises to C'/k, where k = partition coefficient, and the solid has the same composition as the bulk of the liquid.

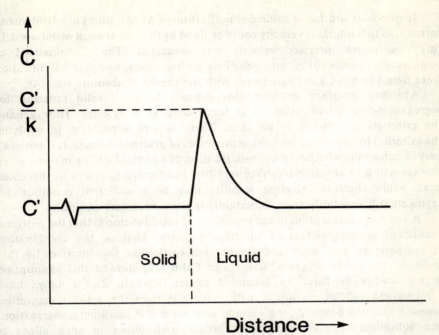

Fig. 7. Steady state segregation-free solidification with aquilibrium solute partitioning. Such solidification is possible only below T_s, the extrapolated solidus.

At very low velocities the interface is stable against breakdown giving cells or dendrites with the associated segregation. For most solidification rates, the interface is unstable, and solute segregation occurs.

Very high growth rates give a condition known as absolute interface stability. At such growth rates the interface is moving too rapidly for solute diffusion in the liquid to give long wavelength instabilities. The solute simply cannot diffuse rapidly enough.

At the same time, short wavelength instabilities which are kinetically possible give very sharp curvatures which are thermodynamically unstable. The interface is therefore stable, and segregation-free solidification occurs. There is, however, still equilibrium solute partitioning at the interface, as shown in Fig. 7.

Midson and Jones [14] proposed absolute interface stability as a criterion of extension of solid solubility. Boettinger, et al. [15, 16] and Juarez-Islas and Jones [17] discussed the role of absolute stability in producing segregation-free solidification. Boettinger, et at. found the velocities needed to give segregation-free solidification in Ag-Cu alloys to be in reasonable agreement with those predicted on the basis of absolute interface stability. They noted that their interface velocities were lower than these usually needed for solute trapping and attributed their result to absolute interface stability.

Juarez-Islas and Jones studied solidification of Al-Mn alloys in a Bridgmann furnace, so that interface velocity could be fixed by the rate of ingot withdrawal. In this experiment interface velocity was measured. The combinations of temperature, composition, and velocities giving segregation-free solidification were found to be in good agreement with the theory of absolute stability.

Absolute interface stability thus appears to be a valid criterion for segregation-free solidification in at least some alloy systems. How general the criterion is remains to be seen. There is also a problem in applying the criterion to alloys and quenching processes of practical interest. It is not at all easy to determine whether or not solidification of a particular alloy in some given process will give absolute interface stability. Furthermore, as will be discussed next, while absolute interface stability may be a sufficient condition for segregation-free solidification, it certainly is not a necessary condition.

A common assumption in the modelling of solidification is that the partition coefficient is independent of interface velocity. That is, the combination of temperature and solid and liquid composition at the interface is that given by the phase diagram. Baker and Cahn [12] showed this assumption to not always be true. As discussed earlier, Zn-rich, Zn-Cd alloys have a retrograde solidus, so alloys with more Cd than the solidus maximum *cannot* be cooled below T_s, the extrapolated solidus. Accordingly, segregation-free solidification with equilibrium solute partitioning in such alloys is impossible. Since Baker and Cahn obtained segregation-free solidification, some mechanism other than absolute interface stability must have been operative. Baker and Cahn appealed to a solute trapping mechanism, whereby Cd atoms are incorporated into the solid, though they undergo an increase in chemical potential in the process. The actual partition coefficient is then unity, resulting in massive solidification.

Boettinger, et al. [15, 16] discussed the various theories which give a velocity-dependent partition coefficient. Figure 8 shows the predictions of these theories. In each case, the partition coefficient increases monotonically from the equilibrium value (0.44) to unity as the interface velocity increases. The dimensionless velocity is $\beta = Va_o/D$, where $V =$ velocity, $a_o =$ lattice constant, and D is a diffusion coefficient, which in the various models ranges from that in the liquid to that in the interface. In all theories the inflection point in k occurs at about $\beta = 1$, where the time for the interface to move one lattice spacing equals the time for an atom to diffuse one lattice spacing in the liquid or in the interface. For higher velocities the interface simply outruns the diffusing atoms and solute segregation is impossible.

The theories for a velocity-dependent partition coefficient are not terribly well developed. In addition, the ancillary data on diffusivities and interface kinetics needed to calculate k are often not available. In addition, as noted by Baker and Cahn, some solidification theories predict that k will decrease with velocity, whereas others predict an increase.

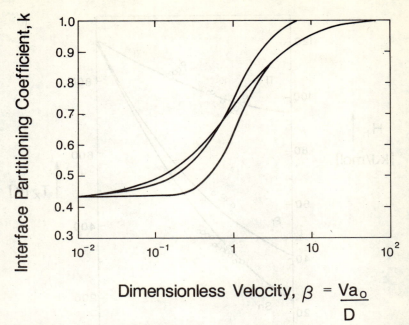

$$\text{Dimensionless Velocity, } \beta = \frac{Va_o}{D}$$

Fig. 8. Dependence of partitioning coefficient on dimensionless interface velocity as predicted by various theories. For $\beta \gtrsim 1$, the solute cannot diffuse rapidly enough to avoid being trapped by the moving interface [15].

Crystallization of amorphous alloys and TSSE by solute trapping are similar in that both are related to the ease of atom movement in the non-crystalline phase. As such, our knowledge of the former may teach us something about the latter.

Buschow [18] and Miedema, et al. [8] related the temperature of crystallization of glassy alloys to the enthalpy to form a hole in the glassy phase the size of the smaller atom. Vacancies in liquids or glassy alloys are thought to have a near-zero enthalpy of migration [19] so the diffusion coefficient is determined by the hole concentration.

Miedema used his macroscopic atom theory to calculate the enthalpy of hole, or vacancy formation in solid or liquid alloys [9]. If the hole concentration has the usual exponential dependence on ΔH_h, the enthalpy of hole formation, then crystallization would occur at a particular ratio of $\Delta H_h / T$. A semi-empirical relation was in fact found to connect T_x, the temperature of crystallization and ΔH_h.

$$T_x = 7.5 \, \Delta H_h$$

where T_x is in Kelvins and ΔH_h is kJ/mol holes.

Figures 9 and 10 plot T_x vs. ΔH_h for a number of amorphous alloys. The data are

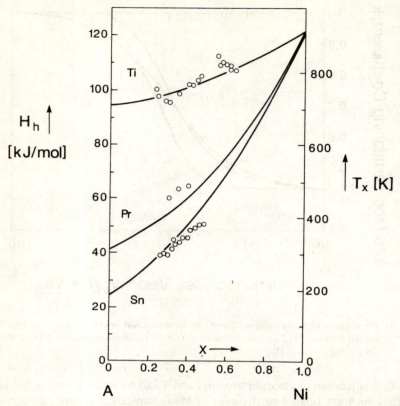

Fig. 9. Dependence of the crystallization temperature T_x in various amorphous alloys $A_{1-x}Ni_x$ on the formation enthalpy of a hole the size of the smaller atom [8].

seen to fit eqn. (8) very well, with some deviation toward crystallization temperatures slightly lower than predicted. The fit is good enough to conclude that the kinetic criterion is adequate in predicting the temperature of crystallization of amorphous alloys.

One might expect that the temperature of crystallization is related in some way to solution thermodynamics, that a stable compound would form even at relatively low temperatures, giving crystallization. Figure 10 also plots the heat of compound formation versus composition for a number of alloy systems. No correlation exists. Miedema, et al. found a similar lack of dependence of T_x on the enthalpy of mixing of the liquid alloy, so solution thermodynamics appear not to play a major role in determining the crystallization temperature of amorphous alloys.

Since solute trapping above T_s requires that solutes *not* diffuse ahead of the solid liquid interface, ΔH_h and the conditions for TSSE may in some cases be related.

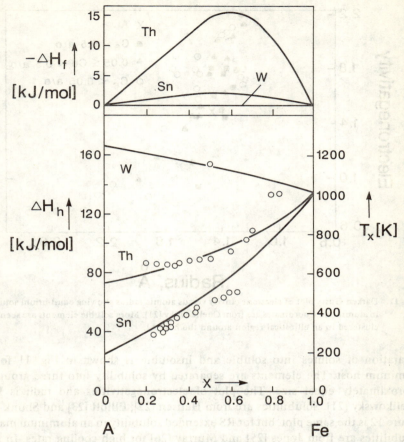

Fig. 10. Plot of experimental crystallization temperatures T_x in various $A_{1-x}Fe_x$ amorphous alloys versus the corresponding heats of hole for formation in the solid state. The enthalpies of compound formation are also shown for comparison [8].

C. Crystal chemistry criteria

Several criteria to predict equilibrium solid solubility have been devised in terms of such elemental crystal properties as Wigner-Seitz or Goldschmidt atomic radius, or electronegativity, or in the case of Miedema, electron density at the boundary of the Wigner-Seitz cell and electron potential. These criteria have some predictive power for equilibrium solubility, and it is reasonable to suspect that they might be useful in predicting non-equilibrium solid solubility under rapid solidification.

Darken and Gurry located the metallic elements on a two-dimensional plot of electronegativity and atomic radius. Elements with significant solubility in a host metal tended to be clustered in an elliptical region around the host. The

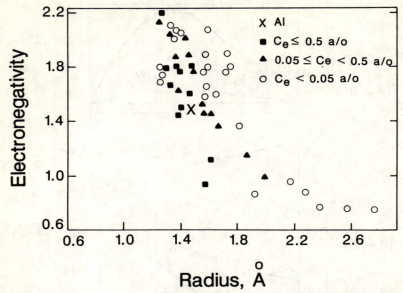

Fig. 11. Darken-Gurry plot of electronegativity versus atomic radius showing equilibrium solubility in aluminum. Coordinates are from Chelikowsky [21]. More soluble elements are seen to be clustered in an ellipitical region around the host Al.

separation of solutes into soluble and insoluble is shown in Fig. 11 for an aluminum host. The elements are separated by solubility into three groups of approximately equal size. The plot of electronegativities and radii is from Chelikowsky [21]. Solubilities are from Hansen [22], Elliott [23] and Shunk [24]. Figure 12 is the same plot, but for RS extended solubility in an aluminum matrix. Solubilities are from Jones [25] and Murray [26] for high cooling rates. In both cases soluble elements are clustered around the Al host, but so are some insoluble elements, while some soluble elements are located well away from the host. In Fig. 11, soluble elements Mn, Cr, and Re are located among the insoluble elements, while highly and moderately soluble elements are well mixed. Much the same kind of mixing is seen in Fig. 12 for RS extended solubility. In both cases the Darken and Gurry coordinates provide valuable guidance as to which elements may or may not be soluble, though leading to the conclusion that coordinates which better characterize solid state solubility are to be desired.

In Figs. 11 and 12 the Pauling atomic radii are used. Other choices of radius would give slightly different plots, and depending on the system, slightly better or worse separation into soluble and insoluble elements.

Jones [27] analysed equilibrium solid solubility in terms of the Wigner-Seitz atomic radius and Miedema's values for the heat of solution of the solute in the matrix. In a study of magnesium as a solvent Jones found a better

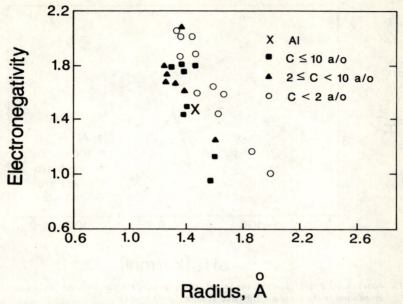

Fig. 12. As for Fig. 11, but non-equilibrium solubility under rapid solidification. Solubilities are from Jones [25] and Murray [26].

division between soluble and insoluble solutes than was obtained by either Darken-Gurry coordinates or the Meidema coordinates of ϕ^* and $n_{ws}^{1/3}$.

Jones also found that the plot gave an excellent separation between elements with high and low equilibrium solubility in aluminum, and proceeded to apply the coordinates to non-equilibrium solidification. Figure 13 shows his results for equilibrium and Fig. 14 for non-equilibrium solidification of Al-based alloys. A reasonable separation between elements on the basis of solubility is effected in both cases. It should be noted, however, that the Darken and Gurry plots and the Miedema plots involve only two crystal chemistry parameters. Jones plots are of atomic size versus enthalpy of mixing, which involves at least two crystal chemistry parameters. Thus the Jones plots involve three parameters to the other plots two, and would be expected to fit the data better.

Chelikowsky [21] compared the effectiveness of Darken-Gurry and slightly modified Miedema plots as predictors of equilibrium solubility in divalent hosts. He found the Miedema coordinates systematically superior, even for Mg-based alloys, where the Darken-Gurry coordinates are most successful.

Figure 15 shows a Miedema plot for equilibrium solubility in aluminum. The solubilities are divided somewhat arbitrarily into three approximately equal groups. It is seen that solutes with 0.5 a/o or greater solubility are clustered nicely around the host Al. Sparingly soluble solutes, with $.05$ a/o $\leq C_e < 0.5$ a/o tend to cluster just farther out, but are mixed in with some insoluble ($C_e < .05$ a/o)

292

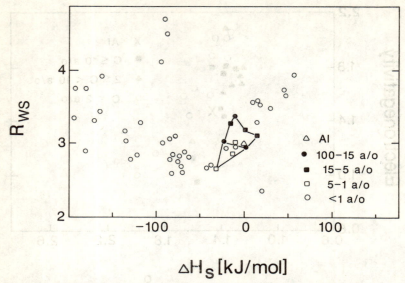

Fig. 13. Jones plot of atomic radius versus heat of mixing for equilibrium solubility in aluminum. The more soluble elements are clustered around the Al-host [27].

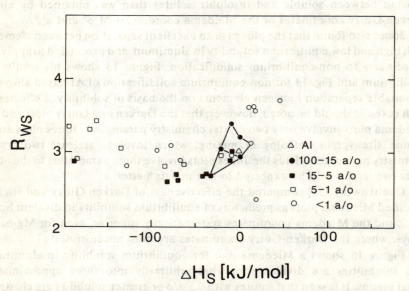

Fig. 14. As Fig. 13, except non-equilibrium solid solubility in Al. The more soluble elements tend to be close to the aluminum host. The separation into soluble and insoluble elements is far from perfect [5].

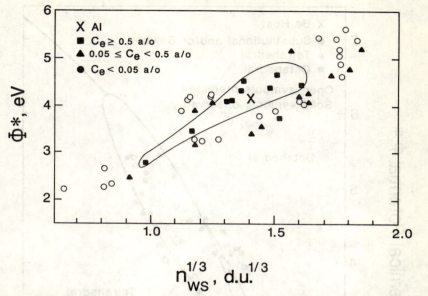

Fig. 15. Miedema plot of equilibrium solid solubility in an aluminum host. Soluble elements are clustered around the host element in a region of convex curvature. Slightly soluble elements are somewhat mixed with insoluble elements. A boundary is drawn around highly soluble elements as an aid to the eye.

elements. The Miedema plot thus gives a good, but not perfect separation into the three solubility groups.

The separation is better than that obtained on a Darken-Gurry plot (Fig. 11) or on a plot of heat of solution in solvent versus Wigner-Seitz radius (Fig. 13). Clearly the Miedema coordinates reflect substantially the fundamental factors governing equilibrium solid solubility.

Chelikowsky [21] found that the Miedema coordinates were remarkably successful in predicting no-equilibrium site preference under ion implantation. Injected ions cannot precipitate out of solution, but may enter solid solution three different ways: in substitutional sites, or in octahedral or tetrahedral interstitial sites. Figure 16 shows the results for ion implantation in a beryllium host. The injected elements separate without exception into three regions separated by smooth convex borders. This remarkable separation is even better than found by Chelikowsky for equilibrium solubility in Mg or as shown in Fig. 15, for equilibrium solubility in Al.

The success of the Miedema plot in predicting equilibrium solubility and predicting site selection in ion implanted solids and crystallization in metallic glasses suggest its extension to TSSE. Figure 17 shows such a plot, again with the solubilities separated into three comparable groups. The ability of the Miedema coordinates to separate the results into three regions is striking, and superior to

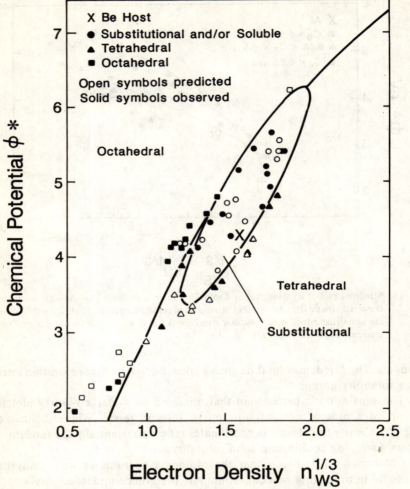

Fig. 16. Miedema plot of site preference of implanted solutes in a beryllium host. The plot gives a perfect separation into three regions with smoothly curved borders [21].

the other schemes just discussed. The separation is perfect except that cerium with an extended solubility of 1.9 a/o is included with solutes in the 2–10 a/o range. Furthermore the boundaries of the solubility regions are convex and smoothly curved, unlike the star-shaped boundaries found using other coordinates. Clearly the two Miedema coordinates do an excellent job of characterizing whatever forces govern TSSE in Al.

We may consider the meaning of the clean separation of the elements on the Miedema plot into three regions by two convex contours. First for illustration consider the heats of mixing of liquids, and ignore the non-metals and pairs of

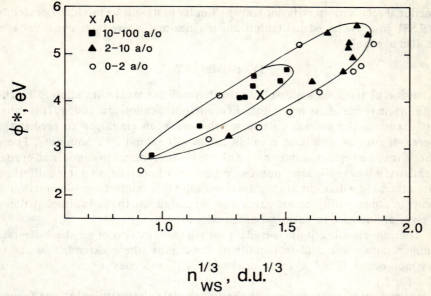

Fig. 17. Miedema plot of terminal solid solubility extension in rapidly solidified Al-based alloys. A near-perfect separation into highly soluble, moderately soluble, and relatively insoluble solutes is effected. Boundaries are drawn as aid to the eye.

metals which have *p-d* or *s-d* hybridization. Then for mixing a particular host element with various other elements.

$$\Delta H\alpha - P(\Delta\phi^*)^2 + Q(\Delta n_{ws}^{1/3})^2 \tag{9}$$

Setting $\Delta H = $ constant one may solve for some function $f(\Delta\phi^*, \Delta n_{ws}^{1/3}) = 0$ which may be plotted on the ϕ^*, $n_{ws}^{1/3}$ plane. For $\Delta H = 0$, the plot is a pair of straight lines through the host element, drawn with the slopes shown in Fig. 4. One could set ΔH at each of a succession of constants and obtain a set of equal ΔH contours in the ϕ^*, $n_{ws}^{1/3}$ plane.

One could not, for example construct such contours for the heats of mixing of solids because $\Delta\phi^*$ and $\Delta n_{ws}^{1/3}$ are not sufficient to determine the solid solubilities. This point is seen in Fig. 14, where only a very tortuous, non-physical contour could separate the high and low solubility elements.

Any attempt to construct more contours would result in intersections of contours of different solubility. Such intersections would have elements with the same values of $\Delta\phi^*$ and $\Delta n_{ws}^{1/3}$ giving different solid solubilities, which proves that those two variables alone do not determine solid solubilities.

The smooth, non-intersecting contours for TSSE indicate strongly that for aluminum at least, the extended solubility under *RS* is governed by $\Delta\phi^*$ and $\Delta n_{ws}^{1/3}$.

Why $\Delta\phi^*$ and $\Delta n_{ws}^{1/3}$ should have this predictive power for TSSE and what the functional relationship is, is not known. Similar plots will be produced presently for TSSE in alloys based on titanium and magnesium, where extensive data bases are also available.

IV. SUMMARY

Extension of terminal solid solubilities beyond the maximum allowed by the equilibrium is one of the major goals of rapid solidification processing. This paper first examines the various criteria which have been employed to predict the degree—if any—by which rapid solidification will extend solid solubility. These criteria may be separated into three groups; kinetic, thermodynamic, and crystal chemistry. All have the same purpose, predicting whether or not non-equilibrium solute trapping will occur at the advancing solid liquid interface. When trapping occurs in a binary alloy, one element is incorporated into the solid although doing so gives an increase in chemical potential.

Thermodynamic criteria, based on enthalpies of mixing or on phase diagram configurations are useful in identifying conditions where extended solubility may not occur $(T > T_o)$, most occur $(T < T_s)$, and may or may not occur $(T_s < T < T_o)$.

Several attempts have been made to calculate the partitioning coefficient, k, in terms of interface velocity and atomic kinetic parameters. Some theories predict that $k \to 1$ for high enough interface velocity. Unfortunately, these theories are not highly developed. In addition, some theories predict that k will decrease with increasing velocity.

Various crystal chemistry parameters have been used to predict equilibrium solid solubility. Darken and Gurry used atomic radius and electronegativity. Jones used coordinates of Wigner-Seitz radius and enthalpy of mixing of the solid solution. Neither set of coordinates was able to effect a complete separation between soluble and insoluble solutes.

We show that the Darken-Gurry and Jones coordinates also effect only a partial separation of soluble and insoluble elements, under rapid solidification conditions.

We propose a criterion for TSSE based on the electronegativity—electron density coordinates of Miedema. These coordinates have been remarkably successful in calculations of enthalpies, including those of mixing and of compound formulation, and of vacancy formation. The Miedema coordinates have also been very useful in predicting equilibrium solubility limits, crystallization temperatures of metallic glasses, and solute site preference in ion implantation. We find that these coordinates effect a near-perfect separation between highly soluble, slightly soluble, and insoluble elements under rapid solidification of Al-based alloys.

We propose that the two coordinates are sufficient to describe the kinetic processes which govern non-equilibrium solute trapping during solidification. Why these coordinates describe this complex process and the form of the functional relationship are not presently known.

ACKNOWLEDGEMENTS

K.C. Russell is grateful to the Air Force Office of Scientific Research for a summer faculty appointment at AFWAL/MLLS where much of this research was performed.

REFERENCES

1. Ansell, G.S. (ed.). *Proc. Second Bolton Landing Conference on Oxide Dispersion Strengthening,* New York, Gordon & Breach (1968).
2. Hehmann, F. and H. Jones. In: *Rapidly Solidified Materials and Their Magnetic and Mechanical Properties,* B.C. Giessen, D.E. Polk and A.I. Taub, eds. Materials Research Society, Pittsburgh, Pa. 259-274 (1986).
3. Jones, H. "Rapid Solidification of Metals and Alloys," The Institution of Metallurgists, London, (1982).
4. Froes, F.H., Y-W. Kim and F.J. Hehmann. "Rapid Solidification of Al, Mg, and Ti," *Journal of Metals,* 14-21 Aug. (1987).
5. Jones, H. In: *Rapidly Solidified Metastable Materials,* B.H. Kear and B.C. Giessen, eds., New York, North Holland, pp. 303-315 (1984).
6. Miedema, A.R., P.F. de Chatel and F.R. de Boer. *Physica* **100B** 1-28 (1980).
7. Niessen, A.K., F.R. de Boer, R. Boom, P.F. de Chatel, W.C.M. Mattens and A.R. Miedema. *Calphad* **7**, 1-70 (1983).
8. Miedema, A.R., A.K. Niesen and K.H.J. Buschow. *Met.* **100**, 71-84 (1984).
9. Miedema, A.R., *Z. Metalk.* **70**. 345-353 (1979).
10. Niessen, A.K. and F.R. de Boer. *J. Less Comm. Met.* **82**, 75-80 (1981).
11. Swalin, R.A. *Thermodynamics of Solids,* John Wiley & Sons (1967).
12. Baker, J.C. and J.W. Cahn. *Acta Metall* **17**, 575-578 (1969).
13. Arantharaman, T.R., P. Ramachandrarao, C. Suryanarayana, S. Lele and K. Chattopadhyay. *Trans. Indian Inst. of Metals* **30**, (6) 1-13 (Dec. 1977).
14. Midson, S.P. and H. Jones, In: *Rapidly Quenched Metals,* T. Masumoto and K. Suzuki, eds., Japanese Institute of Metals, 1539-1544 (1982).
15. Boettinger, W.J., S.R. Coriell and R.F. Sekerka. *Mats. Sci & Eng.* **65**, 27-36 (1984).
16. Boettinger, W.J., D. Shechtman, R.J. Schaefer and F.S. Biancaniello. *Met. Trans.* **15A**, 55-66 (1984).
17. Juarez-Islas, J.A. and H. Jones. *Acta Met.* **35**, 499-507 (1987).
18. Buschow, K.H.J. *Solid State Comm.,* **43**, 171-174 (1982).
19. Faber, T.H. *Introduction to the Theory of Liquid·Metals,* Cambridge, Cambridge University Press, 1972.
20. Darken, L. and R.W. Gurry. *Physical Chemistry of Metals,* New York, McGraw-Hill, 1952.
21. Chelikowsky, J.R. *Phys. Rev. B.* **19**, 686-701 (1979).
22. Hansen, M. *Constitution of Binary Alloys,* McGraw-Hill, New York (1958).
23. Elliott, R.P. *Constitution of Binary Alloys, First Supplement* McGraw-Hill, New York (1965).
24. Shunk, F.A. *Constitution of Binary Alloys, Second Supplement,* McGraw-Hill, New York (1969).
25. Jones, H. *Aluminum* **54**, 274-281 (1978).
26. Murray, J.L. In: *Alloy Phase Diagrams,* L.H. Bennett, T.B. Massalski and B.C. Giessen, eds., New York, North Holland, 249-262 (1983).
27. Jones, H. *Mater. Sci. & Eng.* **57**, L5-L8 (1983).

13

Supercooling and Rapid Solidification Using EM Levitation

S.P. ELDER[1] AND G.J. ABBASCHIAN[2]

[1]*Graduate Research Assistant*
[2]*Chairman and Professor*
Department of Materials Science and Engineering University of Florida, Gainesville, FL 32611, U.S.A.

ABSTRACT

In the past two-and-a-half decades, there has been a great deal of interest in modifying the microstructures, segregation profiles, and properties of alloys by various rapid solidification processes such as power atomization, melt spinning, and laser surface melting. These processes generally increase the rate of heat removal from the solidifying material by improving the heat transfer coefficient to the surroundings and/or by increasing the surface to volume ratio of the material being solidified. It is also possible to obtain high solidification rates if a large supercooling is imposed on the liquid prior to nucleation. The supercooling may also cause diverse modes of solidification such as partitionless (massive) solidification, enhanced supersaturation, grain size refinement, change of the solidification morphology, and/or formation of alternate phases. In this paper, a general review of the use of electromagnetic levitation to obtain bulk supercooling in various alloys and the effects of supercooling on grain size, microstructure, and solute distribution are presented. These are followed by a discussion of the recent experiments on metastable liquid phase separation upon supercooling of copper-iron and copper-cobalt alloys.

INTRODUCTION

Materials scientists and engineers are constantly challenged with the question of how to use materials most effectively in broad ranges of products and applications. To respond to the challenge, they make use of the science of processing-structure relationships in order to engineer new or modified materials with desired properties. An excellent example is rapid solidification processing (RSP). This term refers to a variety of processes which are used to modify structures, and hence properties, of metals and alloys during solidification. Rapid solidification has emerged from the work of Duwez and his co-workers in the early 1960's into an exciting technology. General reviews of various rapid solidification techniques can be found in various articles and proceedings [1–5]. RSP is commonly used to achieve close to equilibrium conditions in some alloy systems while non-equilibrium conditions in others. For the former, rapid solidification is used to obtain homogeneous structures and hence isotropic properties by the reduction or elimination of macro- and microsegregation, elimination of dendritic structures, and/or formation of refined microstructures. For the latter, the process is used to obtain metastable phases, to extend solubility limits, and/or to form metallic glasses.

The major focus of rapid solidification processing work so far has been concentrated on achieving rapid rates of heat removal during solidification. This has been done by improving the contact between the metal and the surroundings, and/or by increasing the contact area between the melt and the cooling medium. Examples are spreading the melt on a substrate, as in splat quenching or melt spinning, and dividing the melt into small particles, as in powder atomization or sputtering. Because of the limitations of the heat transfer coefficients between the melt and surroundings, the cooling rates during melt spinning, splat quenching, or powder atomization techniques are usually 10^6 K/s or less. Another technique of achieving high cooling rates is by melting a localized area on the surface of a sample using a high heat input from a laser or an electron beam. The molten layer is then rapidly quenched on the unmelted substrate. Since the majority of the heat of solidification is removed by the cold substrate, high cooling rates in the range of 10^6–10^9 K/s can be achieved.

It is also possible to obtain high solidification rates by bulk supercooling of the liquid. In a supercooled liquid, solidification initially takes place rapidly since the latent heat of fusion is absorbed by both the supercooled liquid and the forming solid. The initial solidification rate of the supercooled liquid is difficult to measure. However, it is generally believed that the rate during recalescence far exceeds that commonly observed in other rapid solidification processes. For example, recent experiments [6] involving high speed cinematography during solidification of supercooled Ni-Sn alloys show that the dendrite tip velocity depends upon the degree of supercooling and has a value of approximately 16 m/s at 260 K supercooling. A similar tip velocity can also be estimated from the theoretical prediction of Lipton, Kurz and Trivedi [7] which relates the velocity to the total tip undercooling at small and large Peclet numbers. The tip undercooling is the sum of three parts; the thermal undercooling, the solutal undercooling, and the curvature undercooling.

It should be noted that solidification of a supercooled liquid occurs non-isothermally, and the growth rate decreases as the temperature rises upon recalescence. Beyond the maximum recalescence temperature, the growth is similar to other solidification processes, and the rate is limited by the heat removal to the surroundings. In theory, it is possible to supercool the liquid to such an extent that all the solidification is complete before the temperature recalesces to the solidus temperature of the alloy. However, as shown elsewhere [8], the required supercooling far exceeds that calculated from the heat of fusion and specific heat of the material. It is possible to combine bulk supercooling with rapid solidification, such as splat cooling, to overcome the detrimental effects of recalescence. Indeed, under such circumstances, one would expect to observe beneficial effects of combined supercooling followed by rapid solidification.

Among various techniques to obtain high supercoolings, the most widely used are melt emulsification [9-11], melting in molten slag or fused silica [12-24], solidification during free fall in a drop tube [15, 16], and electromagnetic levitation melting [17-19]. The maximum supercoolings obtained for several high temperature metals and alloys [10, 12-15, 20-26] using these techniques are given in Table 1. Data for lower melting point materials can be found in references 20 and 27.

Table 1. Supercoolings data for high temperature metals and alloys

Melt or alloy	Melt emulsification technique	Glass encapsulated technique	Drop tube technique	EM levitation technique
Ni	319[20], 370[10]	–	–	341[21], 340[17]
Fe	295[20], 440[10]	–	–	324[21]
Fe-24 w/o Ni	–	230[12], 300[13]	–	320[22]
Fe-35 w/o Ni	–	–	–	344[21]
Fe-50 w/o Ni	400[10]	–	–	337[21]
Cu	236[20], 245[10]	–	–	266[21], > 400[23]
Ni-32 w/o Sn	–	240[14]	–	–
Ni-25 w/o Cu	–	169[12]	–	–
Ni-30 w/o Cu	278[24]	–	–	300[21]
Nb	–	–	525[15]	460[25]
Nb-58 a/o Si	–	–	–	210[25]
Nb-63 a/o Ti	–	–	–	320[25]
Hf	–	–	530[15]	–
Cr	–	–	–	183[26]

In melt emulsification [9-11], the metal or alloy is finely dispersed in a suspending medium so that the particles do not come in contact with each other. When the metal is divided into many small droplets, the catalytic effects of the active nucleants are restricted to a small number of the droplets therefore allowing for large supercoolings. The suspending medium must be thermally stable and must not react with the metal over the temperature range of the experiment. Because of this, the technique has been limited to low melting temperature and non-reactive materials. More recently, a dispersion of fine particles has been achieved in a ceramic matrix [10] allowing this technique to be applied to higher melting materials.

To obtain large supercoolings in bulk samples, the molten metal may be encapsulated in molten slag or fused silica [12-14]. The slag or glass encapsulant not only isolates the liquid metal from heterogeneous nucleation sites or container walls but also may act as a scavenger removing nucleants from within the volume of the melt. This technique has been successfully used to supercool bulk metals. However, the technique is limited by the choice of the encasing material. As was the case for the melt emulsification technique, the encasing material must not react with the metal. Another disadvantage these two techniques share is the inability to rapidly solidify the supercooled liquid since the solidification rate is limited by the heat transfer through the suspending medium or encasing material.

In the drop tube and electromagnetic levitation techniques [15-19], bulk alloys may be melted and solidified without coming into physical contact with any kind of crucible. In the drop tube technique, a molten sample is dropped and allowed to solidify during free fall (low g) through an enclosed tube. The tube may be evacuated and backfilled with an inert gas. The longest drop tube currently used for solidification studies is the 100 m drop tube at the NASA Marshall Space Flight Center [15]. In electromagnetic levitation, samples are cyclically heated and cooled while they are levitated in a controlled atmosphere using electromagnetic force. Large supercoolings are possible using these two techniques due to the reduction in contamination and the elimination of the presence of container walls. Furthermore, the techniques may be used to melt high temperature and/or reactive materials. With electromagnetic levitation, it is possible to observe *in situ* the onset of solidification, and measure the supercooling directly using a pyrometer. In the drop tube technique, on the other hand, the amount of supercooling is obtained through calculations which require an accurate knowledge of various material constants [28]. However, the greatest advantage in using electromagnetic levitation over the drop tube technique is one of feasibility. The number of drop tube facilities are few, the costs in building one are high, and the space required to house one is great. Nevertheless, the facility makes it possible to study solidification under reduced gravity and to simulate space processing conditions. As discussed later, gravity induced fluid flow can have a major effect on the solidification morphology of supercooled alloys.

In addition to the rapid solidification rates, supercooling of the liquid may cause a drastic reduction in the grain size, a change in the microsegregation profile, a change in the morphology from dendritic to non-dendritic, and/or the formation of metastable phases, as will be discussed later. In this paper the effects of supercooling on the microstructure of Ni, Fe-Ni, Nb-Ti, Fe-Cu, and Co-Cu alloys are discussed. The experiments were conducted using the electromagnetic levitation technique as described here.

ELECTROMAGNETIC LEVITATION

The basic EM levitation system used at the University of Florida is shown

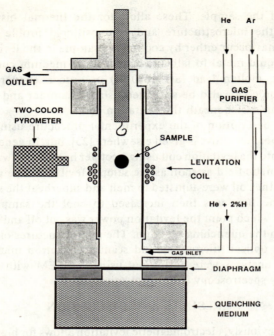

Fig. 1. A schematic representation of the electromagnetic levitation melting apparatus.

schematically in Fig. 1 [19, 29]. It is comprised of a high frequency generator, induction coil, melting chamber, gas delivery system, pyrometer, and chart recorder. Levitation and melting are achieved by using a specially shaped induction coil. The high frequency current supplied by the generator produces an alternating magnetic field in the induction coil, which in turn induces an alternating current in a conductive sample placed in the coil. The induced current, which has the same frequency but is 180 degrees out of phase with the applied field, serves both to heat the sample and to create an opposing magnetic flux that moves the sample from the stronger to weaker part of the field. For levitation to occur there must be a lifting force equal to the weight of the sample; therefore, a strong field gradient that decreases vertically is needed. This is accomplished using a conical induction coil. There are typically one or two oppositely wound turns on top of the coil to provide horizontal stability and to dampen oscillations.

The melting chamber provides a controlled environment in which the sample levitates. The environment is typically one of flowing gas supplied by the gas delivery system. The purpose of the gas is to control the temperature of the sample and to protect against oxidation; therefore, there are facilities for various gases: neutral, reducing, and those of high and low thermal conductivity. A single-color pyrometer and a two-color pyrometer are used to continuously monitor the temperature of the sample. These allow for the thermal history to be later correlated to the microstructure and compositional profile of the sample. Solidification may occur either by cooling the sample in the levitated state or by allowing the liquid metal to fall into a quenching medium consisting of a gas column, water, molten lead (at ~650 K) or onto a copper substrate. The supercooled liquid can also be solidified using a hammer and anvil type splat cooler placed directly beneath the levitation coil.

A detailed description of the experimental procedure using EM levitation to melt and supercool alloys is given elsewhere [23], but, in general, the samples were lowered into the levitation coil using a copper hook which was removed from the electromagnetic field as soon as the sample levitated. The gas flow and the power input to the coil were adjusted to melt and superheat the sample by about 200–300 K. The flow was then increased to cool the sample to a desired supercooling at which point the levitation power was cut off and the supercooled sample fell into the quenching medium. The microstructures of the alloys were analyzed using optical microscopy and scanning electron microscopy (SEM). Compositional analyses were performed using the SEM with electron energy dispersive x-ray spectroscopy (EEDS) or microprobe.

RESULTS AND DISCUSSION

As indicated previously, electromagnetic levitation allows for high supercoolings. Table 1 shows some of the supercoolings achieved using the technique. The supercooling is found to affect the microstructure in one or more of the following ways: (a) refinement of the grain size and/or dendrite arm spacing, (b) change in the solidification morphology from dendritic to nondendritic, (c) change in the segregation profile by producing partitionless solidification, (d) nucleation of alternate phases, and/or (e) in certain systems, metastable melt separation. These effects are discussed here, with specific examples given for Ni, Fe-Ni, Nb-Ti, Fe-Cu, and Co-Cu alloys.

A. Refinement of the microstructure

One effect of supercooling is the refinement of the microstructure, both grain size and dendrite arm spacing. A refinement in the structure is expected because of the increased nucleation density and rapid rate of heat removal with supercooling. An example is given by the recent work of Amaya et al. [17], who utilized EM levitation to study the effect of supercooling on the grain size of high purity Ni. They found that the average grain size of Ni decreased monotonically with increasing supercooling up to 340 K. The relationship between grain size and the degree of supercooling is shown in Fig. 2a. Two photographs of pure Ni at different supercoolings are shown in Fig. 2b and 2c to further illustrate the effect of supercooling. It was further observed that when the supercooling exceeded 225 K, the samples contained a duplex structure consisting of fine-grained regions dispersed in a coarse grained matrix. The average size of the fine grains

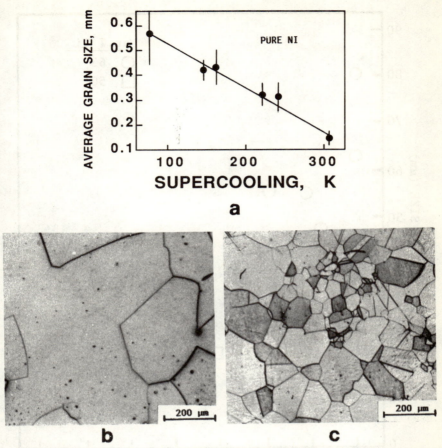

Fig. 2. The average grain size as a function of supercooling for pure nickel (a), and optical micrographs demonstrating the microstructure after supercoolings of 100 K (b) and 305 K (c).

was about one tenth of the matrix. In addition, most of the equiaxed grains contained twins.

Another example of the effect of supercooling on the grain size is seen in recent work by Gokhale and Abbaschian on Nb-Ti alloys [25]. The average grain size as a function of supercooling, plotted in Fig. 3, shows a gradual decrease with increasing supercooling.

In contrast with the above-mentioned studies, there are a number of other investigations [30–34] which show fine grains at small supercoolings, followed by coarser grains at larger supercoolings. The coarse grains may change abruptly to fine grains again when the supercooling exceeds a critical limit, mostly in the range of 150–200 K. Several mechanisms have been proposed to describe the abrupt changes in the grain size at the critical supercooling, such as selective

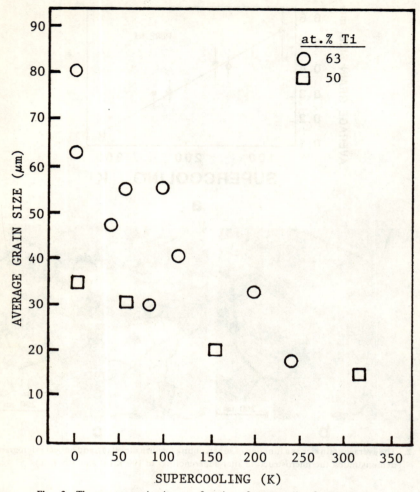

Fig. 3. The average grain size as a function of supercooling for Nb-Ti alloys.

activation of different nucleants [35, 36], shrinkage cavitation [31, 37, 38], and recrystallization [39–41]. Another explanation, and possibly the most likely one, is that the abrupt grain size reduction is due to the shift in the solidification morphology from dendritic to non-dendritic, rather than being directly related to the supercooling. This is evidenced from the work of Abbaschian and Flemings [42] which showed that the grain size of samples with dendritic solidification morphology was about an order of magnitude large than that with non-dendritic spherical morphology. For a given solidification morphology, no abrupt grain size change was observed as a function of the supercooling. The effect of supercooling on the solidification morphology is presented in the next section.

307

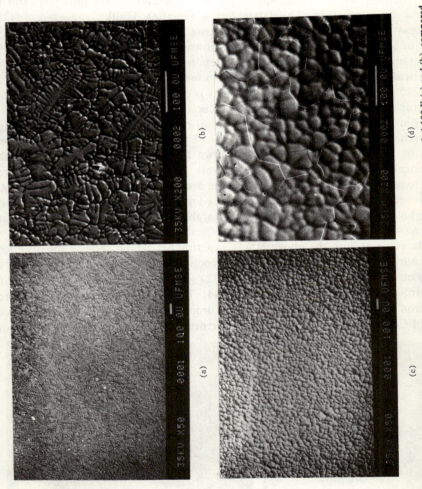

Fig. 4. The dendritic morphology observed in a Nb-63 a/o Ti alloy supercooled 100 K (a) and (b), compared with the spherical morphology observed at 40 K supercooling (c) and (d).

308

B. Solidification morphology

Bulk supercooling may drastically affect the solidification microstructure by changing it from the commonly observed dendritic to non-dendritic morphology. This effect has been observed in a number of systems, including Fe-Ni [42, 43] and Nb-Ti [25]. The different morphologies of Fe-Ni alloys which form at various supercoolings are dendritic, spherical, mixed dendritic and spherical, and a rosette type structure. The latter has a radiating fanlike appearance [43], whereas the spherical morphology consists of equiaxed elements uniformly distributed throughout the sample. It is shown that at supercoolings less than 170 K, the microstructure is predominantly dendritic; however, spherical, and mixed spherical and dendritic morphologies also form. At supercoolings between 170 and 220 K, only the spherical morphology is obtained, which gradually changes towards a rosette type structure at supercoolings greater than 220 K.

Similar types of dendritic and spherical morphologies have been observed upon the supercooling of Nb-50 and 63 a/o Ti alloy systems using EM levitation melting [25]. Typical examples are shown for Nb-63 a/o Ti in Fig. 4. Both spherical and dendritic morphologies were observed in samples which were supercooled less than about 150 K, with the dendritic morphology predominating. However, when the alloys were supercooled more than 150 K, only the spherical morphology was found. The spherical morphology persisted up to the highest supercooling studied, 320 K.

In contrast with the above Nb-Ti experiments, which showed the absence of dendritic structure beyond 150 K supercooling, dendritic structure has recently been observed in Nb-63 a/o Ti alloys at 320 ± 50 K supercooling when the alloy was solidified in a low-g environment [44]. This sample was melted using the electron beam furnace and the 100-meter drop tube at the NASA Marshall Space Flight Center. Figure 5 compares the microstructure of the sample solidified in

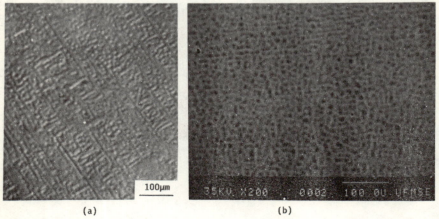

(a)   (b)

Fig. 5. The microstructure of Nb-63 a/o Ti alloy supercooled 320 ± 50 K using the drop tube (a) and supercooled 240 K using EM levitation (b).

the drop tube with that of a sample solidified using EM levitation; the former has dendritic structure, whereas the latter shows spherical cells. The difference in the microstructure clearly indicates the profound effect of gravity induced fluid flow, or its absence, on the microstructure.

The formation of the spherical morphology is believed to be due to the fragmentation of dendrites which form during the early stages of solidification. Once fragmented, each dendrite fragment continues to grow and coarsen independently, leading to the observed spherical morphology. Several reasons have been proposed for the fragmentation of dendrites, including remelting of dendrites upon recalescence above the solidus temperature of the alloy, coarsening and fragmentation caused by surface energy driven forces [6], and fragmentation caused by liquid flow [43]. The comparison of the results obtained using EM levitation with that of the drop tube experiments on one hand, and with those of the molten slag covered experiments on the other, indicate that dendrite fragmentation is most likely caused by the fluid flow. Fluid flow may be driven by either external or internal sources. Among external sources, some depend upon the experimental technique, such as flow induced by the electromagnetic field, the drag force of the coolant gas stream, and interaction of the metal with the quenching medium. Another external source of fluid flow is the gravity induced convection caused by thermal or compositional gradients. In contrast, the fluid flow necessary to feed the solidification shrinkage is independent of the external sources.

C. Partitionless solidification

Another effect of bulk supercooling is on the microsegregation pattern within the dendrites. Normal coring takes place in conventionally solidified dendrites; that is, the center of the dendrite arm is poor in solute (for cases where the distribution coefficient is less than unity) and the solute concentration increases towards the edge, following a Scheil type equation. For highly supercooled dendrites, on the other hand, the central portion of the arms may form without partitioning with a composition equal to that of the bulk alloy. This can be described by considering the solidification sequence of alloy C_o, see Fig. 6, which is supercooled to temperature T_N. Upon nucleation at this temperature, the initial solid which forms has the same composition as the original alloy until the recalescence drives the temperature up to T_p where partitioning occurs. Beyond this temperature, the solute concentration of the solid decreases until the maximum recalescence temperature, T_R, is reached. The temperature then begins to decrease and solidification proceeds in a manner similar to the conventional solidification processes with no supercooling. This traces the solidus curve on the phase diagram until solidification is complete.

An example of the above-mentioned segregation profile is given in Fig. 7 for an Fe-10 w/o Ni alloy solidified with 45 K supercooling. In this photomicrograph, the central Ni-rich regions are seen as dark crosses, surrounded by the Ni-poor regions which appear lighter in color. The crosses represent

310

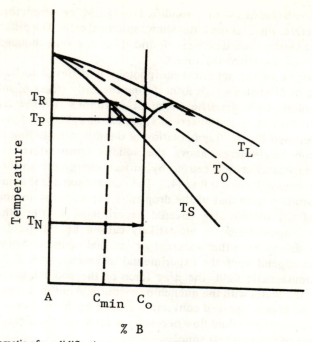

Fig. 6. A schematic of a solidification sequence of alloy C_o, illustrating partitionless solidification.

Fig. 7. Photomicrograph illustrating solute-rich crosses of Fe-10 w/o Ni alloy supercooled to 45 K.

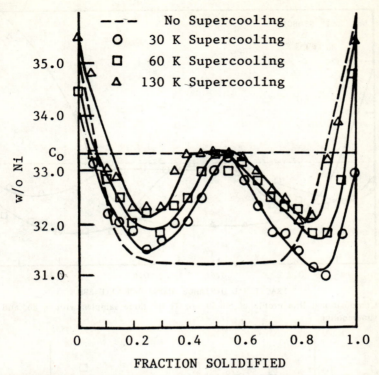

Fig. 8. Compositional line profile of Ni concentration along line AB of Fe-33 w/o Ni for four samples with 0, 30, 60, and 130 K supercooling.

thermal dendrites which form from the supercooled liquid. Compositional line profiles made across the spherical or dendritic elements, which provide quantitative evidence of partitionless solidification, are given for several supercoolings in Fig. 8. The line profiles show that the composition of the central region of spherical or dendritic elements is the same as that of the original alloy. Outside this region, the composition decreases reaching a minimum and then increases towards the edge of the element.

Similar observations are made during supercooling of Nb-Ti alloys, as shown in Fig. 9 for Nb-50 a/o Ti. The figure also includes the compositional profile for a sample which was solidified without any supercooling. As expected, the latter exhibits a Scheil type behavior, with the minimum solute constant at the center of the dendrite arm.

The fraction of the solid forming partitionlessly (the normalized width of thermal dendrites) depends on the amount of supercooling and seems to be independent of the alloy composition. This can be seen from Fig. 10, where the fraction is plotted against the bulk supercooling for two alloy compositions. In addition, as evidenced from Figs. 8 and 9, the level of the

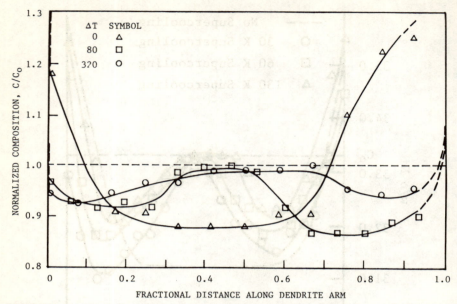

Fig. 9. Compositional line profile of Nb-50 a/o Ti for three samples with 0, 80, and 320 K supercooling.

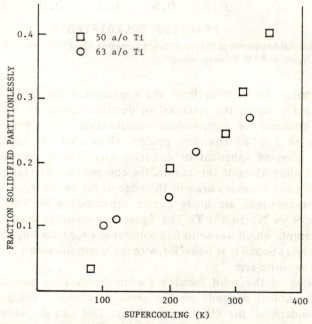

Fig. 10. The fraction of partitionless solidification for Nb-50 a/o Ti and Nb-65 a/o Ti alloys as a function of supercooling.

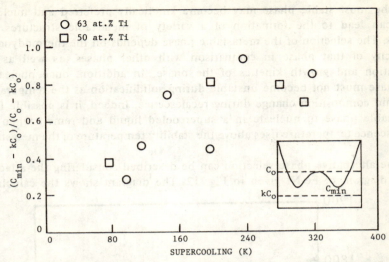

Fig. 11. The normalized composition minima as a function of supercooling for Nb-50 a/o Ti and Nb-63 a/o Ti alloys.

compositional minima depends on the supercooling. The minima generally shift towards the original alloy composition as supercooling is increased (see Fig. 11). This is due to a reduction in the maximum recalescence temperature with increasing supercooling. It may also be due to solute trapping which becomes more pronounced as the solidification rate increases with increased supercooling. This trend together with that of the width of the central plateau shown in Fig. 10 offer strong evidence that supercooled alloys become compositionally more homogeneous as the degree of bulk supercooling increases.

The experimental results show that the amount of partitionless solidification is considerably less than that predicted theoretically from

$$f_s^\circ = \frac{C_p}{H_f}(T_p - T_N)$$

where C_p is the average heat capacity of the solid and liquid, H_f is the heat of fusion, and T_n and T_p as defined earlier. The above relationship is routinely used to calculate the amount of "hypercooling" necessary to achieve complete segregationless solidification. As discussed in detail elsewhere [8, 43], the relationship overestimates the amount of partitionless solidification because of non-uniformity of temperature within the sample during recalescence and also because of the curvature effects at the tip of thermal dendrites.

D. Formation of metastable phases

When an alloy is supercooled, thermodynamically permissible phases other

314

than the most stable phase may become kinetically preferred and nucleate. This can lead to the formation of a variety of metastable structures and phases. The selection of the metastable phase depends on the thermodynamic hierarchy of that phase in comparison with other phases, as well as the nucleation and growth kinetics of the phase. In addition, once nucleated, the phase must not become unstable during solidification as the temperature or liquid composition change during recalescence. Indeed, it is possible for a metastable phase to nucleate in a supercooled liquid and remelt when the recalescence temperature rises above the stability temperature of the metastable phase.

The alternative phase selection can be described considering the attached phase diagram of Fe-Ni given in Fig. 12. The diagram shows the calculated

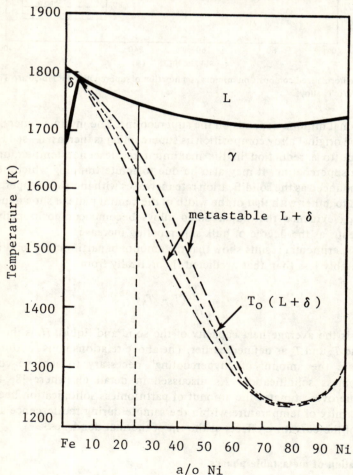

Fig. 12. The Fe-Ni phase diagram with calculated metastable liquidus and solidus extensions of the δ phase. [45]

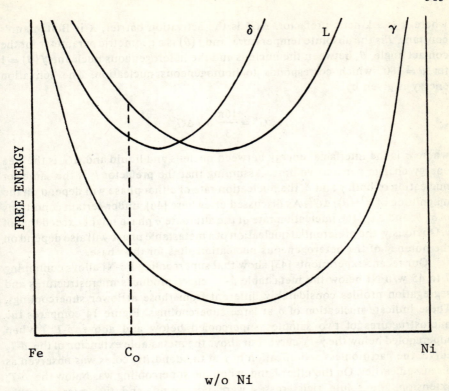

FREE ENERGY

Fe   C_o   Ni

w/o Ni

Fig. 13. A schematic of the free energy curves of the L, δ, and γ phases in supercooled Fe-Ni alloys.

metastable liquidus and solidus extensions for δ [45]. Considering solidification of a supercooled alloy with 25 w/o Ni when the supercooling is less than approximately 100 K, only nucleation of γ is thermodynamically permissible. However, when the supercooling becomes larger, say 250 K, formation of the δ phase becomes thermodynamically possible as well. This is because, as shown in Fig. 13, the liquid to δ transformation at this temperature will result in the reduction of the free energy of the system, albeit not as much as that for the formation of γ.

The above-mentioned thermodynamic consideration gives the necessary but not sufficient condition for the formation of the alternate δ phase. For the phase to form in preference to γ, its nucleation kinetics have to also be favorable. Assuming that nucleation of the γ or δ takes place heterogeneously, the steady state nucleation rate, I, can be expressed by the classical rate theory of Turnbull [11, 20] as

$$I = k \exp\left(\frac{-\Delta G^* f(\theta)}{KT}\right)$$

where k is a kinetic prefactor, ΔG^* is the activation barrier, K is Boltzmann's constant, T is the absolute temperature, and $f(\theta)$ is a geometric parameter for the contact angle, θ, between the nucleus and the heterogeneous nucleant: $f(\theta) = 1$ for $\theta = 180°$ which corresponds to homogeneous nucleation. The activation energy is given by

$$\Delta G^* = \frac{16\pi}{3} \, \sigma^3 / \Delta G_v^2$$

where σ is the interfacial energy between nucleus and liquid and ΔG_v is the free energy change per unit volume. Assuming that the prefactor k is the same for nucleation of both γ and δ, the nucleation rate of either phase will depend on the magnitude of $\sigma^3 f(\theta) / \Delta G_v^2$. As discussed elsewhere [46], under certain conditions of σ, θ, and ΔG_v, the nucleation rate of the alternate δ phase might exceed that of γ. Obviously, the preferential nucleation of a metastable phase will also depend on the potency of the heterogeneous nucleation sites for the phase.

Our recent experiments [43] show that supercooling Fe-Ni alloys containing 7 to 45 w/o Ni below the metastable δ-T_o curve produces microstructures and segregation profiles considerably different from those at lower supercoolings. These indicate nucleation of δ at large supercoolings. Figure 14 compares the microstructures of two samples supercooled below and above δ-T_o. When supercooled below the γ-T_o curve, but above the metastable extension of the δ-T_o curve, the partitionless solidification of γ at the dendritic cores was observed as discussed earlier. On the other hand, when the supercooling was below the δ-T_o extension, the solute rich crosses were not found, and the microstructure consisted of fine martensitic plates with embedded α particles.

Compositional line profiles of the two samples are compared in Fig. 15. For supercoolings above the δ-T_o curve, the central region has the original alloy composition indicative of partitionless solidification. Whereas, for supercoolings below δ-T_o, the compositional profile contains a central plateau with a fluctuating Ni composition which occupies the majority of the element. Beyond this region, the composition gradually increases towards the periphery. It should be noted that although this compositional profile is similar in shape to that of a non-supercooled sample, the minimum concentration level of that supercooled below δ-T_o is much higher. The differences between the two samples are attributed to the nucleation of the δ phase, instead of γ, when the supercooling is below the extended δ-T_o. It should be emphasized that when the recalescence temperature rises above the metastable δ-liquidus, the existing δ phase becomes unstable and will tend to remelt or transform to γ by a solid state massive transformation. It might be also possible for γ to nucleate independently or on the existing δ.

E. Metastable melt separation

Certain alloy systems are completely miscible in the liquid state but exhibit a definite tendency towards metastable liquid immiscibility. This is frequently

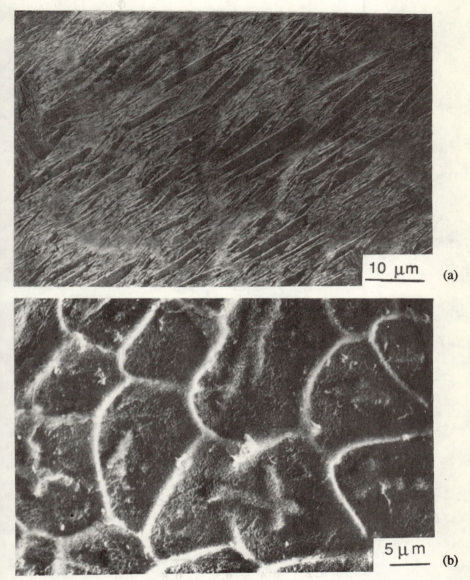

Fig. 14. Photomicrographs of Fe-10 w/o Ni alloy supercooled to 45 K—nucleation temperature above the metastable extension of δ–T_o (a) and to 200 K—nucleation temperature below the metastable extension of δ–T_o (b).

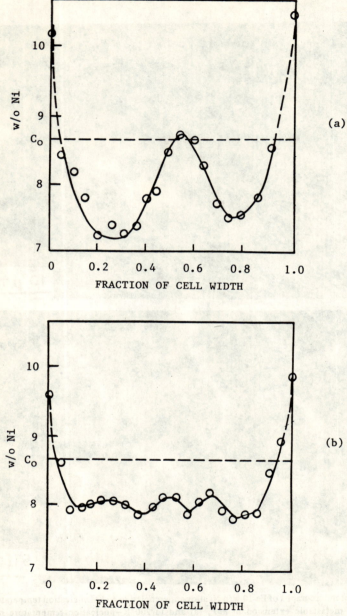

Fig. 15. Ni compositional line profiles for Fe-8.7 w/o Ni supercooled above (a) and below (b) the metastable extension of $\delta\ T_o$

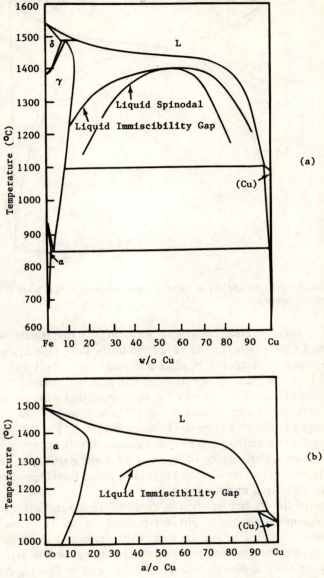

Fig. 16. The Fe-Cu (a) and Co-Cu (b) phase diagrams.

evidenced by a nearly flat liquidus curve and by a large positive deviation of the activities of the components in liquid solution from Raoultian behavior. Such alloys should show a tendency towards liquid phase separation upon supercooling. Fe-Cu and Co-Cu binaries are two examples which fall in this category. The phase diagrams are shown in Fig. 16 with the activities of Fe and Cu in the liquid solution shown in Fig. 17.

320

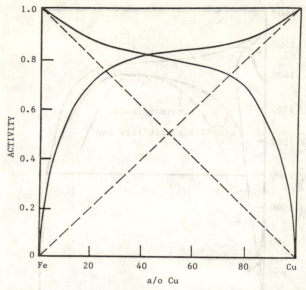

Fig. 17. The activities of Fe and Cu in liquid solution, showing a large positive deviation from Raoultian behavior.

We have utilized electromagnetic levitation to investigate supercooling of Fe-Cu and Co-Cu alloys. The results indicate that both systems become immiscible when supercooled beyond a certain level. This can be seen by comparing Figs. 18 and 19 for two Fe-Cu samples. Figure 18 illustrates the microstructure of an Fe-70 w/o Cu alloy solidified with no supercooling, which consists of Fe-rich dendrites in a Cu-rich matrix. In contrast, the microstructure of the same alloy supercooled by 150 K, Fig. 19, shows Fe-rich particles in a Cu-rich matrix. The Fe-rich particles mostly appeared as randomly distributed spheres with varying sizes; however, some particles were irregularly shaped. This microstructure indicates that the Fe-rich liquid droplets formed by liquid phase separation upon supercooling.

The size of the Fe-rich spheres is expected to depend on the coarsening time and agglomeration in the phase separated region. Once nucleated, the droplets coarsen because of surface energy driven forces. Fluid flow may also bring some droplets in close contact and influence the growth of each other, and/ or impinge, creating a distribution in the size of the Fe-rich droplets. The irregularly shaped droplets are due to collisions between droplets and/or the liquid mixing caused by the electromagnetic field. A cross-section of an Fe-Cu alloy with 60 K supercooling solidified against a copper substrate is shown in Fig. 20. Fine scale phase separation appears on the side which hit the copper substrate and cooled rapidly, whereas dendritic growth appears further away from the substrate. The dendritic structure most likely formed after recalescence.

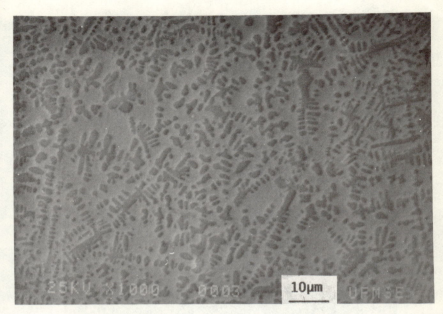

Fig. 18. A photomicrograph of an Fe-70 w/o Cu alloy with no supercooling, showing iron dendrites in copper matrix.

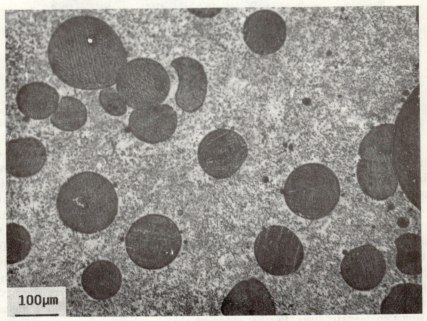

Fig. 19. A photomicrograph of an Fe-70 w/o Cu alloy with 150 K supercooling, showing Fe-rich particles in copper matrix with iron dendrites.

322

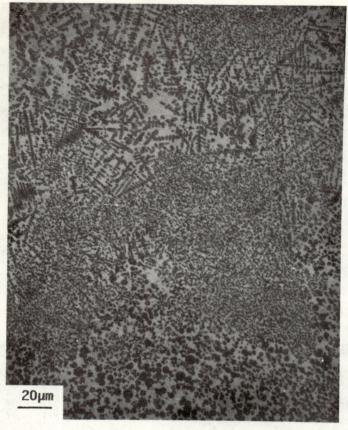

Fig. 20. A photomicrograph of a Fe-70 w/o Cu alloy with 60 K supercooling, solidified against a copper substrate.

The two liquid phases which form upon phase separation experience two different levels of supercooling, and as such, are expected to exhibit different solidification sequences and have different microstructural scales. This is illustrated in Fig. 21 for a Fe-70 w/o Cu alloy with 150 K bulk supercooling. The Fe-rich phase, L_1, has a larger supercooling (approximately 200 K) than that of the overall alloy (150 K) and that of the Cu-rich, L_2 liquid (approximately 75 K). Because of the difference in the supercooling, the L_1 phase is expected to solidify first. Evidence that the L_1 phase is the first to solidify is seen in the fine microstructure of the Fe-rich droplets shown in Fig. 22. This structure is non-dendritic with finely distributed Cu-rich particles. If the matrix were the first to solidify, the heat of fusion would have heated the Fe-rich liquid phase causing a coarser structure, Further evidence that L_1 solidifies first is seen in the preferential nucleation of the Fe from the Cu-rich, L_2, phase on the Fe-rich

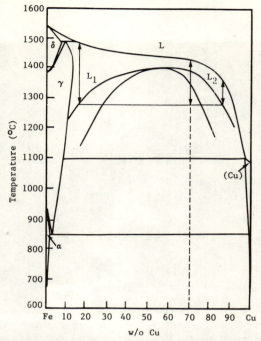

Fig. 21. The Fe-Cu phase diagrams illustrating the difference between the bulk supercooling of Cu-30 w/o Fe (150 K), with those of L_1 (200 K) and L_2 (75 K) formed upon phase separation.

droplets. This can be seen in the photograph in Fig. 23. A slightly coarser dendritic structure was observed in some Fe-rich droplets, shown in Fig. 24, indicating that these Fe-rich droplets may have solidified later in the solidification sequence.

The compositions of the two phases in the phase separated Fe-Cu alloys were measured as a function of supercooling for three alloy compositions supercooled to various degrees. The results are shown in Fig. 25. The data are in good agreement with the theoretical curve calculated by Chuang et al. [47].

Similar metastable liquid phase separation was observed in Co-Cu alloys upon supercooling. A phase separated microstructure, obtained by 75 K supercooling of Co-90 w/o Cu, is compared with the microstructure of a nonsupercooled sample in Fig. 26. The former contains Co-rich spheres within the Cu-rich matrix, while the latter shows conventional Co-rich dendrites in a Cu-rich matrix.

SUMMARY

Electromagnetic levitation has been used to obtain large supercoolings in several different metals and alloys. These supercoolings have been observed to affect the microstructure in different ways. A refinement in the grain size was found to occur

324

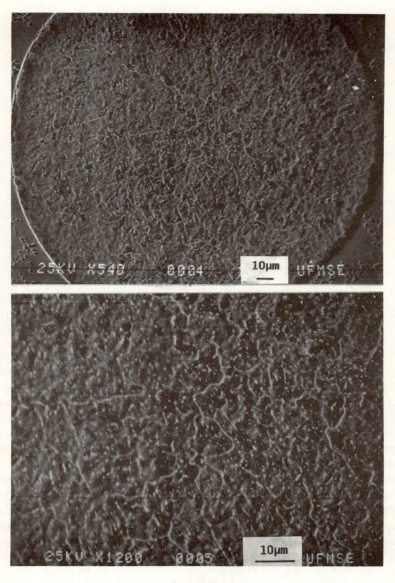

Fig. 22. Photomicrographs of an Fe-rich droplet in a Fe-70 w/o Cu alloy supercooled 105 K, illustrating the fine microstructure.

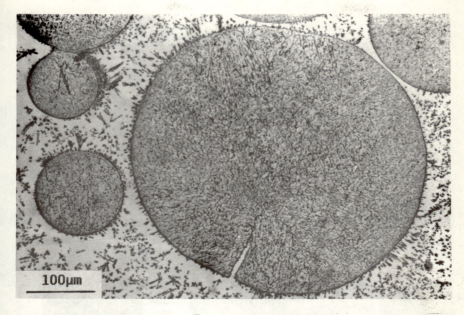

Fig. 23. Photomicrograph of a phase separated Fe-Cu alloy illustrating heterogeneous nucleation of the Fe dendrites in the matrix on Fe-rich particles.

with increasing supercooling in pure Ni. A change in the solidification morphology from dendritic to nondendritic, partitionless solidification, and the formation of a metastable phase were observed in supercooled Fe-Ni and/or Nb-Ti alloys. Metastable melt separation in supercooled Fe-Cu and Co-Cu alloys has been presented.

ACKNOWLEDGEMENTS

The authors wish to acknowledge the support of this research by the NASA-Vanderbilt Center for the Space Processing of Engineering Alloys and Office of Naval Research. Special thanks are due to Professor Robert Bayuzick, Bill Hofmeister, and Gail Whorisky for supplying the drop tube data for Nb-Ti and for enlightening discussions. Special thanks are also due to A. Gokhale for his contributions to Nb-Ti data and to A. Munitz for the Co-Cu results.

326

Fig. 24. Photomicrographs of an Fe-rich droplet in a Fe-70 w/o Cu alloy supercooled 115 K, illustrating coarse structure and shrinkage porosity.

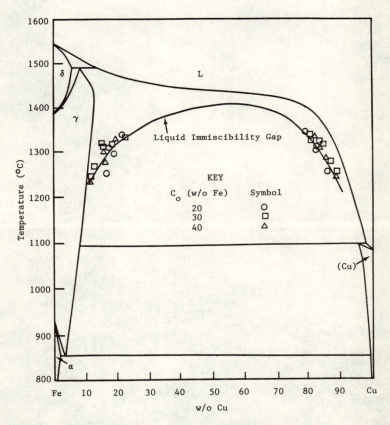

Fig. 25. The average compositions of the Fe-rich and Cu-rich phases for three nominal alloy compositions at several supercoolings plotted on top of the theoretical curve calculated by Chuang et al. [47].

328

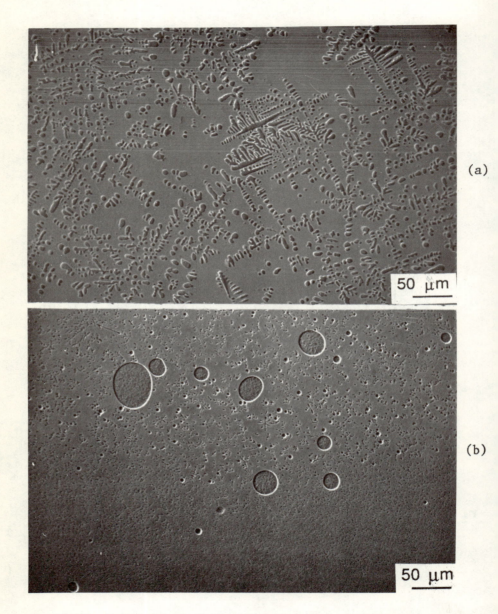

Fig. 26. Microstructure of Co-90 w/o Cu solidified with no supercooling (a), compared with that solidified with 75 K supercooling.

329

REFERENCES

1. Savage, S.J. and F.H. Froes. *Journal of Metals*, **36**, 20 (1984).
2. Lawley, A. In: *Processing of Structural Metals by Rapid Solidification*, F.H. Froes and S.J. Salvage, (eds.), ASM International, p. 31 (1987).
3. Jones, H. In: *Processing of Structural Metals by Rapid Solidification*, Op. cit., p. 77.
4. Mehrabian, R. (Ed.). *Rapid Solidification Processing—Principles and Technologies III*, National Bureau of Standards, Gaithersburg, MD (1982).
5. Berkowitz, B.J. and R.O. Scattergood (Eds.). *Chemistry and Physics of Rapidly Solidified Materials*, The Metallurgical Society of AIME, October 26–29 (1982).
6. Flemings, M.C., Y. Shiohara, Y. Wu and T. Piccone. In: *Undercooled Alloy Phases*, E.W. Collings and C.C. Koch, (Eds.), The Metallurgical Society of AIME, New Orleans, LA, March 2–6, p. 321 (1986).
7. Lipton, J., W. Kurz and R. Trivedi. *Acta Metallurgica*, **35**, 957 (1987).
8. Peteves, S.D., D.D. McDevitt and G.J. Abbaschian. In: *Rapid Solidification Processing—Principles and Technology III*, Op. cit., p. 121.
9. Perepezko, J.H., Y. Shiohara, J.S. Paik and M.C. Flemings. In: *Rapid Solidification Processing—Principles and Technologies III*, Op. cit., p. 28.
10. Rasmussen D.H. and K. Javed. In: *Undercooled Alloy Phases*, Op. cit., p. 79.
11. Turnbull, D. *Journal of Chemical Physics*, **20**, 411 (1952).
12. Kattamis, T.Z. and M.C. Flemings. *Transactions of the American Foundrymen Society*, **75**, 191 (1967).
13. Kattamis, T.Z. and R. Mehrabian. *Journal of Vacuum Science and Technology*, **11**, 1118 (1974).
14. Kattamis, T.Z. and M.C. Flemings. *Metallurgical Transactions*, **1**, 1449 (1970).
15. Bayuzick, R.J., W.H. Hofmeister and M.B. Robinson. In: *Undercooled Alloy Phases*, Op. cit., p. 207.
16. Lacy, L.L., M.B. Robinson, T.J. Rathz, N.D. Evans and R.J. Bayuzick. In: *Materials Processing in the Reduced Gravity Environment of Space*, Guy E. Rindone, (Ed.), Elsevier Science Publishing Co., Inc., New York, p. 87 (1982).
17. Amaya, G.E., J.A. Patchett and G.J. Abbaschian. In: *Grain Refinement in Castings and Welds*, G.J. Abbaschian and S.A. David (Eds.), The Metallurgical Society of AIME, p. 51 (1983).
18. McDevitt, D.D. and G.J. Abbaschian. *Microstructural Science*, **11**, 125 (1983).
19. Ethridge, E.C., J. Theiss, P.A. Curreri and G.J. Abbaschian. NASA Technical Report, TM-82565, November (1983).
20. Turnbull, D. *Journal of Applied Physics*, **21**, 1022 (1950).
21. Willnecker, R., D.M. Herlach and B. Feuerbacher. *Proceedings of the 6th European Symposium on Material Sciences under Microgravity Conditions*, Bordeaux, France, December 2–5, p. 339 (1986).
22. Munitz, A. and G.J. Abbaschian. submitted for publication.
23. Patchett, J.A. and G.J. Abbaschian. *Metallurgical Transactions B*, **16B**, 505 (1985).
24. Cech, R.E. and D. Turnbull. *Journal of Metals*, **191**, 242 (1951).
25. Gokhale, A. and G.J. Abbaschian, to be published.
26. Amaya, German E. *Solidification of Levitation Melted Ni and Ni-Cr Alloys*, Masters Thesis, University of Florida, 1984, p. 58.
27. Perepezko, J.H., B.A. Mueller and K. Ohsaka. In: *Undercooled Alloy Phases*, Op. cit., p. 289.
28. Lacy, Lewis L., Michael B. Robinson, and Thomas J. Rathz. *Journal of Crystal Growth*, **51**, 47 (1981).
29. McDevitt, D.D. and G.J. Abbaschian. In: *Chemistry and Physics of Rapidly Solidified Materials*, Op. cit., p. 49.
30. Kattamis, T.Z. and M.C. Flemings. *Transactions of the Metallurgical Society of AIME*, **236**, 1523 (1966).
31. Walker, J.L. In: *Physical Chemistry of Process Metallurgy*, G.R. St. Pierre, (Ed.), Interscience Publications, New York, p. 845 (1961).
32. Kattamis, T.Z. *Zeitschrift fur Metallkunde*, **61**, 856 (1970).

33. Kattamis, T.Z., W.F. Brower and R. Mehrabian. *Journal of Crystal Growth*, **19**, 229 (1973).
34. Munitz, A. and G.J. Abbaschian. In: *Undercooled Alloy Phases*, Op. cit., p. 23.
35. Mondolfo, L.F. In: *Grain Refinement in Castings and Welds*, Op. cit., p. 3.
36. Crosley, P.B., A.W. Douglas and L.F. Mondolfo. *Solidification of Metals*, Iron and Steel Institute Publication, Number 110, London, p. 10 (1968).
37. Horvay, G. *Proceedings of the 4th International Congress of Applied Mechanisms* (ASME), p. 1315 (1962).
38. Horvay, G. *International Journal of Heat and Mass Transfer*, **8**, 195 (1965).
39. Powell, G.L.F. and L.M. Hogan. *Transactions of the Metallurgical Society of AIME*, **242**, 2133 (1968).
40. Powell, G.L.F. and L.M. Hogan. *Transactions of the Metallurgical Society of AIME*, **245**, 407 (1969).
41. Ovsienko, D.E., B.B. Maslow and B.N. Dneprenko. *Russian Metallurgy*, **4**, 142 (1979).
42. Abbaschian, G.J. and M.C. Flemings. *Metallurgical Transactions A*, **14A**, 1147 (1983).
43. Munitz, A. and G.J. Abbaschian. In: *Undercooled Alloy Phases*, Op. cit., p. 23.
44. Bayuzick, R.J., W.H. Hofmeister and Gail Whoriskey. First Annual Report Submitted to NASA by Vanderbilt University Center for Space Processing of Engineering Materials, p. 80 (1986).
45. Chuang, Y.Y., K.C. Hsieh and Y.A. Chang. *Metallurgical Transactions A*, **17A**, 1373 (1986).
46. Kelly, T.F. and J.B. VanderSande. In: *Chemistry and Physics of Rapidly Solidified Materials*, Op. cit, p. 35.
47. Chuang, Y.Y., R. Schmid and Y.A. Chang. *Metallurgical Transactions A*, **15A**, 1921 (1984).

14

Influence of Surface Oxidation upon Marangoni Convection Effects during Directional Solidification

J.D. VERHOEVEN,[1] M.A. NOACK,[1] W.N. GILL,[2] AND R.M. GINDE[3]

[1]*Department of Materials Science and Engineering and Ames Laboratory, Iowa State University, Ames, Iowa 50011, U.S.A.*
[2]*Department of Chemical Engineering, Rensselaer Polytechnic Institute, Troy, NY 12180-3590, U.S.A.*
[3]*Department of Chemical Engineering, State University of New York at Buffalo, Amherst, NY 14260, U.S.A.*

ABSTRACT

The effects of Marangoni convection upon macrosegregation during plane front solidification have been studied in an ultrahigh vacuum (UHV) system. The purpose of the experiments was to evaluate whether oxide films upon molten Sn surfaces could be made to suppress Marangoni convection. Macrosegregation was evaluated under both an oxidizing atmosphere and under UHV conditions employing ion beam sputter cleaning. Surface cleanliness and oxide film thickness were monitored by *in situ* Auger and electron loss spectroscopies. A cylindrical floating zone geometry was used with the molten zone positioned directly beneath the Auger analyzer. Radioisotope techniques were used to evaluate macrosegregation. Initial experimental results are interpreted in terms of models of the convection velocities calculated for the cylindrical geometry of the experiments.

INTRODUCTION

Liquid convection can be an important factor in the solidification of alloys, particularly in the control of macrosegregation. It is inherent to alloy solidification that both solute and thermal gradients exist in the liquid at the moving solid-liquid interface. These gradients can lead to convection due to the corresponding gradients produced in the density (natural convection) or in the surface tension

(Marangoni convection). Because of the no slip condition in a liquid at container walls, surface tension gradients will not produce convection in the liquid if it is confined by a solid container wall. Therefore Marangoni convection is only important in solidification processes which involve a free liquid surface relatively near to the freezing solid/liquid interface. This situation occurs in welding and in float zone solidification. Recent studies by Heiple et al. [1–2] have shown that Marangoni flow driven by the solute gradient can control the depth to width ratio of stainless steel welds, which are very sensitive to certain impurity element additives, such as S and Se, at less than the 100 ppm level. Interest in Marangoni flow during float zone solidification has been strongly stimulated by the advent of space processing because under microgravity conditions, natural convection can be nearly eliminated and Marangoni convection is left as the dominant convection mechanism. A large literature has developed on theory [3–5] and experiment [6–8] but the experiments are almost exclusively conducted on nonmetallic materials.

An inherent problem in working with liquid metals is control of oxidation at the free surface. However, if a continuous oxide layer were deliberately formed on the free liquid surface, it would eliminate Marangoni flow if the mechanical properties of the oxide layer were such that it could support the shear stress produced by the surface tension gradient and also remain as a continuous layer in spite of volume changes at the solid-liquid interface during solidification. It is possible, therefore, that Marangoni convection could be eliminated by controlled oxidation. The purpose of the present experiments was to develop an experimental technique for evaluating this possibility.

In order to control and evaluate surface cleanliness, the solidification experiments were carried out in an ultrahigh vacuum (UHV) system using the cylindrical mirror analyzer (CMA) type of Auger detector. A cylindrical float zone geometry was utilized in which a horizontal thin disk of Sn was melted outward from its center and then solidified inward. This disk float zone arrangement has the advantage that the molten zone possesses a planar free surface which may be placed directly beneath the CMA for surface analysis. In addition, by utilizing thin disks, the natural convection can be reduced to low levels in earth-based experiments because its magnitude is a strong function of disk thickness. The Sn disk was doped with a small amount of radioactive Bi^{207} which was used to evaluate the radial macrosegregation produced by the solidification process. The plan of the experiments was to solidifiy disks with various oxide thicknesses and from examination of the effect of oxide thickness upon macrosegregation, evaluate the thickness required to suppress Marangoni convection. Although the experiments are not yet completed, this paper presents details of the initial results which demonstrate the potential of the technique and some interesting aspects of surface segregation in these type of experiments.

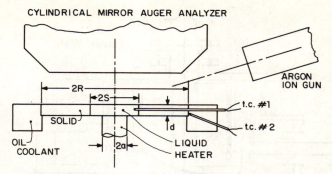

Fig. 1. Schematic representation of the disk float zone apparatus.

EXPERIMENTAL TECHNIQUE

Figure 1 provides a schematic representation of the experimental arrangement, and the actual apparatus is shown in Fig. 2. The Sn disk was 1 mm high ($d = 1$ mm) with a radius, $R = 15$ mm. It was heated in its center by contact from a hot Ta rod of radius, $a = 3$ mm. Figure 2 shows the apparatus without the Sn disk present so that the Ta heater rod is seen protruding up at the center (1). The ring immediately around the Ta rod holds it down on the top of a Cu cylinder which is

Fig. 2. The actual disk float zone apparatus without a tin disk in place.

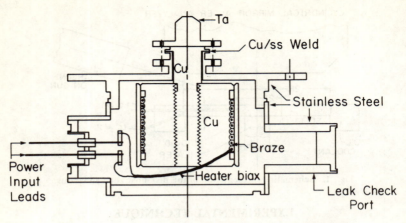

Fig. 3. Cross section of the heater unit shown as (4) on Fig. 2.

enclosed in an evacuated stainless steel can that is shown in Fig. 3. This somewhat elaborate assembly was found to be necessary to avoid outgassing of the heater assembly in the UHV chamber. The biax heater is a 2.4 mm diameter·stainless steel tube containing two heater wires insulated from the stainless tube by a powdered ceramic material. Outgassing from the ceramic insulation at the connector end of the biax heater is contained in the stainless steel vacuum can which is shown in Fig. 2 at (4).

The outer perimeter of the Sn disk was pressed down onto a copper heat sink cylinder with the stainless steel holddown ring shown as (2) on Fig. 2. The temperature of the heat sink cylinder was controlled by hot oil which was pumped from an external constant temperature bath through the Cu tube (labelled (3) in Fig. 2) which was brazed to the heat sink cylinder. The heat sink cylinder was positioned above the heater rod assembly by attachment to a stainless steel plate which was in turn fastened to the heater assembly at (5) in Fig. 2 with ceramic thermal insulating spacers. It was necessary to maintain the entire assembly insulated from ground potential for the Auger analysis and to provide X-Y-Z motion of the sample relative to the CMA. The rod holder (6) was connected to a vacuum feed-through positioner which provided micrometer positioning control and the flexible stainless steel leads (7) were fastened to the vacuum flange through insulators, thereby allowing both insulation from ground and controlled motion of the assembly. The UHV chamber was pumped with both a Ti sublimation pump and a large cryopump which could be sealed off from the chamber through a 6-inch all metal valve. The pressure in the chamber was measured with a nude ionization gauge and was maintained in the 1 to 3×10^{-10} Torr $(1.3$ to 4×10^{-8} Pa$)$ range during the experiments. As shown in Fig. 1, an argon ion sputter gun was positioned to allow the center of the Sn disk to be sputter cleaned. A pressure of 7×10^{-5} Torr $(9.3 \times 10^{-3}$ Pa$)$ Argon was maintained in the chamber during sputter cleaning. In addition, the UHV chamber was fitted with a quadrapole mass analyzer which permitted the residual gas content to be analyzed.

The Bi^{207} radiosotope was obtained in an aqueous carrier solution and a master alloy was made up by electroplating and melting. Final alloys containing 200 ppm by weight of Bi^{207} were vacuum cast and the disks were machined from 1 mm thick rolled sheets of this material. The disks were lightly electropolished to provide good contact with the Ta heater. The initial step of an experiment was to outgas the system with a bakeout of 200–300°C for one week. A programmable dc power supply was used to heat the Ta rod sufficiently to melt the central portion of the disk. The upper surface was cleaned both before and after melting with the argon ion sputter gun, and surface cleanliness monitored with Auger and electron loss spectra collected from the CMA. After the surface was cleaned, the power was turned down at a controlled rate with the programmable supply. Because of the vacuum conditions employed here, the power input from the heater, P, was transferred by conduction through the solid portion of the disk to the heat sink cylinder, so that from a simple heat balance,

$$P = 2\pi k_s d \left[\frac{T(\text{sink}) - 232}{\ln R/S} \right] \qquad (1)$$

where k_s = thermal conductivity of solid Sn, $T(\text{sink})$ = temperature of the oil cooled heat sink in °C and d, R and S are defined in Fig. 1. In the present experiments, $T(\text{sink})$ was held at 160°C and the radius at the solid/liquid interface, S, was predicted by eqn. (1) fairly well. The value of S could be observed *in situ* through a telescope and determined precisely upon removal of the disk from the apparatus because the maximum melt-out interface was clearly visible on the sample surface. The temperature distribution in the melt for pure Sn samples has been analyzed and compared to measured values [9]. The present experiments melted the disk to a maximum radius of 6.5 mm prior to solidification, and then, under computer control, the power was reduced utilizing eqn. (1) to solidify inward at a rate of 1 μm/sec until the S/L interface reached the Ta heater radius, whereupon the power was shut off. Because the time constant of the system was quite long (on order of one hour), the relatively low solidification rate of 1 μm/sec was chosen to insure a constant rate in these initial experiments.

After solidification, the samples were removed and mounted in a lathe on a jig which allowed thin circular slices to be cut out of the sample. The central few millimeters was drilled out and collected and successive circular slices of increasing radii were machined out and collected until a radius greater than the original melt-out radius was achieved. The weight of each cut was measured on a microbalance and radioactive intensity was measured in a scintilation counter. The weights varied from 10 mg near the center to 25 mg at the outer radius and the number of counts in a 1000 sec count interval varied from 7.5 to 9×10^5.

EXPERIMENTAL RESULTS

In previous studies on the oxidation of solid and molten Sn [10, 11], we have found that the CMA detector can be used to measure the low energy electron loss (LEEL) spectrum of Sn and that by utilization of 75 eV beam energies, the LEEL spectrum can detect considerably thinner coverages of oxygen than possible from the Auger spectrum. Figure 4 presents the 75 eV LEEL spectrum taken on the molten Sn zone after it had been sputtered clean. The spectrum is in the derivative mode which means that at the arrowed line labeled (a), for example, a peak occurs in the non-derivative signal and the length of the arrowed line provides a measure of the peak intensity. It is seen that there are several peaks on Fig. 4 and their origin and the effects of oxidation upon their relative intensity have been discussed previously [11]. It was found that the peaks at (a) and (b) provide the most sensitive measue of oxidation of Sn. The 9.8 eV loss peak at (a) is the surface plasmon peak while the 4.8 eV loss peak at (b) is a weak interband transition peak. This interband transition peak was found to be two orders of magnitude larger for both SnO and SnO_2 [11] and consequently the ratio (b)/(a) is a very sensitive measure of the oxygen coverage. The spectrum of Fig. 4 is identical to the cleanest Sn samples achieved by sputter cleaning, and the small (b)/(a) ratio of 0.32 is estimated to correspond to an O_2 coverage of less than 0.01 monolayers.

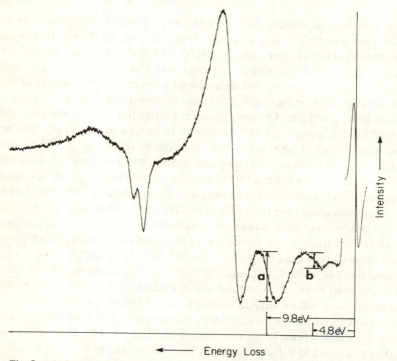

Fig. 4. The first derivative LEEL spectrum of liquid Sn after sputter cleaning. Taken with a 75 eV electron beam energy.

Preliminary experiments on pure Sn samples found two interesting effects. The solid sample was sputter cleaned at a temperature around 20°C below the melting point, then the LEEL spectrum was monitored as the sample was melted. It was found that the (b)/(a) ratio increased from values similar to that shown on Fig. 4 to values of around 2 after melting. The experiment was repeated and found to be reproducible. These results show that a clean solid Sn surface becomes slightly oxidized after melting. Apparently, oxygen dissolved in the solid Sn migrates to the surface on melting. Hence, to achieve a clean molten surface, it is necessary to either sputter clean the surface after melting or hold the molten sample in the UHV chamber for a very long time (on the order of 12 hours) apparently until all the oxygen which had been dissolved in the solid has been removed from the liquid surface into the vacuum.

Preliminary experiments on pure Sn monitored the temperature in the molten Sn with thermocouple #1 (Fig. 1) which consisted of a Chromel-Alumel thermocouple inside a 0.15 mm diameter stainless steel (Type 304) tube welded shut at the end. This thermocouple was not used in the Sn-Bi[207] experiments. After sputter cleaning the molten Sn surface to give values of (b)/(a) < 0.5, it was found by Auger analysis that the surface gave small signals for oxygen, chromium and manganese. Apparently, some of the stainless steel tube was dissolving in the molten Sn and forming oxides of Cr and Mn on the molten surface. This result showed that it was quite important to monitor the Auger spectra as well as the LEEL spectra to be sure that the Sn surfaces were completely clean.

Three Sn-Bi[207] experiments have been carried out; the first experiment under clean conditions and the subsequent two under oxidized conditions. In the clean experiment, the molten zone was formed out to a radius of 5 mm and was then sputter cleaned. The sample was translated under the CMA and it was found that along the solid/liquid interface region, a significant oxygen signal was obtained. It is likely that the oxygen, which comes to the liquid surface on melting, is driven toward the interface by the surface Marangoni flow known to be directed toward the colder interface region. Auger analysis also showed significant levels of Mn, S, Cl, O and C. The sample was therefore sputter cleaned both in the liquid and around the surrounding outer 2 mm of solid until these impurities were absent from the Auger spectra. This procedure was time consuming and took several days. Because of the long hold time, it was decided to melt out from $R = 5$ to $R = 6.5$ mm just prior to solidification to ensure a sharp drop in Bi[207] at the initial S/L interface, which might not be the case otherwise due to thermo-transport of Bi[207] at the interface during the long hold time required for cleaning. Just prior to solidification, Auger and LEEL spectra were taken at five widely spaced locations on the liquid, and Fig. 4 is the LEEL spectrum from one location and Fig. 5 is the Auger spectrum. The spectra were identical over all the locations and indicated an atomically clean surface. The heater power was then increased to drive the S/L interface out to a radius of 6.5 mm and the solidification was immediately carried out at a rate of 1 μm/s.

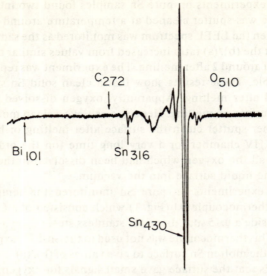

Fig. 5. First derivative Auger spectrum taken on sputter cleaned liquid Sn. The unidentified peaks all come from Sn.

The first oxidation experiment was done in an identical manner except that the surface was not sputter cleaned and an O_2 pressure of 10^{-3} Torr (0.13 Pa) was maintained in the system on freezing. An attempt was not made to evaluate the oxide layer thickness of the first experiment and a second experiment was done to ensure a thicker oxide by utilizing a pressure of 1 Torr (133 Pa) oxygen. This sample, on removal from the chamber, displayed a light blue color on its surface, thus indicating a remnant oxide layer thickness on the order of 4700Å.

The results of the Bi^{207} profile for the oxide free experiment and the first oxide experiment are shown as Fig. 6, where the radioisotope intensity measured as counts/mg-sec has been normalized to 1 in the unmelted region at the left. The abscissa is given as milligrams from the original S/L interface, which was located at a radius of roughly 6.5 mm in both cases. The composition transition at the original meltback interface is seen to be not too sharp, particularly on the clean experiment. This result is almost certainly due to the fact that the original liquid zone was not perfectly circular, which means that several of the circular sample cuts taken on the lathe near the original S/L interface included material from both sides of the original S/L interface. The zero position on Fig. 6 was therefore chosen as an average position. The data for the heavy oxidation experiment lay within the scatter of the data of Fig. 6 and were not included in the figure for the sake of clarity.

The data of Fig. 6 may be analyzed to evaluate the equilibrium distribution coefficient, k_o, by taking the ratio of the extrapolated solid-to-liquid concentrations at the original interface. The results for the data of Fig. 6 give

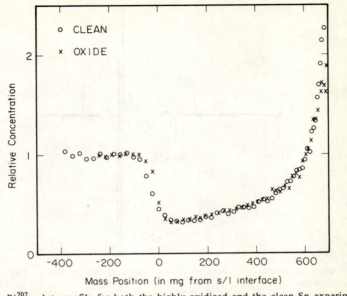

Fig. 6. The Bi^{207} solute profile for both the highly oxidized and the clean Sn experiments. The original ordinate was in counts per mg per sec and has been normalized to 1 for ease of comparison.

$k_o = 0.32$ for both experiments which agrees reasonably well with a previous study which obtained a value of 0.30 [12].

DISCUSSION

Surface cleanliness

The data of Figs. 4 and 5 show conclusively that the original liquid surface was atomically clean. After solidification, the sample was not examined for 72 hours and the Auger spectrum than obtained is shown in Fig. 7. Comparison to Fig. 5 shows that a strong Bi signal has appeared. Auger spectra had also been taken at higher gain and there was clearly no evidence for Bi, O, or C on the molten Sn prior to solidification. The higher gain spectra of the sample after the 72 hours hold showed only a very small buildup of C and O, which is not apparent in Fig. 7. These results indicated that the Bi^{207} appeared to have migrated to the surface subsequent to the solidification. To verify this result, the solid surface was sputtered clean. The Bi peak disappeared after a few minutes of sputtering and then reappeared at roughly the same intensity as shown on Fig. 7 after a 72-hour hold in the vacuum system. The Bi peak was absent from spectra taken near the outer edge of the sample which had never been cleaned. It appears that the Bi^{207} atoms migrate to the surface of unoxidized clean Sn at room temperature. This same type of result has been observed by Frankenthal and

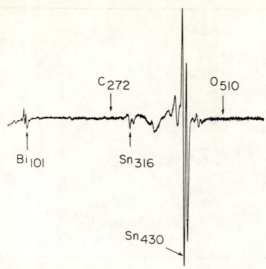

Fig. 7. An auger spectrum taken from the same region as that of Fig. 5, but 72 hrs. after solidification.

Siconolfi who have found that Pb migrates to the surface of clean Sn-0.2% Pb alloys [13]. If surface segregation of Bi or O occurred in the liquid during the time interval of the solidification experiments done here, it would complicate interpretation of the results. Data collected shows that this does not occur in the one-to-two hour time interval of the solidification experiments. Auger spectra collected over a period of one-to-two hours after sputter cleaning liquid Sn showed no evidence of the Bi_{101} peak. A sputter cleaned liquid Sn sample was held in the vacuum system for 12 hours and the LEEL data showed only a very slight increase in oxygen, less than one monolayer. It should be noted that there was no evidence of Ta contamination from the central heater rod.

Macrosegregation

In order to interpret the results of this study, it is necessary to know the composition profile that would be expected if convection were completely eliminated from the liquid. A recent study [14] has obtained solutions to this problem using finite difference techniques for the cylindrical coordinates required here. It was found that the predicted macrosegregation is a function of a dimensionless Peclet number, Pe, defined as the radius of the metal cylinder which is frozen, R, to the diffusion distance, D/V, where D is the mutual diffusion coefficient of the solute and V is the radial solidification velocity. The mutual diffusion coefficient of Bi in Sn has been measured as, $D = 2.33 \times 10^{-4}$ exp $(-1282/T)$ [15]. In the present experiments, $R = 6.5$ mm and $V = 1$ μm/sec. The liquid Sn temperature ranged from 232°C at the initial solid/liquid interface

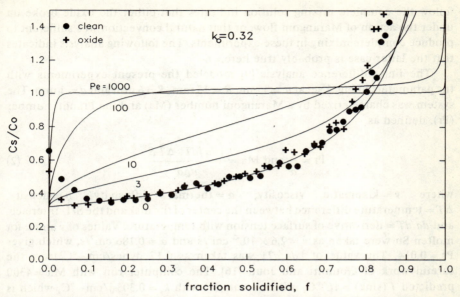

Fig. 8. Comparison of measured solute profiled data to predictions of convection free models [14] for various *Pe* values.

to a value of 370°C at the disk center, see below. Hence the Pe value ranges from values of 3.5 to 2.0 in the liquid so that Pe = 3.0 is a reasonable estimate for these experiments since the largest volume fraction of liquid lies closer to the 3.5 value.

The data of Fig. 1 have been replotted using mass fraction solidified as the ordinant on Fig. 8 and the predicted solute profiles for *Pe* = 0, 3, 10, 100 and 1000 are superimposed on the data. The profile for *Pe* = 0 corresponds to the case of complete mixing in the liquid and it is seen that the experimental results lie on this line within the scatter of the data for the values of *f* where the S/L interface was beyond the heater radius. The no-convection result, *Pe* = 3.0, is clearly above the data and shows that convection, adequate to produce complete mixing in the liquid, was achieved in both the oxide and the oxide-free experiments. The results also show that higher solidification rates would have provided higher *Pe* values and a better test of the convection effects.

Marangoni versus natural convection

The expected flow velocities produced by the Marangoni convection for these experiments has been evaluated using finite difference techniques [9]. The maximum flow velocities which occur at the surface were found to be 20 cm/s. Because this is five orders of magnitude larger than the solidification rates used here, it was clearly expected that the oxide-free experiment should follow the complete mixing (*Pe* = 0) solution. The fact that the oxide experiments also follow the complete mixing solution indicates that either the oxide broke up under the action of Marangoni flow, or that natural convection was sufficient to produce complete mixing in these experiments. The following analysis indicates that the latter case is probably true here.

The finite difference analysis [9] modeled the present experiments with the parameters of Fig. 1 set at $a = 3$ mm, $R = 15$ mm, $S = 6$ mm and $d = 1$ mm. The system was characterized by a Marangoni number (Ma) and the Prandtl number (Pr), defined as,

$$\text{Pr} = \frac{v}{\alpha} \text{ and Ma} = \frac{|d\sigma/dT| \cdot \Delta T \cdot S}{v \varrho \alpha} \tag{2}$$

where $v =$ kinematic viscosity, $\alpha =$ thermal diffusivity, $\varrho =$ density, $\Delta T =$ temperature difference between the center of the disk and the S/L interface, and $d\sigma/dT =$ derivative of surface tension with temperature. Values of v and α for molten Sn were taken as $v = 2.6 \times 10^{-3}$ cm^2/s and $\alpha = 0.186$ cm^2/s, which gives Pr $= 0.014$. The value of $|d\sigma/dT|$ was taken as 0.17 dynes/cm$-$°C from the careful work of Goumiri and Joud [16]. The computer run with Ma $= 4300$ predicted $T(\text{sink}) = 167$°C, utilizing eqn. (1) with $k_s = 0.305$ j/cm$-$°C, which is close to the 160°C value used here; the predicted values for the surface velocity and the surface temperature are plotted in Fig. 9. The general shapes of these curves have been discussed previously for the case of a much larger heater radius and the results here are quite similar. It is seen that surface velocities on the order of 20 cm/s are predicted to occur in the liquid beyond the 3 mm heater radius position and extend nearly to the S/L interface position at 6 mm. As mentioned above, this value of 20 cm/s is larger than the solidification rate by a factor of 2×10^5 and would be expected to produce complete mixing in the liquid in these solidification experiments.

The case of pure natural convection has not been modeled by finite difference techniques, but we may obtain an order of magnitude analysis from the treatment of Birikh [17]. His analysis considers a horizontal trough of height d and infinite extent with a horizontal temperature gradient G, and both Marangoni and natural convection are included. The velocities predicted by his model will be higher than obtained here because of the finite extent of the present system, but a correction factor may be obtained as follows. The surface velocity predicted by Birikh is,

$$v = \frac{Gd|d\sigma/dT|}{4v\varrho} \left[1 + \frac{Bo}{12} \right]; \ Bo = \frac{\varrho \beta g d^2}{|d\sigma/dT|} \tag{3}$$

The $Bo/12$ term represents the contribution from natural convection and taking $\varrho = 7$ gms/cm^3 and $\beta = 9.5 \times 10^{-5}$ °C^{-1} the value of the $Bo/12$ term is found to be quite small, 0.0032. This shows that the surface tension contribution to the flow velocity is some 300 times greater than the gravity term.

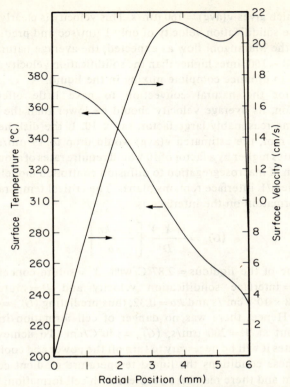

Fig. 9. Finite difference model predictions of surface temperature and velocity for conditions of $Ma = 4300$ with $Pr = 0.014$, for conditions of $d = 1$ mm, $a = 3$ mm, $S = 6$ mm, $R = 15$ mm and $T(\text{sink}) = 167°C$.

If the surface oxide eliminated the Marangoni flow, and only the natural convection velocity remained, an estimate of its magnitude can be made from the Birikh analysis. The average natural convection velocity is predicted to be,

$$v(\text{avg}) = \frac{Gg\beta d^3}{192\, v} \tag{4}$$

To estimate a value for comparison to these experiments, the temperature gradient in the nearly liner region of the temperature profile of Fig. 9 was determined to be 350°C/cm and this was used in eqn. (4) to give $v(\text{avg}) = 620$ µm/s. As mentioned above, this value is an overestimate because of the infinite extent approximation of the Birikh analysis. From eqn. (3), the analysis also predicts a Marangoni surface velocity of 81.2 cm/sec for the present experiments which is considerably higher than the finite difference result which averages 20.5 cm/sec. It is reasonable, therefore, to estimate the effect of the constraining wall of the S/L interface upon natural convection by multiplying 620 µm/s by the ratio of 20.5/81.2 which gives $v(\text{avg}) = 160$ μm/s. This velocity is clearly considerably greater than the solidification velocity of only 1 μm/sec and predicts that if the oxide stopped the Marangoni flow as expected, the average natural convection velocity was still ~ 160 times higher than the solidification velocity and should be quite adequate to produce complete mixing in the liquid.

In order for the natural convection to have little effect upon the macrosegregation, its average velocity should be lower than the solidification velocity by some reasonably large factor, say x 10. If the disk height, d, were reduced to 0.5 mm, the estimated $v(\text{avg})$ would drop to 19 μm/s. To achieve solidification rates higher by a factor of 10 would require rates of around 200 μm/s. The comparison of macrosegregation to diffusion controlled models, as in Fig. 8, requires that the S/L interface remains planar. The critical temperature gradient to avoid cell formation on the interface is

$$(G)_{cr} = \frac{m \, X_s \, V}{D} \left[\frac{1 - ko}{ko} \right] \tag{5}$$

where m = slope of the liquidus = 2.87°C/wt%, X_s = solute concentration = .02 wt% Bi^{207}, V = interface solidification velocity and D = mutual diffusion coefficient $\approx 1.8 \times 10^{-5}$ cm²/s and $ko = 0.32$, thus predicting $(G)_{cr} = 0.68$°C/cm at $V = 1$ μm/sec. Hence, there was no danger of cell formation in the present experiments, but at $V = 200$ μm/s, $(G)_{cr} = 140$°C/cm. To achieve 200 μm/s solidification rates it will be necessary to turn off the power and cool the heat sink block; under these conditions the liquid temperature gradient could well fall below 140°C/cm and there may be a problem with cell formation. Therefore, to carry out the present experiments with negligible natural convection, it appears that it will be necessary to reduce d to less than 0.5 mm. Alternately, if UHV conditions could be achieved in a space flight, the experiments could be done at low g conditions where the natural convection could be reduced dramatically.

SUMMARY AND CONCLUSIONS

When liquid convection occurs during plane front solidification of alloys, the amount of segregation occurring over the length solidified (the macrosegregation) increases relative to the no-convection diffusion controlled models. When the liquid convection velocity becomes appreciably larger than the solidification velocity, the macrosegregation follows the complete mixing model. Hence, measurement of macrosegregation provides a qualitative means of evaluating liquid convection effects in directional freezing experiments, and this approach has been used here. The macrosegregation was evaluated using 200 ppm of radioactive Bi^{207} doped into Sn disks. A floating zone in the center of 1 mm high disks was solidified inward within a UHV chamber. The surface oxide condition of the Sn was controlled by both deliberate oxidation and sputter cleaning. It was demonstrated with Auger and low energy electron less spectroscopies that sputter cleaned solid Sn reoxidized after melting and required additional sputter cleaning to maintain atomically clean surfaces. It was also demonstrated that strong surface segregation of Bi occurs on clean solid Sn at room temperature, but not on liquid Sn over the one-to-two hours time required for the solidification experiments. Macrosegregation was evaluated on cleaned Sn and Sn oxidized to thicknesses on the order of 5000 Å utilizing solidification rates of 1 μm/s. Comparison of the Bi^{207} profiles to calculated convection free and complete mixing profiles showed that complete mixing occurred both with and without the oxide present. Analysis shows that the original purpose of the experiments, to evaluate the thicknesses of oxide required to suppress Marangoni convection, was not possible because the natural convection velocities were large relative to the solidification velocity so that complete mixing is expected both with and without an oxide film. The experiments do demonstrate, however, that controlled UHV directional solidification experiments can be done with *in situ* Auger analysis to evaluate surface cleanliness, and that corresponding macrosegregation profiles can be obtained quite accurately using radioactive isotope techniques. Additional experiments utilizing thinner disks and higher solidification rates should be able to reduce the ratio of the natural convection velocity to solidification velocity adequately to allow oxide film effects upon Marangoni flow to be evaluated.

ACKNOWLEDGMENTS

Discussions with A.J. Bevolo concerning surface analysis were very helpful. Initial work was supported by NASA, Materials Processing in Space Program, Marshall Space Flight Center. The final work was jointly sponsored, in part by the Ames Laboratory, which is operated for the U.S. Department of Energy by Iowa State University, under contract no. W-7405-ENG-82, supported by the Director of Energy Research, Office of Basic Energy Sciences; and the National Science Foundation, Grant no. CBT 85 13606.

REFERENCES

1. Heiple, C.R. and J.R. Roper. *Welding J.* (Research Supplement) **62,** 975 (1982).
2. Heiple, C.R., P. Burghardt, J.R. Roper and J.L. Long. *Proc. Conf. on Effects of Residual, Trace and Microalloying Elements on Weldability and Weld Penetration;* The Welding Institute, Cambridge, U.K., Nov. (1983).
3. Ostrach, S. *Physiochem. Hydrody.*: V.G. Levich Festschrift **2,** 571 (1977).
4. Smith, M.K. and S.H. Davis. *J. Fluid Mech.* **132,** 119 and 145 (1983).
5. Gill, W.N., N.D. Kazarinoff and J.D. Verhoeven. *Amer. Chem. Soc. Symp. Series* No. 292, 47 (1985).
6. Chun, Ch.-H. *J. Cry. Growth,* **48,** 600 (1980).
7. Preisser, F., D. Schwabe and A. Scharmann. *J. Fl. Mech.* **126,** 545 (1983).
8. Kamotani, Y. and S. Ostrach. *J. Thermophys. and Heat Trans.* **1,** 1 (1987).
9. Gill, W.N., N.D. Kazarinoff, C.C. Hsu, M.A. Noack and J.D. Verhoeven. *Adv. Space Res.* **4,** 15 (1984).
10. Bevolo, A.J., J.D. Verhoeven and M. Noack. *J. Vac. Sci. Tech.* **20,** 943 (1982).

346

11. Bevolo, A.J., J.D. Verhoeven and M. Noack. *Surf. Sci.* **134**, 499 (1983).
12. Verhoeven, J.D., E.D. Gibson and R.I. Griffith. *Met Trans.* **6B**, 475 (1975).
13. Frankenthal, R.P. and D.J. Siconolfi. *Surface Sci.* **119**, 331 (1982).
14. Verhoeven, J.D., W.N. Gill, J.A. Puszynski and R.M. Ginde. submitted, *J. Crys. Growth* (1988).
15. Verhoeven, J.D., E.D. Gibson and B. Beardsley. *Met. Trans.* **6B**, 349 (1975).
16. Goumiri, L. and J.C. Joud. *Acta Met.* **30**, 1397 (1982).
17. Birikh, R.V. *J. Appl. Mech. Tech. Phys.* **3**, 43 (1966).

15

Channel Formation during Alloy Solidification

A. HELLAWELL

Michigan Technological University
Houghton, MI 49931, U. S. A.

ABSTRACT

During the solidification of alloys which have a significant freezing range there develops a 'mushy zone' consisting of a dendritic array within a loosely connected body of interdendritic liquid. Generally, such interdendritic liquid is of different density to the remaining bulk liquid so that there can arise a buoyancy force for convection, sometimes evident in the formation of local segregation channels, variously known as 'A' segregates or 'freckles' in the foundry industry. Different geometrical situations are discussed with particular reference to that where growth occurs vertically upwards. Reasons are presented for believing that such channels originate by liquid perturbations close to the dendritic growth front and, assuming such a model, an analysis of the situation is examined. The phenomenon is regarded as an example of thermo-solutal convection and its wider application to other systems is discussed.

I. INTRODUCTION

Among the many microscopic and macroscopic defects to be found in ingot castings are those which appear as long, nearly vertical, pencils of solute rich material, within the columnar regions: these may be of the order of 10 interdendritic spacings in width and up to 1 m in height. In conventional steel billets, with predominantly horizontal heat flow, they have long been recognized as 'A' segregates (Fig. 1), and are rich in less dense components, carbon, sulphur, phosphorus, etc. The incidence of such defects and the conditions which aggravate their occurrence have been documented and recorded in some detail over the past 50 to 60 years and an account of this type of information may be found in a recent review by Moore [1]. More recently, the same type of defects became recognized in other situations, notably in ESR ingots and directionally

348

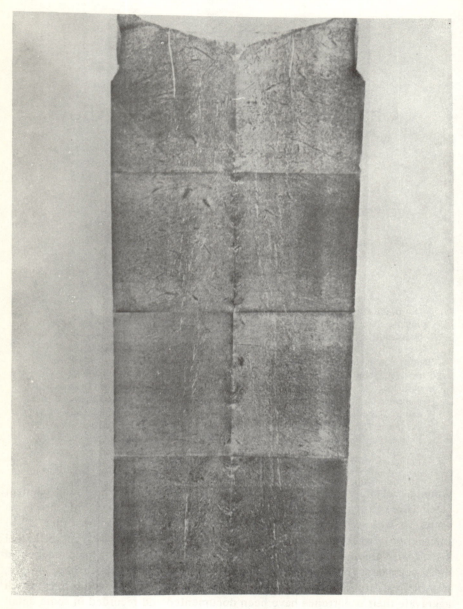

Fig. 1. 'A' segregate channels in the upper part of a high sulphur, high phosphorus medium steel billet.

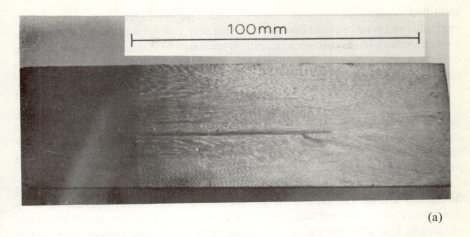

(a)

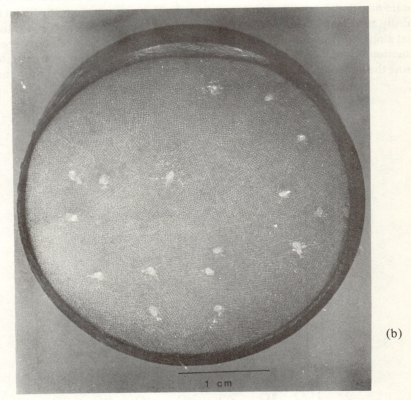

(b)

Fig. 2. (a) Vertical section through a base chilled Pb-10 wt.% Sn ingot, quenched to show dendritic growth front and channel, (b) horizontal section through similar ingot to show channels and dendritic structure.

solidified superalloys where they are termed 'freckle'—Fig. 2 shows a laboratory example produced in a lead base sample. In a very general way, for a given alloy in a given system, the incidence of channels increases with decreasing growth rates or temperature gradients, but of course, the latitude of these conditions is often restricted in a practical casting operation.

It will be readily appreciated that the occurrence of such channels, rich in certain solutes, can represent a serious problem during subsequent processing, such as forging, particularly if the solute-rich regions are the more brittle. Under the more stringent requirements of superalloys, used in aeronautical/engine applications, the presence or absence of freckles can make the difference between rejection or acceptance of an ingot.

It is now recognized that channel segregation arises because interdendritic liquid has a different density to the remaining bulk liquid, because it has a different composition and the density-composition dependence, $d\rho/dc$, generally outweighs that with temperature, $d\rho/dT$. Thus, in an iron-carbon alloy, solid-liquid distribution coefficient, $k_o < 1$, liquidus slope, m, negative, the liquid in the mushy zone may be cooler than the bulk liquid, but it is also richer in solute, so that along the liquidus line, the density actually decreases with temperature. This situation was recognized by Hunt some 20 years ago and simply demonstrated [2] using the aqueous analogue system, NH_4Cl-H_2O (Fig. 3). This is a simple eutectic

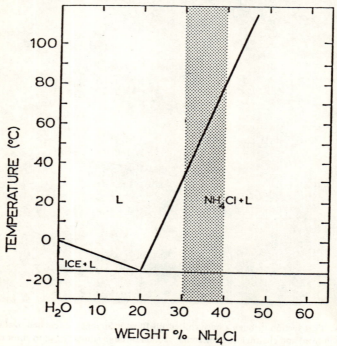

Fig. 3. H_2O-NH_4Cl phase diagram—shaded region corresponds to range of interest.

system with the point at ~ 20 wt.% NH_4Cl and − 16°C, compositions of interest are hypereutectic, up to a maximum of $\gtrsim$ 40 wt.% NH_4Cl where the liquidus reaches the boiling point. In the limited hypereutectic range, the primary solid is the more dense NH_4Cl ($\varrho \sim 1450$ kg m^{-3}) and water is rejected as the less dense solute; NH_4Cl solidifies with a dendritic morphology, but in the composition range available, the fraction of solid in the mushy zone, f_s is very small—e.g. at 35 wt.% NH_4Cl, $f_s \gtrsim 10\%$ at the eutectic temperature—and it yields a very open and permeable mushy zone, by comparison with that found in a typical metallic alloy. Nevertheless, this analogue served to demonstrate, clearly, the development of channels in molds with lateral heat flow and it was further shown, that if the density of the liquid was adjusted by the addition of a third component, $ZnCl_2$, then channels could be prevented or reversed in direction above certain composition levels.

This work was then reinforced by studies at United Technologies [3], working with a base chilled model to simulate ESR ingots, with freckle formation. It was

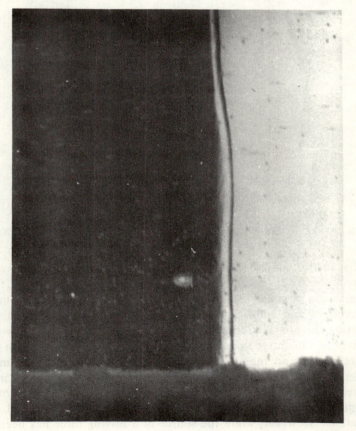

Fig. 4. (a) Side view of water rich solute plume rising from channel in NH_4Cl

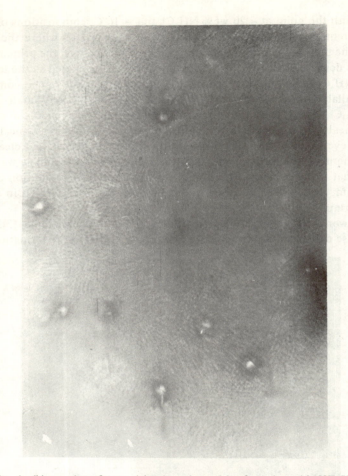

Fig. 4. (b), top view of water rich solute plume rising from channel in NH_4Cl.

shown that in this configuration, vertical growth, anti-parallel to gravity, that channels rose vertically (despite any mold inclination) and, in particular, that each channel ejected a plume of solute-rich liquid, rising through the bulk liquid (Fig. 4), i.e. channel and bulk liquid plume flow was (is) continuous. In this configuration the propagation of a channel involves an almost steady-state re-entrainment of bulk liquid to maintain streamline vertical flow through the bulk liquid—activity around the channel mouth is shown schematically in Fig. 5. Subsequently, there have been parallel studies with metallic systems, involving various geometrical configurations and density changes (e.g. 4–6) confirming the general picture of convection driven by compositional rather than thermal considerations. However, while the source of a driving force has been clearly demonstrated, there remains the question of where and how such perturbations

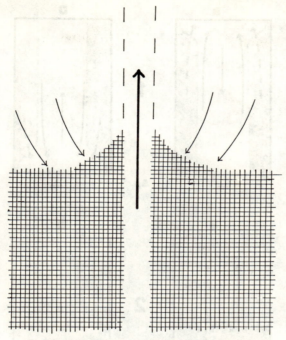

Fig. 5. Schematic re-entrainment at channel exit.

can develop, and, in order better to understand the phenomenon, this is the critical problem requiring experiment and analysis. Before considering this 'where' and 'how', it may be helpful to put in perspective the alternative situations which can arise during the solidification of a material which passes through a significant temperature/composition range.

Figure 6 summarizes the possible geometrical configurations, 1–3, which a materials scientist would wish to consider. In each case, 1–3, we should consider the alternative possibilities, namely, that the liquid density with temperature, *along the liquidus line, $k_o < 1$*, has a density gradient which is either negative (left column) or positive (right column). To bring these home as familiar examples, the left hand would include base alloys in: Fe–C, Pb-Sn, Pb-Sb and NH_4Cl-H_2O, etc., or the right hand side, such as Fe-Mo, Cu-Pb, Al-Cu, etc. Case #1, we recognize as the conventional billet casting: fluid convection patterns in the bulk liquid are much more complex than those indicated, but, notwithstanding, the flow tendencies within the mushy zones (shaded) are quite obviously up or down, and solute accumulation, with attendant depression of the freezing point at top or bottom of the ingot is as expected and as observed: always, what goes up must come down, somehow. Case #2, would apply to any based chilled configuration, it will be recognized, in a metallurgical context, as representative of the typical ESR/VAR configuration or any directional solidification geometry, such as that

354

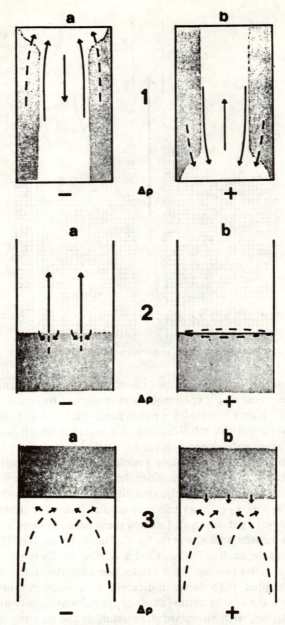

Fig. 6. 1–3, alternative configurations for (a) density inversion within the mushy zones (shaded) and (b) with more dense solute.

for turbine blade production. With $d\rho/dc \times d\rho/dT$ of negative sign, left hand, we have the typical possibility of freckle formation, but, if the gradients are reversed, right hand, only lateral temperature gradients can create a convective problem—the growers of device material will also recognize this configuration. Case #3, will be most familiar to crystal growers using the Czockralski technique, but founders of big castings, or those involved with continuous castings will also recognize this situation. Here, any convective driving forces from mushy-zone versus bulk liquid density differences are either, 3(a), additive or in conflict, 3(b), with those in the bulk liquid. The metallurgist will find no difficulty in relating situation #2(a) to the top region of #1(a), or that of #3(b), to the bottom region of #1(b).

In what follows, we will focus upon the configuration of #2(a), partly because it lends itself to relatively controlled experiment, but also because it is free of the complications of bulk liquid convection and is the more easily adaptable to mathematical analysis. Therefore, let us consider this situation, where the liquid in the mushy zone is less dense than the supernatant, thermally stable bulk liquid. While this is a classical situation of density inversion, it should be noted that the density gradients are a consequence of both heat and composition variations (coefficients of thermal expansion, α, and of solutal expansion, β) and will also require consideration of the respective diffusivities, k and D_L. We will proceed with further experimental evidence.

II. MORE RECENT OBSERVATIONS

The following concerns the base chill geometry of Fig. 6.2 with liquid in the mushy zone less dense than the thermally stabilized, supernatant liquid. In the aqueous system with NH_4Cl, near steady state channels and plumes become established after $\sim 10^3$ s in a sample containing ~ 35 wt. % NH_4Cl (refer to Fig. 3). Direct observation with colored dyes shows that re-entrainment is rapid (a few seconds) and obviously restricted to feed volumes around each channel outlet. As noted, the channel and plume flow is continuous, flow in a plume is streamlined and measured flow rates, < 10 mm s^{-1}, are sufficiently slow for the liquid to attain thermal equilibrium as it rises: this is an important requirement for streamlined, steady state flow—the rates are, however, much too great to allow any possibility of solute mixing by diffusion. The question arises as to where channels originate, whether within the mushy zone at some depth, or at, or close to, the dendritic growth front. While direct observation of the transparent system tends to support the latter view, there are some further experimental observations of interest and relevance.

a) Mould movements

Since this is a gravity-driven phenomenon, it was thought of interest to change the direction of the gravitational vector with respect to the solidifying specimen by slow and systematic movement. To do this, Sample and Hellawell [7] caused a base chilled mould to rotate about an inclined axis (Fig. 7), in such a way that the

356

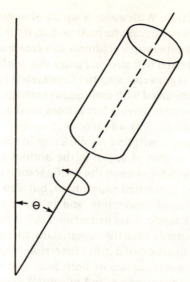

Fig. 7. To illustrate precessional movement caused by rotation about an inclined axis.

gravitational vector precessed around the surface of a cone, at a rate determined by the rotation rate (typically < 10 rpm and angle of inclination (< 30° to the vertical). The results of such experiments showed that channels were indeed prevented, as long as the movements were about an inclined, as opposed to a vertical axis, and that channels would be arrested or restarted by alternating the precessional cycles. Subsequently [8-10], it was also demonstrated that the same type of movement—at slower rates and smaller angles of inclination—produced the same results in lead-tin alloy ingots.

Originally, it was supposed that movements of this type would diffuse the gravitational vector and so prevent interdendritic fluid flow in any given direction, thereby preventing a single channel from developing. However, it should also be remembered that the bulk liquid is simultaneously disturbed and it is necessary to consider the consequences of precessional movements more closely. In moving a container, slowly, in a given sense, a liquid of low viscosity does not move with the container but remains static—i.e., the vessel walls and base (in this case represented by a dendritic growth front) actually move with respect to the bulk liquid. Referring to Fig. 8, a simple rocking movement about a single axis, through an angle θ, causes the bulk liquid to alternately rise and fall, up and down the container walls and across the base (growth front) by a rate determined by the angle θ; actually, Tan θ, and by the width of the vessel. In a narrow tube the fluid movement is obviously less than in a wide, open dish. If now the rotational axis is itself rotated, continuously, as by precession, the result then is that the growth front moves continuously under the bulk liquid in a circular manner, and, quite apart from any other effects, this type of movement will shear

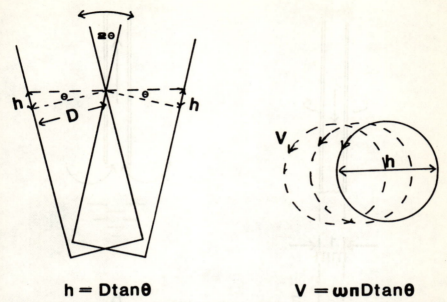

$$h = Dtan\theta \qquad\qquad V = \omega nDtan\theta$$

Fig. 8. To illustrate liquid movements with respect to a mould of diameter, D, rotated at an angular rate, w, about an axis inclined at angle, θ.

off any solute plumes which might otherwise have developed—the activity is across the growth front itself and the result is to inhibit the escape of solute plumes—much as a high wind prevents the escape of smoke from a chimney stack. Experiment shows that the effect is reduced in narrower containers and also, that it is much easier to prevent channels in a metallic system than an aqueous model—almost an order of magnitude in the necessary shear velocity—a point requiring further discussion.

b) Induced channel formation

Two simple experiments with the transparent system have been instructive, referring to Fig. 9, depict an upward growing dendritic front, at which channels were about to form. In one series of experiments, Fig. 9(a), a fine glass tube of diameter $\lesssim$ 1 mm was used to gently drill out artificial channels to depths of $\sim$ 20 mm, after which it was observed that these channels did *not* continue to propagate and were overgrown by the dendritic front. In a second series, Fig. 9(b), the fine tube was lowered to within $\sim$ 2 mm of the growth front and an artificial plume was created by slowly sucking the liquid up the tube—in every such case, a channel rapidly developed beneath the plume. It must be concluded from these simple experiments, that in this geometrical configuration at least, plume flow in the bulk liquid precedes channel formation, and a model involving liquid perturbation close to the growth front would seem to be realistic

358

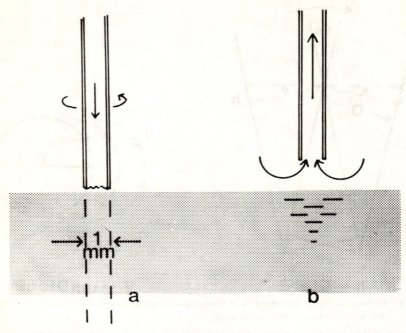

Fig. 9. Artificial formation of channels, (a), caused by drilling into the mushy zone (shaded) and (b) by sucking up liquid and creating an artificial plume in the bulk liquid.

(Fig. 10). The analogy has been drawn here between the problem of releasing a body of stagnant liquid, as in a marsh or swamp: drainage is not accomplished by the creation of internal dykes, but rather by creating a breach at the edge of the swamp, thereby allowing the liquid to escape, leading subsequently to a drainage channel which develops *backwards* from the opening breach.

c) Systematic measurements

In the course of the above experiments [11] and subsequently, working with lead base alloys in the systems Pb-Sn, Pb-Sb and Pb-Sn-Sb [12–13], the conditions of alloy composition, growth rate, temperature profiles and primary dendrite spacings have been carefully recorded and documented. In summary, the general picture indicates that channels form in metallic alloys at much lower liquid fractions, i.e. in less permeable mushy zones than in the aqueous system. It seems, that channels (and therefore plumes) are much more easily formed in the better conducting material, and, cf. above, that such fluid flow is also the more easily inhibited by mold movements.

III. A POSSIBLE ANALYSIS OF CHANNEL "NUCLEATION"

Based upon the observations described above, it seems reasonable to consider the

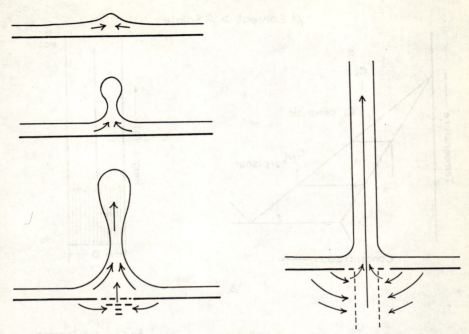

Fig. 10. Schematic model for channel formation by liquid perturbation at the dendritic growth front.

formation of channels in terms of the model shown in Fig. 10 which involves liquid perturbation close to the growth front. What happens in other configurations, as with lateral heat flow (Fig. 6.1), may be more complicated, although the author is of the opinion that the mechanism will not differ in principle. In what follows the general picture is beyond reasonable doubt, but the reader should be warned that there will be some assumptions which are not universally accepted, particularly relating the critical conditions for liquid perturbations and the onset of convective flow.

We refer now to Fig. 11, for a typical alloy diagram and upward growing dendritic front advancing into a positive temperature gradient, dT/dz The bulk liquid is composition, Co. At the growth front, the interdendritic liquid in the mushy zone, has a concentration which rises along the liquidus line, at lower temperature. Correspondingly, the density gradient above the front, $d\varrho/dz$, has a negative slope and is stable against convection, while below the plane of the front, the density gradient, $d\varrho/dz$ is now positive, exhibits density inversion and is unstable. In a simple-minded way [14], it might be supposed that when the solutal density gradient exceeds the thermal density gradient, i.e. $d\varrho/dz$ solutal $> - d\varrho/dz$ thermal, that convection might begin. However, this is not a spontaneous result, because the fluid has a finite viscosity and, if it is translated vertically, it also moves into a region of different temperature and needs to reach thermal equilibrium. This situation was identified originally by Bênard [15] in a purely

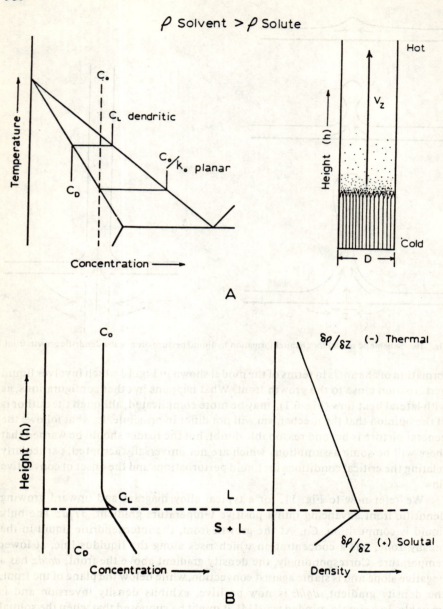

Fig. 11. Composition and density gradients above/below the dendritic front in a vertical sample and positive temperature gradient.

thermal context (liquid heated from below) and analyzed by Rayleigh [16] in what is probably the first example of a perturbation analysis. The analysis has three parts: (1) identification of a buoyancy pressure, (2) translation against viscous drag, and (3) simultaneous attainment of thermal equilibrium. Summation of these contributions is now expressed by a Rayleigh number, above a critical value of which convection may be expected to occur.

In a purely thermal environment, the Rayleigh number is expressed as a dimensionless quantity:

$$R_T = g a \rho \; dT/dz h^4 / k \; \eta,$$

where, g, is the gravitational acceleration
α, the coefficient of thermal expansion
ϱ, the liquid density
dT/dz, the thermal gradient
k, the thermal diffusivity
η, the dynamic viscosity coefficient
h, a characteristic dimension for the system.

An equivalent expression may also be written for a solutal Rayleigh number, R_s, in which a solutal coefficient of expansion, β, replaces the thermal coefficient, α, and solute diffusivity, D_L, replaces the thermal diffusivity. The two numbers may also be combined [17–19] to given an effective Rayleigh number:

$$R_E = R_T - R_s/\tau$$

where τ is the thermal:solutal diffusivity ratio, k/D_L, a so-called Lewis number.

In all such discussions, the characteristic dimension, h, clearly dominates the Rayleigh number, being present, as it is, to the fourth power. Various values for h have been assumed in the literature, from the physical dimensions of the system (height, width) to the characteristic diffusion distance (D_L/V) and the Rayleigh number quoted for a given situation, can obviously differ by many orders of magnitude. It is at this point that the present analysis is not in conformity with previous treatments. Sarazin and Hellawell [12, 13] assumed, from a simple consideration of the dimensions within a Rayleigh number, that convection should begin, in any systems, when the buoyancy pressure exceeds the restraining requirements of viscosity and heat flow, i.e. they assumed that a critical value of $R_{Effective}$ should be unity. A value for the dimension, h, was not preselected, but rather derived from the Rayleigh numbers by inserting all the other available or measured quantities. This may be arbitrary, but what is remarkable, is that the dimension so extracted, both for metallic and aqueous systems, comes out close to that of the measured interdendritic spacings in either case. This result may be a coincidence, but if the problem is one of liquid perturbation (Fig. 10), then it can be argued that the value of h relates to a critical wave length for successful perturbation, and, at a growth front having a characteristic periodicity, that wavelength ought to relate closely to such periodicity—i.e. to the primary dendritic spacing. We do not know if this is physically correct, but the coincidence

362

is difficult to overlook. Clearly, more work will be needed, notably to study how independent control of the dendrite spacings will influence the onset of channel formation. Provisionally, in this connection, it may be worth noting, cf. in the present introduction, that those growth variables which reduce the dendrite spacings, growth rate and temperature gradient, are also those known to reduce the incidence of channel formation.

IV. OTHER SYSTEMS AND SITUATIONS

So far, only solidification has been considered, as it affects heat and solute gradients and their relative diffusion rates, and channels have been regarded as a special case of thermo-solutal convection. It will be readily appreciated, however, that there must be many other systems and materials where parallel opposing density gradients will arise, for example during melting and in any fluid. Thus it should come as no surprise to discover that the same general problem has been recognized by oceanographers before materials scientists/metallurgists became aware of it, in a context where relatively fresh water is surmounted by a body of thermally stable fluid which is more saline. Convection is not spontaneous, but when it does occur, large liquid plumes erupt to give salt fingers or fountains [20, 21]. Another oceanographic example is to be found during the melting of icebergs when water is produced at the melting surface [22], and there may be geological examples in molten magma chambers.

The generality of this type of convective situation has been elegantly illustrated by Schmitt [23] (Fig. 12). This figure has as coordinates, logarithms of

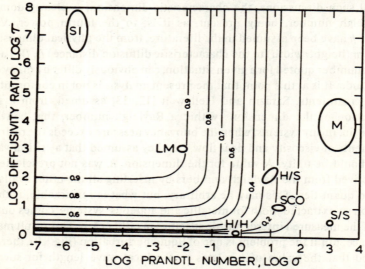

Fig. 12. Plot of Log τ (Lewis number) vs. Log σ (Prandtl number) after Schmitt [23]. Contours correspond to normalized flow rate, various systems and materials are indicated: LM—liquid metals, H/S—heat/salt in aqueous systems, M—molten magmas, SCO—molten semi-conductors, S/S—salt/sugar combinations, H/H—heat and humidity, SI—stellar interiors.

the dimensionless quantities, τ, thermal:solutal diffusivity ratio, and the Prandtl number, σ (ratio of the kinematic viscosity coefficient, v, to the thermal diffusivity, k), and indicated upon it are regimes corresponding to various materials and systems. Fluids of high viscosity, low density and low thermal conductivity fall to the lower right hand of this figure, those with high thermal conductivity and low kinematic viscosity to the upper left. The contours correspond to a dimensionless growth rate, a high value representing ease of perturbation to produce streamlined flow. Individual examples fall within quite small regions on this figure: liquid metals, *LM,* have good thermal conductivities, relatively high Lewis numbers and low Prandtl numbers, perturb readily and flow rapidly. Salt water combinations have somewhat lower Lewis numbers and higher Prandtl numbers and convect more slowly. Molten silicates fall across a wider range and are more sluggish. Salt water solutions with sugar syrup are very sluggish, *S/S.* Even heat and water vapor can be represented on this figure, of possible meteorological interest, and stellar interiors, S.I., of astrological relevance. This is a most enlightening diagram and places whole ranges of natural and laboratory studies onto a common reference framework, where diffusion of two species compete, thus providing a valuable perspective.

V. SUMMARY

a) Channel formation is a consequence of density inversion resulting from liquid composition gradients during solidification. Channel flow is continuous with solute plumes in bulk liquid and is essentially streamlined.

b) Channels are thought to develop at, or close to a dendritic growth front, rather than within the body of the mushy zone. A liquid perturbation model suggests that a characteristic dimension or wavelength corresponds to the primary interdendritic spacing.

c) The phenomenon is regarded as an example of double diffusive convection and, as such, may be regarded as common to similar effects in a variety of other materials systems.

ACKNOWLEDGMENT

This article is based upon a research program supported by NASA through the NASA-Lewis Research Center, Grant #NAG-3-560.

REFERENCES

1. Moore, J.J. and N.A. Shah. *International Metals Reviews,* **28,** 338 (1983).
2. McDonald, R.J. and J.D. Hunt. *Trans. TMS-AIME,* **245,** 1993 (1969) and *Metall. Trans. A.,* **1A,** 1787 (1970).
3. Copley, S.M., A.F. Giamei, S.M. Johnson and M.F. Hornbecker. *Metall. Trans. A,* 1970, **1A,** 2193 (1970).
4. Fisher, K.M. and J.D. Hunt. *Sheffield International Conference on Solidification and Casting,* Metals Society, Book **193,** 235 (1977).

364

5. Bridge, M.R., M.P. Stephanson and J. Beech. *Metals Technology,* **9,** 429 (1982).
6. Mehrabian, R., M. Keane and M.C. Flemings. *Metall. Trans. A,* **1A,** 1209 (1970).
7. Sample, A.K. and A. Hellawell. *Metall. Trans. B,* **13B,** 495 (1982).
8. Sample, A.K. and A. Hellawell. *Modelling of Casting and Welding Processes, II,* Metals Soc. AIME. 119, (1983).
9. Hellawell, A. *Structure and Dynamics of Partially Solidified Systems,* D.E. Loper, ed., NATO ASI Series E, 125, p. 3 (1987).
10. Sample, A.K. and A. Hellawell. *Metall. Trans. A,* **15A,** 2163 (1989).
11. Sample, A.K. M.S. Thesis, Michigan Technological University (1984).
12. Sarazin, J.R. M.S. Thesis, Michigan Technological University (1986).
13. Sarazin, J.R. and A. Hellawell. *Metall. Trans. A,* **19A,** 1861 (1988).
14. Sharp, R.M. and A. Hellawell. *J. Crystal Growth,* **12,** 261 (1972).
15. Benard, H. *Revue Generale des Sciences Pures et Appliquees,* **11,** 1261 and 1309 (1901).
16. Rayleigh, Lord. *Phil. Mag.,* 1916, **32,** 529 (1916).
17. Coriell, S.R., M.R. Cordes, W.J. Boettinger and R.F. Sekerka. *J. Crystal Growth,* 1980, **49,** 13 (1980).
18. Hurle, D.T.J., E. Jakeman and A.A. Wheeler. *J. Crystal Growth,* **58,** 163 (1982).
19. McFaddan, G.B., R.G. Rehm, S.R. Coriell, W. Chuck and K.A. Morrish. *Metall. Trans. A,* **15A,** 2121 (1984).
20. Stern, M.E. *Tellus,* **12,** 172 (1960).
21. Turner, J.S. *Rev. Fluid Mechanics,* **17,** 11 (1985).
22. Huppert, H.E. and J.S. Turner. *J. Fluid Mechanics,* **100,** 367 (1980).
23. Schmitt, R.W. *Phys. Fluids,* **26,** 2373 (1983).

16

Convective and Morphological Instabilities during Directional Solidification

S.R. CORIELL AND G.B. McFADDEN

National Bureau of Standards, Gaithersburg, MD 20899, U.S.A.

ABSTRACT

During the directional solidification of a binary alloy at constant velocity, solute segregation may arise due to either interface instability or fluid flow in the melt. Recent calculations of cellular shapes in the absence of fluid flow and thermosolutal convection caused by the rejection of a lighter solute during growth vertically upwards are presented. Three-dimensional steady-state solutions for nonplanar interface morphologies are computed numerically by finite differences. A linear temperature field is assumed, and the solute field in the melt and the unknown crystal-melt interface position are obtained self-consistently. For a model of an aluminum-chromium alloy with a distribution coefficient greater than unity, the calculations predict hexagonal nodes near the onset of instability. Numerical results for the solute segregation caused by thermosolutal convection are obtained using finite differences in a two-dimensional, time-dependent model that assumes a planar-crystal melt interface. The system is assumed periodic in the horizontal direction, and the possibility of multiple flow states sharing the same period is examined. As the solutal Rayleigh number is varied, multiple steady states, time-periodic states, and quasi-periodic states occur.

1. INTRODUCTION

During solidification and crystal growth from the melt, the crystal-melt interface is subject to morphological instability [1–8]. Under conditions for which unstable interfaces occur, the interface morphology is sometimes cellular, but dendritic (tree-like) growth usually occurs when the degree of instability is large. The solute distribution in the crystal influences the properties of the crystal, and since solid state diffusion is usually very slow, this solute distribution is determined primarily by the solute distribution in the melt at the crystal-melt interface. Thus, the interface morphology and fluid flow in the melt play a central role in determining the properties of the solidified material.

During directional solidification of a single phase solid at constant velocity from a binary liquid melt, there is uniform relative motion of the sample and its thermal environment. Temperature gradients normal to the crystal-melt interface typically of the order of 100 K/cm are used in order to remove the latent heat of fusion; large temperature gradients also help to prevent morphological instability of the crystal-melt interface. Even in vertical growth it is difficult to avoid horizontal temperature gradients since the container walls are not perfectly adiabatic; the release of latent heat and the different thermal conductivities of the container, melt, and crystal give rise to horizontal temperature gradients which drive fluid flow [9]. Even in the absence of horizontal gradients during vertical growth, thermosolutal convection may occur in a binary alloy. Since the equilibrium solute concentration in the melt is different from that in the crystal, the solidification process causes a solute gradient in the melt in front of the crystal-melt interface. For solidification at constant velocity V, the solute gradient is exponential in distance from the interface with a decay distance of D/V, where D is the solute diffusion coefficient in the liquid. The temperature gradient is also exponential with a characteristic distance $\varkappa/V$, where $\varkappa$ is the thermal diffusivity. However, the thermal diffusivity is usually so large compared to the solute diffusivity that the temperature gradient is essentially constant over the solute distance D/V. For growth vertically upward, the temperature field provides a stabilizing influence on convection for a normal liquid, which expands on heating. The solute field can be either stabilizing or destabilizing depending on whether solute is rejected from or preferentially incorporated in the crystal and depending on whether it increases or decreases the density of the melt. The convective instability during directional solidification differs from conventional analyses of double diffusive convection [10] in that the solute gradient is exponential rather than linear and the crystal-melt interface is a free boundary.

In the following, we shall discuss recent three-dimensional, steady-state calculations of cellular morphologies for processing conditions near the onset of morphological instability in the absence of fluid flow in the melt. Such calculations predict the amount of solute segregation and provide information about the pattern of the segregation, e.g., whether hexagonal cells or two-dimensional rolls occur. We then discuss the effect of fluid flow in the melt on solute segregation with particular emphasis on double diffusive convection caused by the rejection of a light solute during direction solidification vertically upwards. Numerical results have been obtained using finite differences in a two-dimensional time-dependent model that assumes a planar crystal-melt interface. As the driving force for convection is varied, multiple steady states, time-periodic states, and quasi-periodic states occur.

2. SOLIDIFICATION BOUNDARY CONDITIONS

The boundary conditions (see, for example [4, 6, 8]) at the crystal-melt interface are

$$u.t_1 = u.t_2 = 0 \tag{1}$$

$$v.n\,(\varrho_s - \varrho_L) = (u.n)\varrho_L \tag{2}$$

$$(v.n)\,L_v = (-k_L\nabla T + k_s\nabla T_s).n \tag{3}$$

$$(v.n)\,(c_s - c) = (\varrho_L/\varrho_S)D\nabla c \cdot n \tag{4}$$

$$T_s = T = T_M + mc - T_m\Gamma K \tag{5}$$

$$c_S = kc \tag{6}$$

Here, v and u are the local velocities, measured with respect to the crystal, of the crystal-melt interface and the fluid, respectively; t_1, t_2, and n are unit tangent vectors and the unit normal vector to the crystal-melt interface, respectively; ϱ_s and ϱ_L are densities of crystal and melt, respectively; L_v is the latent heat of fusion per unit volume of solid; T_s and T are the temperature in the crystal and melt, respectively; k_s and k_L are the thermal conductivities of crystal and melt, respectively; c_S and c are the solute concentrations in the crystal and melt, respectively; T_M is melting point of pure material with a planar interface; m is the change of melting point with solute concentration; Γ is a capillary constant; K is the mean curvature of the crystal-melt interface; and k is the distribution coefficient. The first four boundary conditions follow from conservation laws. The last two boundary conditions assume local equilibrium at the crystal-melt interface; this assumption appears to be a good approximation for most metals at low growth velocities. For faceted materials and for high growth velocities, the last two boundary conditions require modification (see, for example, [11-13]). In general, the capillary constant and non-equilibrium effects will depend on crystallographic orientation, and the liquidus slope, m, and the distribution coefficient, k, will depend on concentration. We have also neglected diffusion in the crystal and cross-coupling of the heat and solute fluxes.

3. MORPHOLOGICAL STABILITY IN THE ABSENCE OF FLOW

It is well known that, for sufficiently high solute concentrations, a planar crystal-melt interface is morphologically unstable [1] and will develop a cellular or dendritic morphology with concomitant solute segregation (microsegregation). Linear stability analysis shows that the planar interface is stable if

$$G^*/mG_c > S(A, k) \tag{7}$$

with

$$G^* = \frac{k_S G_S + k_L G_L}{k_S + k_L} \tag{8}$$

and

$$A = \frac{k^2 T_M \Gamma V}{(k-1)\,mc_\infty D} \tag{9}$$

Here $G_c = Vc_\infty (k-1)/(Dk)$ is the unperturbed solute gradient at the planar interface in the melt, c_∞ is the solute concentration in the melt far from the interface, and G_L and $G_s = (k_L G_L + VL_v)/k_s$ are unperturbed temperature gradients at the interface in the melt and crystal, respectively. G_L is positive if the temperature increases with distance into the melt. The function $S(A, k)$ depends on the two dimensionless parameters A and k and can be calculated from the roots of a cubic polynomial [14]. For small solidification velocities, $A \ll 1$ and $S(A, k) \approx 1$, so that the approximate stability criterion is $G^*/mG_c > 1$, which is called the modified constitutional supercooling criterion. For large solidification velocities, $A > 1$ and $S = 0$, and the interface is stable provided $G^* > 0$. With local equilibrium at the crystal-melt interface, the onset of instability is non-oscillatory in time and the wavelength of the instability decreases with solidification velocity. For example, for the solidification of aluminum containing copper with a temperature gradient in the melt of 200 K/cm, the wavelength at the onset of instability ranges from 0.06 cm to 0.0002 cm as the solidification velocity varies from 10^{-4} cm/s to 10 cm/s [7]. Morphological stability theory has recently been extended to the solidification of multicomponent alloys for a simple model in which each solute acts independently [15], and it was shown that under fairly general conditions there are no linearly unstable modes that are oscillatory in time.

The onset of morphological instability can be oscillatory in time due to anisotropy of interface kinetics [16], velocity dependence of the distribution coefficient [17], or fluid flow in the melt [3, 6, 8]. Oscillatory instability in the form of travelling waves has been observed during the growth of gallium-doped germanium [18].

4. CELLULAR GROWTH

Recently there has been renewed theoretical interest in the highly nonplanar morphologies that occur during crystal growth and alloy solidification. We will discuss some aspects of cellular growth during directional solidification of a binary alloy in the absence of fluid flow. For interface shapes that do not differ too much from planarity, the free boundary problem may be treated by nonlinear expansion techniques [4, 7, 19–22]. Numerical calculations of interface shapes that differ significantly from planarity have also been carried out [23–31].

Three-dimensional steady-state solutions for nonplanar interface morphologies have been computed numerically by using finite differences [31]. The solute field in the melt and the unknown crystal-melt interface position are obtained self-consistently. An example of a steady state cellular interface and the resulting solute distribution is shown in Fig. 1 for an aluminum-chromium alloy. In these calculations, the thermal properties of crystal and melt are assumed to be equal and the latent heat is neglected so that the temperature field is linear. The distribution coefficient k for chromium in aluminum is 1.8 so that the solute distribution $c(x,y)$ at the crystal-melt interface and the interface shape $H(x,y)$ are

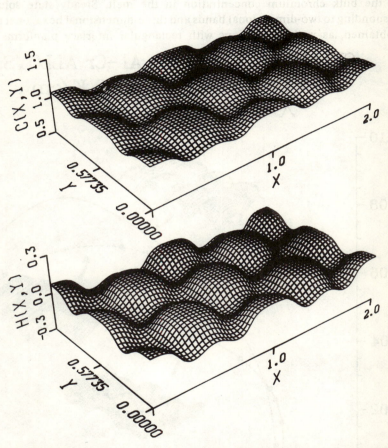

CELLULAR GROWTH Al–Cr ALLOYS
$V = 0.001$ cm/s $G_L = 10$ K/cm
$C_\infty = 1.03C^* = 0.275$ wt.%
$\lambda = \lambda^* = 0.0184$ cm

Fig. 1. The solute distribution (upper plot) and interface shape (lower plot) during the directional solidification of an aluminum alloy containing chromium. The growth velocity is 0.001 cm/s, the temperature gradient in the liquid is 10 K/cm, the bulk concentration of chromium is 0.275 wt. percent, and the wavelength of the cellular interface is 0.0184 cm. From [31].

similar. The computational domain was chosen so that hexagonal symmetry is possible, and, in fact, occurs for the case studied in Fig. 1. Note that if the appropriate solute maxima are connected to form hexagons, there is also a solute maxima at the center of each hexagon; we denote this pattern as hexagonal nodes.

370

A hexagonal pattern with solute maxima only at the corners of the hexagon will be denoted as hexagonal cells.

The results of a large number of calculations for aluminum-chromium alloys at a solidification velocity of 0.001 cm/s, a temperature gradient in the melt of 10 K/cm, and a cellular spacing corresponding to the wavelength at the onset of linear morphological instability are summarized by the bifurcation diagram in Fig. 2 in which $|d|$ is a measure of the amplitude of the interface deformation and c_∞ is the bulk chromium concentration in the melt. Steady state solutions corresponding to two-dimensional bands and three-dimensional hexagonal nodes are obtained, as well as solutions with rectangular interface planforms. The

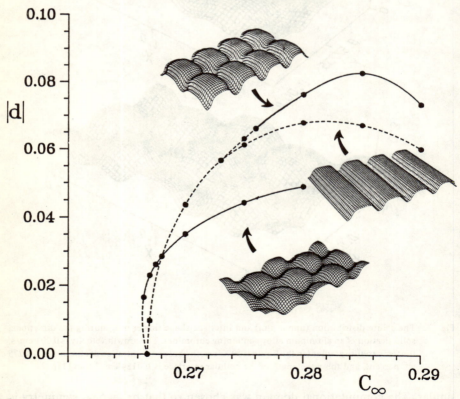

CELLULAR GROWTH Al–Cr ALLOYS
V = 0.001 cm/s G_L = 10 K/cm
$\lambda = \lambda^* = 0.0184$ cm

Fig. 2. The interface deformation as a function of the bulk concentration (in wt.%) of chromium in aluminum for hexagonal solutions, two-dimensional solutions, and rectangular solutions. Unstable steady states are indicated by dashed curves, and stable steady states are indicated by solid curves. From [31].

stability of the computed steady-state solutions is suggested by the behavior of the iterative scheme used in the calculation; successive iterates may be viewed as evolving in a time-like manner.

For sufficiently low concentrations, the only steady state is the planar interface. As the concentration increases, the planar interface becomes unstable, and there is a finite amplitude hexagonal solution. There are no infinitesimal amplitude stable solutions, and the lowest amplitude stable solution occurs at a bulk concentration for which linear theory would predict stability. Near the onset of instability, the stable steady-state solution is three-dimensional and (in terms of the solute distribution) corresponds to nodes rather than cells. As the degree of instability increases (increasing solute concentration), the amplitude of the interface deformation of the node solution increases, until the node solution becomes unstable. For a small range of concentrations, there are two stable steady-state solutions, the node solution and a rectangular solution, which arises from a secondary bifurcation from the unstable two-dimensional roll solutions. The rectangular solutions become stable at a bulk concentration that is 3.6% above the critical concentration for linear instability and remain stable for the remaining range covered by our calculations (to 8.9% above the critical concentration). For the parameter range of the calculations, two-dimensional and cell-like (rather than node-like) solutions are not stable.

The results near the onset of instability could have been anticipated from weakly nonlinear theory. In Rayleigh-Benard convection, it is necessary to introduce a temperature-dependent viscosity to produce vertical asymmetry and produce stable hexagonal solutions [32]. The exponential solute field in the directional solidification problem has inherent vertical asymmetry, and the weakly nonlinear theory suggests stable node-like solutions for all values of the distribution coefficient under the assumption of a linear temperature field.

Since cellular spacings are generally very much smaller than sample dimensions, a computational domain was used with periodic boundary conditions corresponding to the wavelengths of the anticipated cell spacings. In general, it is possible to find a family of steady state solutions (interface shape, temperature and concentration fields) for a range of wavelengths. The selection problem is magnified in three-dimensional calculations where the two wavelengths in the directions transverse to the growth direction both need to be specified. Many of our calculations have been carried out at the critical wave number from linear theory with the ratio of the transverse wavelengths chosen to allow hexagonal solutions. We have also calculated a few solutions on a square planform, and it seems likely that there are stable steady-state solutions for a wide range of the two transverse wavelengths. The mechanism of wavelength and pattern selection is an unsolved fundamental problem.

Young et al. [33] have treated nonlinear effects for anisotropic interface kinetics in the limit of very small segregation coefficients. The presence of anisotropy causes cells to propagate at an angle to the growth direction.

372

5. COMBINED DOUBLE DIFFUSIVE AND MORPHOLOGICAL INSTABILITIES

The onset of coupled morphological and double diffusive instabilities of the quiescent base state during directional solidification of a binary alloy at constant velocity has been determined by linear stability analysis [34–39]. Specific calculations have been carried out for dilute concentrations of tin in lead and ethanol in succinonitrile; for both systems, the solute is rejected at the crystal-melt interface and decreases the density of the melt so that the solute field is destabilizing with respect to convection. The concentration of tin in lead at the onset of instability as a function of increasing growth velocity exhibits a minimum, followed by a maximum, and then another minimum (see Fig. 2 of reference [8]). For small solidification velocities, the onset of instability is via a convective mode, while for large solidification velocities, it is via a morphological mode. The qualitative behavior of the morphological mode can be understood by recognizing that the onset of instability is determined by a balance between the destabilizing solute gradient and the combined stabilizing effect of the positive temperature gradient and the crystal-melt surface tension. The solute gradient is proportional to the interface velocity, and this destabilizing influence increases with velocity. However, at velocities greater than about 1 cm/s, the wavelength at the onset of instability becomes sufficiently small that the stabilizing role of the crystal-melt surface tension becomes important. In other words, for small velocities the temperature gradient is the dominant stabilizing influence while for large velocities the surface tension is dominant. Similarly, the qualitative behavior of the convective mode can be understood by recognizing that the relevant length scale is D/V, that the solute gradient is destabilizing, and that viscosity and the temperature gradient are stabilizing. At low velocities, stability is determined by a balance of the solute gradient and the temperature gradient, and at large velocities by a balance of the solute gradient and viscous forces. The convective mode is obviously a strong function of the gravitational field, whereas the morphological mode is relatively insensitive to gravitational field. Very small solute concentrations give rise to convection, for example, at a solidification velocity of 10^{-4} cm/s, the critical concentration is $3.0\,(10^{-4})$ wt.% tin. For the lead-tin and succinonitrile-ethanol alloys, the wavelength of the convective instability is usually an order of magnitude greater than that of the morphological instability, and the coupling between them is relatively weak. In general, the onset of instability is non-oscillatory in time; however, near the crossover point where the critical concentrations for convective and morphological instabilities are comparable, the onset of instability can be oscillatory [38, 39].

The effect of a forced Couette flow, parallel to the horizontal crystal-melt interface, on the onset has also been treated [40]. Such a flow does not affect perturbations with wave vectors perpendicular to the flow. For perturbations with wave vectors parallel to the flow, the onset of morphological instability is somewhat suppressed and double diffusive instability is greatly suppressed. When instabilities occur, they are oscillatory and correspond to travelling waves. For values of the crystal growth velocity for which coupled morphological and convective modes occur, the presence of a forced flow produces sufficient decoupling to allow otherwise degenerate branches to be identified.

Young and Davis [41] have recently studied the effect of buoyancy on morphological stability for alloys with very small distribution coefficients. A weakly nonlinear analysis of steady solutal convection has also been carried out [42].

The fact that morphological and convective length scales can differ by orders of magnitude has so far precluded numerical calculations of the nonlinear aspects of the coupled problem. Useful results may sometimes be obtained by assuming that the morphological and convective modes are uncoupled. Experiments of Glicksman and colleagues, combined with linear stability analysis, provide an example in which this decoupling is not valid [8, 43–45].

6. DOUBLE DIFFUSIVE CONVECTION DURING DIRECTIONAL SOLIDIFICATION

Numerical calculations of the temperature, solute, and fluid flow fields in the melt and the temperature field in the solid have been carried out by using finite differences in a two-dimensional, time-dependent model that assumes a small Prandtl number and a planar crystal-melt interface [46, 47]. The small Prandtl number of liquid metals and semiconductors, viz., of the order of 0.01, allows the equation for the temperature variation from the base state to be approximated by a linear diffusion equation. The assumption of a planar interface precludes morphological instability, and requires relaxation of the usual solidification boundary condition; in particular we do not require that the interface temperature equal the equilibrium temperature. Linear stability calculations have verified the range of validity of this approximation with respect to the onset of convective instability. An example of these linear stability calculations are shown in Fig. 3 for the lead-tin system at a growth velocity of 2.0 (10^{-4}) cm/s and a temperature gradient in the melt of 200 K/cm. The tin concentration (in units of the concentration at the onset of convection) is shown as a function of the wavelength λ (in units of the diffusion length D/V) of a sinusoidal perturbation. As the concentration increases, the horizontal wavelength of the fastest growing mode decreases rapidly. A higher eigenmode is also shown in the figure.

Typical values of the Schmidt number for metals and semiconductors are of the order of 100 and 10, respectively; calculations have been carried out for Schmidt number of 81 (corresponding to lead), 10, and unity. The flow is assumed to be periodic in the horizontal direction with a given period. To save computer time the calculations are performed in one-half of a period in the horizontal direction, with planes of symmetry imposed at each lateral boundary. These symmetry planes admit an alternative interpretation, in that they may also be considered to represent idealized, perfectly insulating, stress-free side walls.

LINEAR THEORY FOR LEAD–TIN SYSTEM

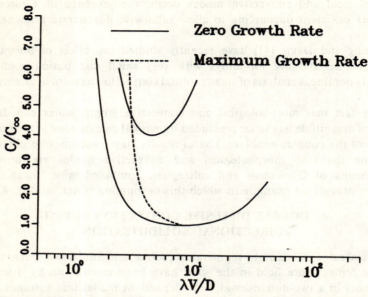

Fig. 3. The normalized concentration of tin in lead at the onset of instability during directional solidification at 2.0 μm/s and a temperature gradient in the liquid of 200 K/cm as a function of the dimensionless wavelength of a sinusoidal perturbation. Two eigenmodes and the maximum growth rate are shown.

With the assumption of a planar interface, linear stability calculations for a destabilizing solute field and a stabilizing temperature field (the salt finger regime of double diffusive convection [10]) indicate that the onset of convection is not oscillatory in time, and this is in agreement with the finite difference calculations in which steady-state convection is found. However, as the solutal Rayleigh number increases, the flow becomes periodic in time. We have carried out a detailed study of the flow transitions which occur for a Schmidt number equal to unity; a more limited number of calculations for Schmidt numbers of 10 and 81 have shown the same qualitative behaviour.

A bifurcation diagram for steady-state solutions for a Schmidt number of unity is shown in Fig. 4. A measure of the intensity of the flow (the maximum of the stream function over all spatial positions) is plotted as a function of the driving force for convection, viz the solutal Rayleigh number, Rs. All other parameters are kept constant in these calculations. For $Rs < Rs^*$, the motionless state is stable, and the stream function vanishes. At $Rs = Rs^*$, the motionless state is unstable to infinitesimal perturbations, and as Rs increases, the maximum value of the stream function increases. The dots along the Rs axis indicate eigenvalues from linear stability calculations; for example, at $Rs = 1.67 \, Rs^*$, the motionless

Time–Independent States

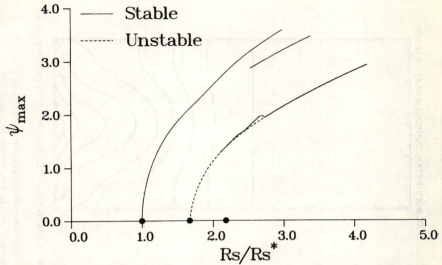

Fig. 4. The maximum value of the stream function as a function of the solutal Rayleigh number for time-independent flow states. The points indicated along the Rs/Rs^* axis are eigenvalues from linear stability theory Stable solutions are shown as solid curves and unstable solutions are shown as dashed curves. From [47].

state becomes unstable to perturbations with one-half the basic wavelength. However, the nonlinear flow state which bifurcates from $Rs = 1.67\ Rs^*$ is unstable with respect to perturbations of the basic wavelength. The unstable (dashed) portion of the curve was calculated from our time dependent code by not allowing such perturbations. At $Rs = 2.1\ Rs^*$, the nonlinear flow state with one-half the basic wavelength becomes stable. Above $2.5\ Rs^*$ another steady-state solution becomes stable; this solution has the basic wavelength, but a different structure in the vertical direction and is presumably related to the second eigenmode of the basic wavelength, which for the motionless state occurs at $2.18\ Rs^*$. Above the Rs values indicated by the solid curves, the flow becomes periodic in time. For $Rs \approx 6\ Rs^*$, we have calculated quasi-periodic flows with two dominant periods.

When the horizontal wavelength of the convective flow is comparable to the lateral dimensions of the sample, boundary conditions for rigid walls are more appropriate. We have carried out such calculations for lateral walls that are perfectly insulating with respect to solute and perfectly conducting with respect to heat. One of the interesting results is that perfectly conducting walls can actually destabilize the system, i.e., a decrease in the separation of the sidewalls may lower the solutal Rayleigh number at which convection first occurs [48]. The qualitative behavior of convection is similar to that for periodic boundary conditions on the lateral walls in that multiple steady states and time-periodic

376

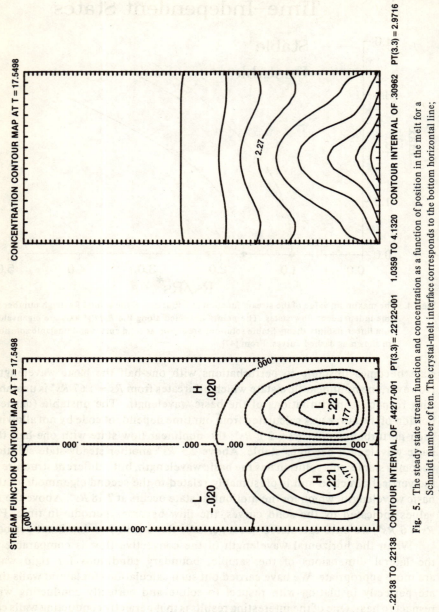

Fig. 5. The steady state stream function and concentration as a function of position in the melt for a Schmidt number of ten. The crystal-melt interface corresponds to the bottom horizontal line; the horizontal and vertical tick marks indicate the mesh used in the calculations.

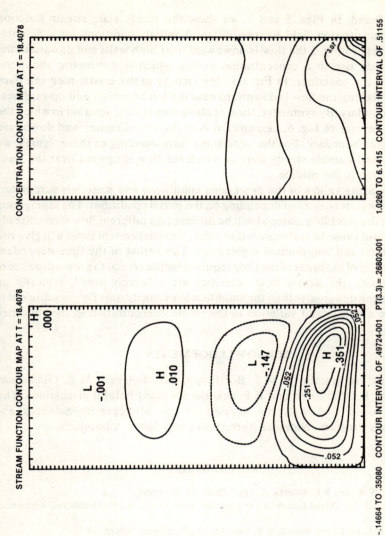

Fig. 6. The steady state stream function and concentration as a function of position in the melt for a Schmidt number of ten. The values of processing conditions and thermophysical properties are identical to Fig. 5; only the initial conditions are different from those of Fig. 5.

states are found. In Figs. 5 and 6, we show the steady state stream function and the concentration field for two different initial conditions for a Schmidt number of ten. In Fig. 5 the flow is downward near both walls and upward in the middle, giving rise to a concentration profile which is symmetric about the midplane of the container. In Fig. 6 in the vicinity of the crystal-melt interface (bottom of figure), the flow is downward near the left boundary and upward near the right boundary. By symmetry, there is also a steady state solution in which the flow is the reverse of Fig. 6, i.e., upward near the left boundary and downward near the right boundary. For the conditions corresponding to these figures, we have not found a stable steady state in which the flow is upward near the walls and downward in the middle.

For the same values of the processing conditions and materials parameters different flow states can exist depending on the initial conditions. The segregation of solute in the solidified material will be different for different flow states. Steady state flows will cause lateral segregation, while time dependent flows will give rise to both lateral and longitudinal segregation. The period of the time-dependent flows is usually of the order of the time required for the crystal to grow a distance of D/V. Although the actual flow velocities are relatively small, typically an order of magnitude greater than the solidification velocity, this flow is sufficient to give rise to a 50% lateral variation in the solute concentration in the solidified material.

ACKNOWLEDGEMENTS

The authors are grateful to W.J. Boettinger, R.F. Boisvert, M.E. Glicksman, R.G. Rehm, R.J. Schaefer, and R.F. Sekerka for many helpful discussions. This work was conducted with the support of the Microgravity Sciences and Applications Program, National Aeronautics and Space Administration.

REFERENCES

1. Mullins, W.W. and R.F. Sekerka. *J. Appl. Phys.*, **34**, 323 (1964).
2. Sekerka, R.F. In: *Crystal Growth: An Introduction*, P. Hartman, ed., North-Holland, Amsterdam (1973).
3. Delves, R.T. In: *Crystal Growth*, B.R. Pamplin, ed., Pergamon, Oxford (1974).
4. Wollkind, D.J. In: *Preparation and Properties of Solid State Materials*, W.R. Wilcox, ed., Vol 4, Marcel Dekker, New York (1979).
5. Langer, J.S. *Rev. Mod. Phys.* **52**, 1 (1980).
6. Coriell, S.R. and R.F. Sekerka. *Physico Chem. Hydrodyn.* **2**, 281 (1981).
7. Coriell, S.R., G.B. McFadden and R.F. Sekerka. *Ann. Rev. Mater. Sci.* **15**, 119 (1985).
8. Glicksman, M.E., S.R. Coriell and G.B. McFadden. *Ann. Rev. Fluid Mech.* **18**, 307 (1986).
9. Adornato, P.M. and R.A. Brown. *J. Crystal Growth*, **80**, 155 (1987).
10. Turner, J.S. *Buoyancy Effects in Fluids*, Cambridge University Press, Cambridge (1973).
11. Boettinger, W.J., S.R. Coriell and R.F. Sekerka. *Mat. Sci. Eng.* **65**, 27 (1984).
12. Parker, R.L. *Solid State Phys.* **25**, 151 (1970).
13. Caroli, B., C. Caroli and B. Roulet. *J. Crystal Growth* **66**, 575 (1984).
14. Sekerka, R.F. *J. Appl. Phys.* **36**, 264 (1965).

15. Coriell, S.R., G.B. McFadden, P.W. Voorhees and R.F. Sekerka. *J. Crystal Growth* **82**, 295 (1987).
16. Coriell, S.R. and R.F. Sekerka, *J. Crystal Growth* **34**, 157 (1976).
17. Coriell, S.R. and R.F. Sekerka. *J. Crystal Growth* **61**, 499 (1983).
18. Singh, R., A.F. Witt and H.C. Gatos. *J. Electrochem. Soc.* **121**, 380 (1974).
19. Sriranganathan, R., D.J. Wollkind and D.B. Oulton. *J. Crystal Growth* **62**, 265 (1983).
20. Caroli, B., C. Caroli, and B. Roulet. *J. Crystal Growth* **68**, 677 (1984).
21. Wheeler, A.A. *IMA J. Appl. Math.* **35**, 131 (1985).
22. Alexander, J.I.D., D.J. Wollkind and R.F. Sekerka. *J. Crystal Growth* **79**, 849 (1986).
23. Ungar, L.H. and R.A. Brown. *Phys. Rev.* **B29**, 1367 (1984).
24. McFadden, G.B. and S.R. Coriell. *Physica* **12D**, 253 (1984).
25. Ungar, L.H. and R.A. Brown. *Phys. Rev.* **B30**, 3993 (1984).
26. Ungar, L.H., M.J. Bennett and R.A. Brown. *Phys. Rev.* **B31**, 5923 (1985).
27. Ungar, L.H. and R.A. Brown. *Phys. Rev.* **B31**, 5931 (1985).
28. Wheeler, A.A. *Quart J. Mech. Appl. Math.* **39**, 381 (1986).
29. Kelly, F.X. and L.H. Ungar. *Phys. Rev.* **B34**, 1746 (1986).
30. Hunt, J.D. and D.G. McCartney. *Acta Metall,* **35**, 89 (1987).
31. McFadden, G.B., R.F. Boisvert and S.R. Coriell. *J. Crystal Growth* **84**, 371 (1987).
32. Segel, L.A. In: *Non-equilibrium Thermodynamics Variational Techniques and Stability,* R.J. Donnelly, R. Herman and I. Prigogine, eds., University of Chicago Press, Chicago (1966).
33. Young, G.W., S.H. Davis and K. Brattkus. *J. Crystal Growth* **83**, 560 (1987).
34. Coriell, S.R., M.R. Cordes, W.J. Boettinger and R.F. Sekerka. *J. Crystal Growth* **49**, 13 (1980).
35. Coriell, S.R., M.R. Cordes, W.J. Boettinger and R.F. Sekerka. *Adv. Space Res.* **1**, 5 (1981).
36. Hurle, D.T.J., E. Jakeman and A.A. Wheeler. *J. Crystal Growth* **58**, 163 (1982).
37. Hurle, D.T.J., E. Jakeman, and A.A. Wheeler. *Phys. Fluids,* **26**, 624 (1983).
38. Schaefer, R.J. and S.R. Coriell. *Metall. Trans.* **15A**, 2109 (1984).
39. Caroli, B., C. Caroli, C. Misbah and B. Roulet. *J. Phys.* (Paris) **46**, 401 (1985).
40. Coriell, S.R., G.B. McFadden, R.F. Boisvert and R.F. Sekerka. *J. Crystal Growth,* **69**, 15 (1984).
41. Young, G.W. and S.H. Davis. *Phys. Rev.* **B34**, 3388 (1986).
42. Jenkins, D.R. *PhysicoChem. Hydrodyn.* **6**, 521 (1985).
43. Coriell, S.R., G.B. McFadden, R.F. Boisvert, M.E. Glicksman and Q.T. Fang. *J. Crystal Growth* **66**, 514 (1984).
44. McFadden, G.B., S.R. Coriell, R.F. Boisvert, M.E. Glicksman and Q.T. Fang. *Metall. Trans.* **15A**, 2117 (1984).
45. Fang, Q.T., M.E. Glicksman, S.R. Coriell, G.B. McFadden and R.F. Boisvert. *J. Fluid Mech.* **151**, 121 (1985).
46. McFadden, G.B., R.G. Rehm, S.R. Coriell, W. Chuck and K.A. Morrish. *Metall. Trans.* **15A**, 2125 (1984).
47. McFadden G.B. and S.R. Coriell. *Phys. Fluids* **30**, 659 (1987).
48. McFadden, G.B., S.R. Coriell and R.F. Boisvert. *Phys. Fluids* **28**, 2716 (1985).

17

Phase Selection in Non-equilibrium Processing

R.J. SCHAEFER

National Bureau of Standards, Gaithersburg, MD 20899, U. S. A.

ABSTRACT

When an alloy is solidified rapidly, it is often thermodynamically possible for several different phases to form. Which of these phases dominates the final microstructure depends on the nucleation and growth kinetics of the several possible phases, and also on the detailed thermal conditions of the specific solidification process. These effects are discussed with reference to the aluminum-manganese system, where the results of several types of experiments can be understood in terms of the strikingly different nucleation and growth behavior of several stabale and metastabale phases.

1. INTRODUCTION

The first stage in understanding the microstructure resulting from a solidification process is to know which phases are present. In most alloy systems more than one phase will be present when solidification is complete, and to understand the development of the microstructure one needs to know which phase forms first and how it affects the subsequent formation of other phases. This paper will address these problems without discussing the detailed morphology of the phases.

While the phase diagram shows us which phases are present at equilibrium, even moderate rates of solidification lead to substantial deviations from equilibrium, and during rapid solidification processes these deviations become extreme. One must therefore consider not only the equilibria but also the kinetics of the many different phase transformation processes which lead to the solidification microstructure.

A convenient representation of an alloy phase selection process would be one such as that shown in Fig. 1, where the phases produced by solidification are mapped on a plot of alloy composition versus a processing parameter

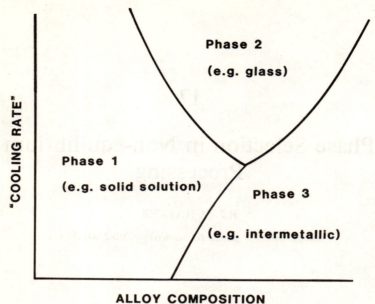

ALLOY COMPOSITION

Fig. 1. A schematic phase selection map.

such as the cooling rate. Although this type of plot may provide some general guidelines, it cannot apply to the many different types of solidification processes to which an alloy might be subjected. A major obstacle to the development of this type of representation is the general impossibility of describing the processing parameter by a single variable as indicated in Fig. 1. In real processes the temperature within a liquid metal normally varies with time and space in some complex manner. Moreover, the response of the liquid to this variation will depend on the presence or absence of solid phases which may act as substrates or nucleation sites for crystal growth, and will also depend on the geometry of the sample which may or may not allow nucleated crystals to propagate over large distances.

Nucleation, growth, and polyphase reactions all play roles in phase selection processes. Most (but not all) rapid solidification processes start with a nucleation event which may be either homogeneous or heterogeneous. The first phase of the alloy·system to nucleate will frequently serve as a heterogeneous nucleant for other phases.

Growth of a crystalline phase in a rapidly cooling alloy will usually occur at substantial supercooling, strongly dependent upon the crystallographic character of the phase and its composition relative to that of the melt. As the temperature drops, the solidification of other phases may become possible. Peritectic or eutectic reactions, however, become less likely at rapid solidification rates. Solid state transformations may also occur if the cooling rate following solidification is not large enough to suppress them.

Thus when considering the phases formed during rapid solidification, the supercooling is the relevant parameter in determining nucleation rates, the velocity with which the isotherms move can determine the type of solidification front which becomes established, and the cooling rate can determine how far one solidification process progresses before another becomes possible.

Three major types of rapid solidification processes may be considered. In atomization processes, droplets of the alloy, typically $< 100\ \mu$m in diameter, are isolated from one another and a separate nucleation event is required to start crystal growth in each particle. After nucleation, the particles may undergo substantial recalescence before continuing to cool.

During resolidification following surface melting by lasers or electron beams, the presence of a crystalline substrate eliminates the need for an initial nucleation event. Under these circumstances, the resolidification of the molten layer is controlled largely by the rate at which the isotherms move toward the surface (the solidification velocity) as the heat source is removed. If this rate becomes large, it is possible for large supercoolings to develop within the liquid.

In melt spinning or other processes in which liquid metal impinges onto a cold surface, the thermal conditions are more difficult to determine but both large supercooling and a high rate of isotherm motion may be involved.

2. NUCLEATION PROCESSES

Nucleation of a crystal phase in a supercooled melt is almost always a heterogeneous process, i.e., it occurs on a substrate of some other crystalline phase, at a supercooling smaller than that which would be required in the absence of the substrate. The identity of these substrates is rarely known, but oxide films or small "motes" of refractory compounds are frequently suspected. Whatever their nature, these nucleants are finite in number and their action can sometimes be ameliorated by atomization of the liquid into small droplets. By this process the nucleants are isolated into a small fraction of the droplets and the remaining majority may then be supercooled to a higher level until nucleation occurs by the action of some less potent nucleant, or homogeneously. In a classic series of experiments, Turnbull [1] used this method to obtain high supercoolings in pure liquid metals. More recently, Perepezko and his students [2] have made extensive studies of supercooled alloy droplets and have been able to measure nucleation temperatures of metastable as well as stable alloy phases. In these experiments, very high levels of supercooling were obtained even in slowly cooled alloys by emulsifying the melt in a suitable liquid medium.

Although very careful experiments are able to measure nucleation rate as a function of temperature [3], in most cases the nucleation rate varies so rapidly with temperature that it appears to be zero above some temperature T_n and extremely high below this temperature, with the result that one can effectively consider T_n to be a fixed temperature at which nucleation occurs. Moreover, it is sometimes found that the phase with the highest nucleation temperature is not the phase with the highest equilibrium melting temperature. Thus the phase which nucleated in supercooled bismuth droplets was found to correspond to a phase which at equilibrium exists only at high pressure [4].

Experiments such as those of Turnbull and Perepezko employing slow cooling can be used to make quite precise measurements of nucleation temperatures in supercooled alloy droplets, and it is usually found that the smaller the size of the droplets, the larger the degree of supercooling attained. Because the cooling rate in these experiments is independent of droplet size, the larger supercooling of the smaller droplets must be attributed to more effective isolation of the heterogeneous nucleants. In more rapid metal atomization processes, the cooling rates are much higher and it is not possible to determine nucleation temperatures, but the effects of similar size-dependent nucleation phenomena can be seen. Thus Kelly et al. [5] found that in 303 stainless steel powders there were both fcc and bcc particles, with the metastable bcc particles having a smaller average size.

Nucleation is also required for the formation of crystalline phases in melt-spun ribbons. Whether this nucleation will occur only directly adjacent to the chill surface or throughout the thickness of the ribbon will depend on the growth characteristics as well as on the nucleation characteristics of the crystallizing phase. In surface resolidification, nucleation is not required, but it may occur if the growth characteristics of the phases in the substrates allow significant supercooling to develop.

In many alloy systems we have enough thermodynamic information to tell us when the nucleation of different phases will become possible, but we will rarely know much about the possible effects of heterogeneous nucleants. Nonetheless, experience gained with one type of solidification process can often help us to predict or at least explain the results of other types of processes.

3. GROWTH

Interfacial supercooling is needed to drive crystallization processes at a finite rate, and during rapid solidification processes the solid-liquid interface can fall hundreds of degrees below the equilibrium liquidus temperature. Moreover, even when a phase is growing with an interface temperature only a few degrees below its own liquidus temperature, it may be much more highly supercooled with respect to the liquidus of another phase. Because this may make it possible for the other phase to nucleate in the melt and replace the first phase, it is worth examining in more detail the factors which affect the amount of supercooling required to drive crystallization.

In alloy solidification, the two primary contributors to interfacial supercooling are solute redistribution and interfacial kinetics. Different phases can vary greatly in the amount of supercooling of these types required to drive their growth at high velocity. Much of the theory of coupled eutectic growth and the way in which the composition for coupled growth varies with solidification velocity (see Kurz and Fisher [6] for a review) is based on considerations of the supercoolings required to drive solidification of the different phases.

Many intermetallic compounds appear to solidify with effectively fixed compositions, even during rapid solidification processes. When such compounds form in melts which differ significantly from their fixed compositions, large levels of supercooling are required to drive the solute diffusion process, and a high degree of structural refinement of the solidifying phase may occur to minimize the diffusion distances. For such phases, the growth process is so greatly slowed by solute diffusion that local heat flow plays an insignificant role. The solidification process may then be regarded as isothermal growth from a supersaturated solid solution.

At the opposite extreme are phases which can grow with a wide range of compositions close to the melt composition, for example a terminal phase solid solution with a solute partition coefficient close to unity. Such phases can grow with a composition equal to that of the melt when the interface temperature is only a few degrees below the equilibrium liquidus, and there is little potential for interfacial supercooling to develop.

The appearance of crystallographic facets on the solid-liquid interface of a growing crystal is an indication of a large interfacial kinetic effect. The facets, being a manifestation of the difficulty of attaching atoms to certain crystallographic planes on the growing surface of the crystal, are an indication that a significant undercooling is required to drive the crystal growth. Jackson [7] derived a factor proportional to the entropy of fusion which gives a fairly reliable indication of the tendency of crystal phases to form facets when growing from the melt. High entropies of fusion are thus correlated to facet formation, slow growth rates, and high levels of kinetic undercooling.

Thus in general we can expect that large levels of supercooling may develop during solidification of faceted, high-entropy-of-fusion phases having narrow ranges of composition whereas much lower lower levels of supercooling are required to drive rapid solidification of low-entropy-of-fusion phases having wide composition ranges.

In evaluating the possibilities for a liquid to transform into different solid phases, it is instructive to consider the T_o curves for these phases. These are curves on the phase diagram which show where the liquid and solid phases of the same composition would have the same free energy, and they always lie between the solidus and liquidus curves. Perepezko and Boettinger [8] have shown for several different types of phase diagrams how the T_o curves can be used to estimate which phases can form from the melt without change of composition. Alloy compositions in which no phase can form without large composition changes are prime candidates for metallic glass formation at high cooling rates.

4. POLYPHASE REACTIONS

Eutectic and peritectic reactions require diffusion of solute and, therefore, become more difficult at high solidification rates. In a study of the Ag-Cu system [9] it was found that the maximum growth rate for coupled eutectic growth was only about 2.5 cm/sec, and above this velocity a different, single-phase growth form occurred. Observation of a eutectic structure is generally an indication that the local solidification rate was not very high.

In the idealized peritectic reaction, solid of one phase reacts with liquid to form a second phase. The second phase frequently forms a continuous coating on the first phase, thereby forming a barrier to the solute diffusion required for the reaction to proceed. Therefore peritectic reactions cannot go to completion in rapid solidification processes. Instead, the high temperature phase may survive and become coated with a layer of the low temperature phase which forms directly from the melt.

5. THE ALUMINUM-MANGANESE SYSTEM

The rapid solidification of aluminum-manganese alloys has been intensively studied in the past several years, largely because of the discovery of the quasicrystalline phases [10, 11] in this system. Because this system contains so many equilibrium and metastable phases, and because it has been so thoroughly studied, it provides some excellent examples of phase selection which can be explained in terms of the different nucleation and growth kinetics of the various phases. Solidification from the melt in aluminum-manganese has been studied by surface melting and resolidification, by melt spinning, and by solidification of isolated droplets. Although the different methods give quite different results, they can all be understood to result from a consistent sequence of nucleation and growth phenomena.

A. The phase diagram

Figure 2 represents the Al-rich end of the equilibrium phase diagram of the Al-Mn system as determined by Murray et al. [12]. The dotted lines show the approximate position of the metastable solidus and liquidus of the icosahedral quasicrystal phase, as determined by Knapp and Follstaedt [13] at high temperatures and as estimated from observations of eutectic structures at lower Mn contents [14]. The decagonal phase is slightly more stable than the icosahedral phase and its liquidus curve is slightly above that of the icosahedral phase [13].

B. Surface melting

Surface melting experiments have been carried out by scanning an electron beam across the surface of alloy bars containing up to 30 wt% Mn [15, 16]. In such experiments, the velocity of resolidification at the trailing side of the melt zone is less than the scan velocity by a simple geometric factor, but near the center of the melt trail this factor is relatively insignificant compared to the wide range of velocities used, except at very high scan rates.

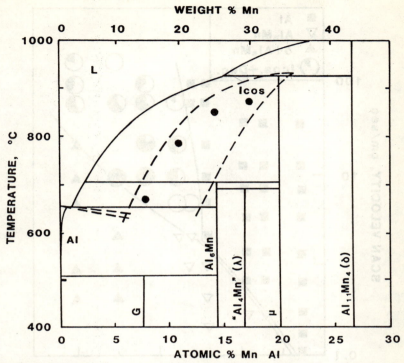

WEIGHT % Mn

Fig. 2. Part of the Al-Mn phase diagram, with dashed lines showing the approximate positions of the metastable solidus and liquidus of the icosahedral phase and of supersaturated fcc Al.

Figure 3 shows the phases which dominate resolidified melts in samples which have been scanned by electron beams, as a function of Mn content of the alloy. The figure has been divided into three regions; a low Mn, high velocity region in which cellular Al grows from the substrate; and a high Mn, low velocity region in which intermetallic phases grow from the substrate; a high Mn, high velocity region in which the quasicrystalline icosahedral and decagonal phases nucleate within the melt zone. Between the first two regions is a very narrow band in which coupled growth of Al/Al$_6$Mn eutectic occurs.

Eady et al. [17] carried out a series of directional solidification experiments at velocities up to 0.2 cm/sec and Mn contents up to 5.5 wt%. They showed that with increasing growth velocity the Mn concentration for coupled growth of the Al-Al$_6$Mn eutectic increased rapidly. Identical results were recently obtained in more detailed measurements by Juarez-Islas and Jones [18]. This is the trend expected according to principles described by several authors such as Kurz and Fisher [6], for a eutectic between a faceted intermetallic phase and an unfaceted phase with a partition coefficient close to unity. The electron beam results are consistent with an extrapolation of the trends established by Eady et al. It was

388

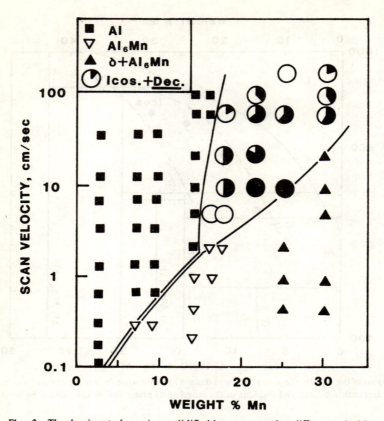

Fig. 3. The dominant phases in resolidified layers scanned at different velocities.

found that Al-Al$_6$Mn eutectic was observed at Mn contents as high as 16 wt%, but at the highest concentrations it occurred only as a transient stage when the growth of primary Al$_6$Mn dendrites ended; in alloys with 14 or less wt% Mn, it was then soon followed by cellular Al.

The transition from primary Al$_6$Mn to eutectic to cellular Al with increasing solidification velocity is thus consistent with well-established concepts of the velocity dependence of the composition for coupled eutectic growth. The phase diagram in Fig. 2 suggests that primary crystals of a higher Mn phase such as the hexagonal λ or μ could form in alloys containing more than 5 wt% Mn. These phases, however, apparently grow only quite slowly, and they never dominated the melt zone. A few scattered hexagonal crystals were seen in the slowest melts of the 25 wt%Mn alloy, but they never propagated over a large part of the melt zone.

Contrastingly, the δ(Al$_{11}$Mn$_4$) phase (triclinic) was able to grow rapidly in alloys of 25 to 30 wt% Mn. It grew in the form of thin plates which were always coated with an epitaxial coating of Al$_6$Mn. The orientation relationship between the two phases [19] is such that an almost perfect match exists between their

Fig. 4. Al-16 wt% Mn, scanned left to right (top view). The dendritic crystals growing in from the sides are Al₆Mn, and the crystals nucleated within the melt zone are the icosahedral phase.

structures on the interface plane. It is concluded that this structural matching leads to the easy nucleation of Al_6Mn on the rapidly growing plates of δ phase, with the result that the μ and λ phases are bypassed.

Quasicrystalline phases nucleate within the melt zone of alloys containing more than about 15 wt% Mn when the scanning velocity is sufficiently large. Figure 4 shows an alloy scanned at a velocity just above the upper limit at which the Al_6Mn dendrites are able to propagate throughout the melt zone. The extent of propagation of Al_6Mn dendrites in such cases is rather erratic because of their different orientations resulting from growth from different substrate grains. In the center of the melt zone, quasicrystal phase grains have nucleated. At scan velocities only slightly higher than that shown in Fig. 5, the growth of Al_6Mn from the substrate is almost completely eliminated and the quasicrystal phases fill the melt zone. In many cases a mixture of the icosahedral

390

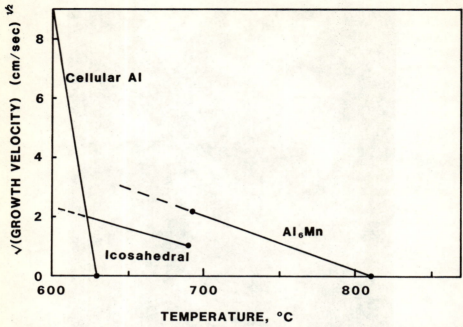

Fig. 5. Semi-schematic representation of growth rates of different phases in an Al-18 wt% Mn alloy.

phase, with six five-fold rotation axes, and the decagonal phase, with a single ten-fold rotation axis, is present. Studies [20] of 25 wt% Mn alloys scanned at speeds from 10 to 100 cm/sec revealed that the icosahedral phase nucleates within the melt zone and that the decagonal phase is then nucleated epitaxially on the icosahedral phase. The decagonal phase nucleates with its ten-fold axis parallel to one of the five-fold axes of the icosahedral phase, so that six orientational variants of the decagonal phase are produced by each icosahedral phase dendrite. The decagonal phase is more stable than the icosahedral phase, and replaces it completely in melts scanned at 10 cm/sec. In melts scanned at 50 cm/sec, the icosahedral to decagonal transformation is only partial, whereas in melts scanned at 100 cm/sec the formation of the decagonal phase is completely suppressed, and the icosahedral crystals remain untransformed.

The transformation from icosahedral to decagonal phase is one of the few processes in which the cooling rate appears to be the controlling parameter. On the basis of a simple heat flow model it was estimated that in the 25 wt% Mn alloy this transformation was complete at cooling rates of approximately 10^5 K/sec and was suppressed at cooling rates of 10^6 K/sec.

The liquidus of the icosahedral phase [13] is only slightly below the equilibrium liquidus, and the nucleation temperature of the icosahedral phase [21] is only slightly lower. Thus when rapid melt scanning causes supercooling to develop in the melt zone ahead of the advancing crystals of the equilibrium phases, it is the icosahedral phase which nucleates. In alloys containing < 20 wt% Mn, it has a distinctive morphology which immediately reveals its unusual symmetry. In the 16–18 wt% Mn region where the transition from Al_6Mn to icosahedral phase occurs, one can observe that the Al_6Mn dendrites or the Al_6Mn/Al eutectic grows around and engulfs some of the icosahedral crystals. From the shape of the engulfed crystals one can estimate that the Al_6Mn phase is growing about 5 times as fast as the icosahedral phase; nonetheless, the icosahedral phase displaces the Al_6Mn because of its much larger nucleation rate.

In alloys containing 16–18 wt% Mn, at scanning speeds above 50 cm/sec the icosahedral crystals in the resolidified layer become progressively smaller as the scan speed increases. They are engulfed by a rapidly advancing solidification front of cellular Al before they are able to grow to the large sizes seen in the slower melts. For the cellular Al to form, the temperature of the advancing solidification front must have dropped to approximately the metastable solidus temperature of aluminum at this composition. Once this occurs, however, further increases in velocity may result in relatively little further increase in interface supercooling, as the cellular aluminum can grow with little diffusional redistribution of Mn and interfacial kinetic effects are probably small. Figure 5 is a semi-schematic representation of dendrite growth rates as a function of temperature for Al_6Mn, icosahedral phase and cellular Al in an alloy containing 18 wt% Mn. Points which are known or can be estimated with reasonable precision are: the velocities of the Al_6Mn dendrites (5 cm/sec) and the icosahedral dendrites (1 cm/sec) at the (unknown) temperature where the icosahedral phase nucleation rate becomes large; a lower limit to the growth velocity of the icosahedral phase in rapidly scanned melts (to reach the observed size in the brief time available for growth at least 2.5 cm/sec is required); the metastable liquidus temperature of Al_6Mn [12]; and the temperature of the advancing cellular Al front (~ 640°C, from extrapolation of the equilibrium solidus and liquidus). Using these values, the curves in Fig. 5 have been sketched in: note that the curve for the growth velocity of the icosahedral phase starts at a finite value because supercooling is needed to produce nucleation within the melt, whereas no supercooling is needed to start growth of the other phases, which propagate from the substrate.

C. Powders

Mueller et al. [21] studied the solidification of Al-Mn powders as a function of Mn content up to 30 wt% and cooling rates from 0.5 to 500°C/sec. They found phases quite different from those produced by surface melting, but the results are not inconsistent.

The liquid alloys were emulsified in a molten eutectic salt, which promoted formation of a stable, noncatalytic coating on the surface of the alloy droplets. When the salt is dissolved away with distilled water, the powder particles remain isolated from one another by the coating. The cooling rates employed were 0.5° C/sec in a DTA system, 25°C/sec by air cooling, and 500°C/sec by water quenching.

In addition to the thermal analysis of the cooling powders, the solidified product was examined by SEM and x-ray diffraction.

The powders cooled at 0.5°C/sec contained Al and λ with the volume fraction of the latter increasing as the Mn content of the alloy increased. The formation of the Al_6Mn phase was bypassed, and a metastable eutectic between Al and λ was observed, just 3°C below the stable Al/Al_6Mn eutectic temperature. The failure of Al_6Mn to form was attributed to the difficulty of nucleating this phase.

In the powders cooled at 25°C/sec, the decagonal phase was observed in addition to λ and aluminum, especially at higher Mn concentrations. The powders cooled at 500°C/sec contained in addition a small amount of the icosahedral phase. On the basis of microstructural observations, it appeared that the decagonal phase in these droplets formed by replacing the icosahedral phase through the same process of epitaxial nucleation observed in the surface melting experiments. The measured nucleation temperatures in the 25°C/sec experiments were thus concluded to represent the initial formation of the icosahedral phase. These temperatures are shown as dots on Fig. 2, where it can be seen that they are consistent with the proposed metastable liquidus curve for the icosahedral phase.

The most dramatic contrast between the powders and the surface melted alloys is the prominent appearance of λ phase in the former at concentrations and cooling rates which would have produced Al_6Mn in the latter. This contrast was clearly due to the phases in the powders being dominated by nucleation processes (faster for λ phase) whereas the phases in the resolidified melts were dominated by growth from the substrate (faster for Al_6Mn).

Ridder and Bendersky [22] examined Al-Mn powders made by an electrohydrodynamic atomization process which produces droplets with diameters of less than one micrometer. They found that these droplets solidified as the icosahedral phase with an extremely fine grain size, indicating that the nucleation rate for this phase can become very high.

D. Melt-spun ribbons

The phase constituency of melt-spun ribbons of Al-Mn is relatively simple; in ribbons containing < 40 wt% Mn, the equilibrium intermetallic phases fail to form. The solid solubility of Mn in Al is greatly extended beyond its equilibrium value, and the icosahedral phase starts to appear at 12–16 wt% Mn, depending on the cooling rate. The volume fraction of icosahedral phase then increases with increasing Mn concentration, approaching unity at approximately 34 wt% Mn in the most rapidly solidified ribbons. In ribbons which solidify less rapidly, the icosahedral phase is replaced by the decagonal phase as the Mn concentration increases above about 27 wt% [23].

The supersaturated Al formed in melt-spun ribbons takes the form of columnar grains growing from the chill surface to the top surface of the ribbons, whereas the icosahedral grains have an equiaxed structure indicating that they nucleated within the melt. When the aluminum grows upward from the chill surface, its solid-liquid interface must lie below the metastable extension of the aluminum liquidus as shown in Fig. 2. The liquid ahead of the advancing aluminum interface in the more concentrated alloys is thus well below the temperature at which the icosahedral phase can nucleate.

6. DISCUSSION

Based on the different solidification experiments described above, one can construct Table 1 representing the nucleation and growth behavior of several of the phases in the Al-Mn system.

Table 1. Solidification behavior of Al-Mn phases

Phase	Maximum Growth Rate (cm/sec)	Nucleation in Melt	Heterogeneously Nucleated by
Al	> 200	*	several
Al$_6$Mn	2.5	*	δ
λ	very low	moderate	
μ	very low	*	
δ	20	*	
icosahedral	1	high	
decagonal	< 1	*	icosahedral

*Not detected in these experiments.

Aluminum can grow very rapidly and appears to be nucleated by several intermetallic phases, but only when the temperature has dropped to below its liquidus. It can form metastable eutectics with λ [21] or the icosahedral phase [14], in addition to its stable eutectic with Al$_6$Mn. The other phases which can grow fast enough to keep up with a moving surface melt at velocities of several centimeters per second are the δ and Al$_6$Mn phases, with the former sometimes acting as a heterogeneous nucleant for the latter. Contrastingly, the λ and μ phases appear to grow only very slowly.

The icosahedral phase grows at only a moderate rate, but nucleates at a very high rate when the melt becomes sufficiently supercooled. This can occur in surface melts when the temperature of the crystals growing upward from the substrate drops as the scan velocity increases. It occurs in melt-spun ribbons where the only other phase present is aluminum, which can grow only at a temperature below its metastable liquidus. Sufficient supercooling for nucleation of the icosahedral phase develops in alloy droplets if the cooling rate is high enough. The λ phase can nucleate at a somewhat higher temperature than the icosahedral phase, but apparently its slow growth kinetics prevent it from attaining a detectable size before the faster growing icosahedral phase nucleates, except in slowly cooled (0.5°C/sec) droplets.

394

We have thus seen that a representation such as Fig. 1 is an oversimplification, because it does not consider the different roles of nucleation and growth phenomena in the various solidification processes. Figure 3 is a reasonable representation for one type of process, although even there the exact location of the boundaries would depend somewhat on details, such as the diameter of the melt pool. Nonetheless, for a system which has been examined in great detail we have found that we can understand the phases produced by several different rapid solidification processes in terms of a consistent set of nucleation and growth characteristics for the different phases.

Although we can make some reasonable assumptions about the growth behavior of crystalline phases on the basis of information about their solubility range, crystallography, entropy of fusion or growth forms, we usually do not have many clues to their nucleation properties. The work with Al-Mn demonstrates the importance of using several different rapid solidification techniques to gain a full understanding of phase selection in a complex system.

REFERENCES

1. Turnbull, D. and E. Cech. *J. Appl. Phys.* **21**, 804 (1950).
2. Perepezko, J.H. *Mat. Sci. and Eng.* **65**, 125 (1984).
3. Turnbull, D. *J. Chem. Phys.* **20**, 411 (1952).
4. Yoon, W., J.S. Paik, D. LaCourt and J.H. Perepezko. *J. Appl. Phys.* **60**, 3489 (1986).
5. Kelly, T., M. Cohen and J. VanderSande. *Met. Trans.* **15A**, 819 (1984).
6. Kurz, W. and D.J. Fisher. *Int. Metall. Rev.* **24**, 177 (1987).
7. Jackson, K.A. In: *Liquid Metals and Solidification*. American Society for Metals, Cleveland. p. 174 (1958).
8. Perepezko, J.H. and W.J. Boettinger. *Mat. Res. Soc. Symp. Proc.* **19**, 223 (1983).
9. Boettinger, W.J., D. Shechtman, R.J. Schaefer and F.S. Biancaniello. *Met. Trans.* **15A**, 55 (1984).
10. Shechtman, D., I. Blech, D. Gratias and J.W. Cahn. *Phys. Rev. Lett.* **53**, 1951 (1984).
11. Bendersky, L.A. *Phys. Rev. Lett.* **55**, 1461 (1985).
12. Murray, J.L., A.J. McAlister, R.J. Schaefer, L.A. Bendersky, F.S. Biancaniello and D.L. Moffat. *Met. Trans.* **18A**, 385 (1987).
13. Knapp, J.A. and D.M. Follstaedt. *Phys. Rev. Lett.* **58**, 2454 (1987).
14. Yu-Zhang, K., J. Bigot, G. Martin, R. Portier and D. Gratias. *Proc. XIth Cong. on Elect. Microscopy,* Kyoto, p. 167 (1986).
15. Schaefer, R.J., L.A. Bendersky, D. Shechtman, W.J. Boettinger and F.S. Biancaniello. *Met. Trans.* **17A**, 2117 (1986).
16. Schaefer, R.J. and D. Shechtman. to be published.
17. Eady, J.A., L.M. Hogan and P.G. Davies. *J. Aust. Inst. Metals* **20**, 23 (1975).
18. Juarez-Islas, J.A. and H. Jones. *Acta Met.* **35**, 499 (1987).
19. Schaefer, R.J. and L.A. Bendersky. *Mat. Res. Soc. Symp. Proc.* **58**, 217 (1986).
20. Schaefer, R.J. and L.A. Bendersky. *Scripta Met.* **20**, 745 (1986).
21. Mueller, B.A., R.J. Schaefer and J. Perepezko. *J. Mater. Res.* **2**, 809 (1987).
22. Bendersky, L.A. and S.D. Ridder. *J. Mater. Res.* **1**, 405 (1986).
23. Schaefer, R.J. *Scripta Met.* **20**, 1187 (1986).

18

Thin Film Sensors

K. RAJANNA, S. MOHAN AND E.S.R. GOPAL

Instrumentation and Services Unit,
Indian Institute of Science,
Bangalore 560 012, India.

ABSTRACT

The paper is concerned with two types of sensors, namely thin film strain gauges and thin film thermocouples, which are of extensive use not only for conventional applications, but also for measurements under special conditions such as material processing and characterization. The versatility of thin film technology to prepare these types of sensors is highlighted. After explaining the basic features, the general class of strain gauges and the sensing material aspects for strain gauges, and the utility of cermet materials for preparing thin film strain gauges are described. Also, some preliminary attempts to prepare and study the specialized thin film strain gauges are presented. A review is also given of thin film thermocouples, their relative merits over the conventional wire type thermocouples with specific examples and the recent ongoing development work in this area.

1. INTRODUCTION

In most of the material processing and characterization applications, it is necessary to measure quantities such as temperature, pressure, flow rate, strain etc. Sensors are needed to detect and measure these physical parameters and to convert them into electrical signals. The present day technology demands sensors of high sensitivity and resolution.

With the advent of thin film technology, it is possible to prepare sensors with characteristics that cannot be achieved easily with bulk materials. In brief it is possible to make sensors in thin film form with tailor-made properties. In this article, two types of sensors i.e., strain gauges and thermocouples, which are of extensive use for measuring the parameters in material processing and characterization, are discussed. The importance of these two sensors in thin film form, their relative merits over their bulk counterparts, fabrication technology, material selection and applications are discussed with specific examples (in case of thermocouples) as well as with a discussion of the state of art of the technology.

2. THIN FILM STRAIN GAUGES

Basically, strain gauge is a device used to measure the linear deformation (mechanical surface strain) occurring in the material during loading. The discovery of the basic operating principle of strain gauge dates back to 1856, when Lord Kelvin reported that certain metallic conductors when subjected to mechanical strain exhibited a corresponding change in electrical resistance. In general all electrically conducting materials possess a strain sensitivity. The dimensionless number F which is variously termed as the electrical resistance-strain coefficient, the strain sensitivity factor or the gauge factor is expressed mathematically [1] as,

$$F = \frac{\Delta R/R}{\Delta L/L}$$

or since $\Delta L/L = \epsilon$, the strain

$$F = \frac{\Delta R/R}{\epsilon}$$

where R and L represent respectively, the initial resistance and initial length, while ΔR and ΔL represent the small changes in resistance and length which occur as the material is strained.

The conventional types of strain gauges developed over the years are metallic wire, metallic foil and semiconductor strain gauges. A major new advance to strain gauge technology is the introduction of thin film strain gauges [2]. Because of the versatility of thin film technology, thin film strain gauges can be tailor-made to have desired properties so that they are useful not only for conventional applications but also for measurements under special conditions such as material processing studies. Figure 1 shows the broad classification of strain gauges. Some of the important advantages of thin film strain gauges compared to conventional bulk type strain gauges are that the former can:

a) be made to have high gauge factor and low temperature coefficient of resistance (TCR),

b) be made of any metal, metal alloy, semiconductor or cermet films,

c) possess good linearity and low hysteresis, and

d) be mass produced with considerable cost reduction.

In the case of metal/metal alloy film strain gauges, the gauge factor value is not appreciable. Even though, discontinuous metal films can offer high gauge factor, they are not suitable because of the associated problems of stability and reproducibility [3]. On the other hand semiconductor films appear to be very attractive because of their high strain sensitivity [hence high gauge factor]. But

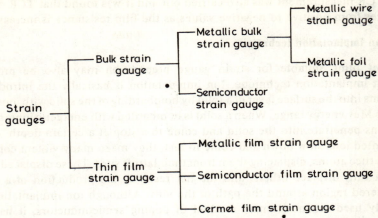

Fig. 1. Broad classification of strain gauges.

they exhibit non-linear resistance-strain characteristics and are greatly affected by temperature changes [4]. Strain gauges made of cermet films can have higher gauge factor coupled with good stability and controllable temperature coefficient of resistance [3]. Cermet films are basically metal-dielectric composites [5]. Here the conducting film is diluted by dispersing it in a matrix of insulating material. The electrical conductance of the cermet film structure is similar in many ways to that found in discontinuous metal films and amorphous solids. Depending on the materials used, their relative amounts and the mean particle size and spacing, the electrical transport character can be changed from an activated type with a negative temperature coefficient of resistance (TCR) to a metallic one with a positive TCR, allowing a continuous range of properties to be obtained [6].

Meiksin et al. [7] have studied chromium-silicon monoxide cermet films with respect to strain gauge application. In their study commercially available 70%-30% Cr-SiO pellets were deposited on glass substrates by flash evaporation. They stabilized the films by post-deposition heat treatment. The sheet resistance of the films after baking ranged from 18k ohm to 670 k ohm/ □ . The gauge factor of the films was found to be in the range 12–50 with very good stability. They also studied gold-silicon monoxide cermet films. It is important to note that gold does not react with SiO, which enhances the ideal structure of metallic islands embedded in an insulating matrix. Films were deposited by the two-source method with compositions varying from 60–40% to 90–10% Au-SiO by volume. Film thickness in their study ranged from 730 to 2360Å. The optimum composition was determined as 80–20% with a gauge factor of 38.

Witt [8] prepared gold-glass cermet films by *rf* sputtering. The substrates used were 7059 glass slides with evaporated Cr/Au contacts. The sheet resistance of cermet films ranged from 100Ω to 1 MΩ/□. Following the deposition, the films were stabilized in vacuum. The maximum gauge factor value obtained was 42.

The TCR measurement was also carried out and it was found that TCR of films changes from positive to negative values as the film resistance is increased.

2.1 Ion implantation techniques

Cermet films suitable for strain gauge preparation may also be produced by ion implantation technique. Ion implantation is basically the introduction of atoms into the surface layer of a solid by bombarding of the solid with ions in the Kev to Mev energy range. When a solid is bombarded with energetic ions, most of the ions penetrate into the solid and come to a stop at a certain depth. As the implanted ions slow down and come to rest, they make many violent collisions with lattice atoms, displacing them from their lattice sites. These displaced atoms can in turn displace others, and the net result is the production of a highly disordered region around the path of the ions. Although ion implantation was initially used as an alternate method of doping semiconductors, it has now become an accepted method of modifying near surface and properties of thin films in general [9, 10]. Using ion implantation technique cermet films can be produced by bombarding metal films with gas (like argon, oxygen and nitrogen) ions [9, 11]. Alternatively, a cermet film may also be produced by implanting metal to a dielectric film. The advantage of ion implantation technique over the other methods (evaporation and sputtering) of cermet film preparation is that the former method is controllable and reproducible and does not suffer from the problems of fluctuating evaporation or sputtering rates [9].

Stroud [12] studied the effect of strain on the resistance of oxygen ion implanted titanium thin films. The films deposited by evaporation were bombarded with 30 Kev oxygen ions to such dose levels that sheet resistances between 1 KΩ/□ and 100 KΩ/□ were obtained. The background pressure during bombardment was less than 10^{-6} torr. The maximum gauge factor value obtained was around 10.

Robinson et al. [13] have studied the strain sensitive property of tantalum thin films bombarded with argon ions. Tantalum films on glass substrates were prepared by electron beam evaporation and bombarded with collimated beam of 40 Kev argon ions to doses of up to 3×10^{17} ions per cm^2. The maximum gauge factor reported is about 6 and the TCR value ranged from −150 ppm/°C to +400 ppm/°C.

The effect of argon ion bombardment on the strain gauge factor of thin gold films was studied by Robinson et al. [14]. Evaporated gold films of approximately 200 Å in thickness were bombarded with argon ions to increase the sheet resistance to a maximum of 40 KΩ/□. The beam was scanned horizontally and vertically to ensure uniformity of dose. The strain gauge coefficient of resistance was measured for films with a wide range of sheet resistance, and was found to be almost invariant with an average value of 2.6. The temperature coefficient of the strain gauge factor was found to be similar in magnitude but opposite in sign to the TCR, which was measured as $+12 \times 10^4$ °C^{-1}.

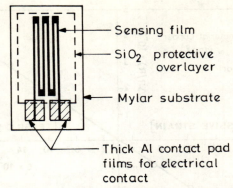

Fig. 2. Schematic of the thin film strain gauge (copper film of thickness about 120 nm deposited on mylar substrate).

2.2 Preliminary work on thin film strain gauge

Even though it is ideal to make the strain gauge using cermet film, we describe here the study made on metal film as a preliminary attempt.

2.2.1 Preparation of thin film strain gauge

Copper film (thickness $t \approx 1200$ Å) was deposited on to mylar substrate by resistive evaporation from a molybdenum boat. The evaporation was carried out at a pressure of about 3×10^{-5} torr. Prior to film deposition the substrate was subjected to glow-discharge cleaning for about 10 minutes. Pattern formation of the film as shown schematically in Fig. 2 was done using photolithography technique [15]. In order to attach the electrical leads to the sensing element (gauge), thick aluminium contact pad films were deposited at the two ends of the sensing film pattern using mechanical mask. To provide protection to the thin film strain gauge against atmospheric attack, fingerprints etc; an over layer (thickness $\approx 2,000$ Å) of SiO_2 was deposited on the sensing film pattern by electron beam evaporation.

2.2.2 Response of the gauge

Before, studying the response of the thin film strain gauge, it was subjected to repeated vacuum heat treatment for stabilization.

The cantilever method has been adopted to study the response of the gauge to tensile and compressive strains. The details of the experimental arrangement and calculation of strain ($\in$) value are given elsewhere [16]. The variation of relative change in resistance $\Delta R/R$ with strain ($\in$) is shown in Fig. 3. It is evident that the response of the gauge is linear in tensile as well as compressive modes. Also, after subjecting the gauge to several loading and unloading cycles, no noticeable hysteresis effect was observed in its resistance–strain characteristics.

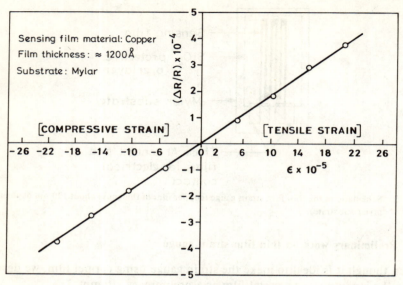

Fig. 3. Response of the thin film strain gauge.

3. THIN FILM THERMOCOUPLES

The conventional thermocouples are made up of wires. The selection of the material is done depending on the range in which they are supposed to be used. However, for special applications (like material processing) they suffer from several disadvantages. The shortcomings are due to: (a) heat capacity, (b) response time, (c) environmental effects, (d) stability, (e) control over characteristics, and (f) compatibility with the components.

Especially, in situations like temperature measurement on critical areas of turbine blades, wire thermocouples embedded in machine grooves in the blade lead to: (i) distortion and disturbances in the flow of gas streams, and (ii) uncertainty in measurement due to heat transfer characteristics of blade or vane. On the other hand thin film thermocouple mounted directly on the critical component possess excellent heat transfer characteristics to define precise component temperatures without disturbing the gas stream.

Among the metallurgical applications the feasibility of fabricating thin film thermocouples on internal combustion hardware has been extensively investigated [17, 18]. The specific aim was to measure the surface temperature of valves, valve sealings, cylinder walls, combustion chamber surface and Piston heads during engine operations. This is a real test case for the application of thin film thermocouples; as one has to think of the following problems in the selection of materials and in the fabrication processes. The thermocouples are exposed to very high temperatures (as high as 1300°K) and to a variety of corrosive gases. They have to be fixed on to uneven surfaces of irregular shapes. It has to be a

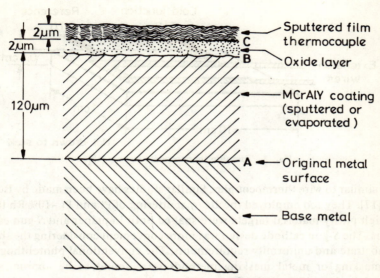

Fig. 4. Multilayer design of thin film thermocouple [18].

part of the process technology. This has been studied in detail by a number of investigators. Kreider [18] has designed a multilayer stock as shown in Fig. 4. The base metal is an alloy coated with MCrAly of thickness 150 microns by sputtering technique.

Thermal oxidation of this coating was performed at temperatures ranging from 1075 to 1350°K. This is covered by an Al_2O_3 layer of thickness 2 microns prepared by reactive sputtering technique. The temperature sensor, a Pt and Pt-10% Rh thermocouple in this case was deposited as a thin film of thickness 2 microns over the oxide layer by rf magnetron technique. The major problems that were encountered are the following.

   i) electrical shorting through the insulation
  ii) delamination in service
 iii) open circuits developed in thermal cycling
 iv) oxide layer integrity
  v) adhesion of thermocouple material to oxide
 vi) oxidation resistance of thermocouple material
vii) durability at high temperature.

For studying the thermocouple output performance, 10 cm long thin film thermocouple test bars shown in Fig. 5 have been fabricated. The test bars were placed through the wall of the furnace with the hot and cold junctions monitored with reference thermocouples. The thermocouples were highly adherent and withstood the repeated thermal cycling in air up to 1100°K without insulation breakdown. The output characteristics of the thin film thermocouples were very

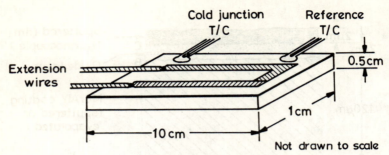

Fig. 5. Thin film thermocouple test bar [18].

much similar to wire thermocouples. Similar studies have been made by Budhani et al. [17]. They too employed the *DC/rf* sputtering of Pt and Pt –10% Rh targets. The high purity material targets were shaped both in planar and *S*-gun cathode designs. The *S*-gun cathode design was more convenient considering the shape of the substrate and uniformity requirements. The conventional photolithography, tape masking or metal masking methods were found to be appropriate for sequential deposition on flat test samples. Only tape masking and photolithography techniques were used on curved blades and vanes. The deposition parameters and the technique have been optimized for improved adhesion. Contamination of the thermocouple elements by metals diffusing from the Al_2O_3 base and the products of combustion were found to be major constraints. Stress-free stochiometric Si_3N_4 coatings on top of the thermocouples reduced the corrosion of thermocouple elements. The problems that still needed attention were, thermocouple lead connections, strain relief and uniform coatings on large area curved surfaces.

Tong et al. [19] developed thin film Pt-Ir thermocouple which exhibited bulk like thermoelectric behaviour up to 790°C, the highest temperature observed with thin film pure element thermocouples. This thermocouple was fabricated by sequential evaporation of iridium (99.99% pure) and platinum (99.99% pure) on alumina substrates at a pressure of 2.1 and 1.7×10^{-6} torr respectively. It was observed that high substrate temperatures were required to obtain good adhesion between iridium and alumina substrates. The thickness of the Ir and Pt thin films were 0.29 and 0.51 micron respectively. A protective layer of aluminium oxide (1.72 micron thickness) was deposited on the thermocouple films by reactive evaporation. This oxide layer provided adequate protection from environmental corrosion or abrasion and improved the life of iridium at elevated temperatures. The thermal response time of the thermocouple measured with a Q-switch Nd:YAG laser and a signal averaging oscilloscope was found to be 60 ns. The ease of implementation with other thin film devices and the highly reproducible temperature calibration made this thin film thermocouple ideally suited for monitoring instantaneous temperatures during device processing. Pt-Ir thermocouples find application in:

i) studies of laser annealing of thin films,

ii) measuring optical absorption of thin films, and

iii) to measure thermal conductivity of thin films.

Marshall et al. [20] prepared thin film thermocouples with different combinations from nickel, iron, copper, constantan, chromel and alumel. The simple metal elements were evaporated by resistive heating method whereas alloys were deposited by flash evaporation. They observed that maximum thermal e.m.f of a thin film thermocouple was obtained when the film thickness of each element was greater than 2500 Å.

Bolker and Sidles [21] prepared the thin film platinum thermometers by *rf* sputter deposition and used them to measure film temperature *in situ* during the sputtering process. A serpentine resistor pattern of platinum was *rf* sputtered on to corning 7059 glass substrates through a molybdenum shadow mask. Molybdenum pads [sputtered films of thickness 7000 Å] were provided for end contacts. Sputtered SiO_2 coating served as a protecting layer. These thermometers were found to have a sensitivity of 2°C per ohm.

A thin film germanium resistor was prepared by Regelsberger et al. [22] and was used for fast and accurate thermometry in non-adiabatic microcalorimeter. Optically polished sapphire single crystals were used as the substrates. A thin film germanium resistor was deposited (thickness: 65 nm) on to these substrates. Gold contacts with an undercoating of chromium were used as electrodes. The good mechanical property and stable temperature characteristics of this sensor allowed easy handling and reliable thermometry at low temperatures.

Of late, superconducting materials have also been used as thermal sensors. The temperature dependence of the magnetic penetration depth in a superconductor, in the Kinetic inductance limit, is proposed as a basis for a sensitive thermometer by McDonald [23]. The development of high T_c superconductors coupled with the development of the technology of depositing them as thin films can be a future prospectus for developing highly sensitive thin film thermocouples.

ACKNOWLEDGEMENT

We gratefully acknowledge the support of different funding agencies for initiating the developmental work in the area of thin film sensors. We thank all our colleagues for contributing to the vacuum and thin film instrumentation needed in this work.

REFERENCES

1. Perry, C.C. and H.R. Lissner. *Strain Gauge Primer,* 2nd Edition, Mc Graw-Hill, New York, p. 158 (1962).
2. Perino, P.R. *Instruments and Control Systems,* p 119, December (1965).
3. Chopra, K.L. and Inderjeet Kaur. *Thin Film Device Applications*, Plenum Press, New York, p. 169 (1983).

4. Baker, M.A. In: *Strain Gauge Technology,* A.L. Window and G.S. Holister, eds., Applied Science Publishers, London, Ch. 6 (1982).
5. Chopra, K.L. *Thin Film Phenomena*, McGraw-Hill, New York, p. 390 (1979).
6. Devenyi, A., W. Theiner, S.K. Sharma, J. Aalfeld and R. Manaila-Devenyi. *Thin Solid Films*, 15, 39 (1973).
7. Meiksin, Z.H., E.J. Stolinski, H.B. Kuo, R.A. Mirchandani and K.J. Shah. *Thin Solid Films* 12, 85 (1972).
8. Witt, G.R. *Thin Solid Films*, 13, 109 (1972).
9. Stroud, P.T. *Thin Solid Films* 11, 1 (1972).
10. Stephens, K.G. and I.H. Wilson. *Thin Solid Films*, 50, 325 (1978).
11. Wilson, I.H., K.H. Goh and K.G. Stephens. *Thin Solid Films*, 33, 205 (1976).
12. Stroud, P.T. *Thin Solid Films*, 10, 205 (1972).
13. Robinson, R.G.R., K.G. Stephens and I.H. Wilson. *Thin Solid Films*, 68, 305 (1980).
14. Robinson, R.G.R., K.G. Stephens and I.H. Wilson. *Thin Solid Films*, 27, 251 (1975).
15. Glang, R. and L.V. Gregor. In: *Handbook of Thin Film Technology*, L.I. Maissel and R. Glang, eds., McGraw-Hill, New York, Ch. 7 (1970).
16. Rajanna, K. and S. Mohan. *J. Mater. Sci. Lett.* 6, 1027 (1987).
17. Budhani, R.C., S. Prakash and R.F. Bunshah. *J. Vac. Sci. Technol.* A4(6), 2609 Nov/Dec (1986).
18. Kreider, K.G. *J. Vac. Sci. Technol.* A4(6), 2618, Nov/Dec (1986).
19. Tong, H.M., G. Arjavalingam, R.D. Haynes, G.N. Hyer and J.J. Ritsko. *Rev. Sci. Instrum.* 58(5), 875 May (1987).
20. Marshall, R., L. Atlas and T. Putner. *J. Sci. Instrum.* 43, 144 (1966).
21. Bolker, B.F.T. and P.H. Sidles. *J. Vac. Sci. Technol.* 14(1), 205 Jan/Feb (1977).
22. Regelsberger, M., R. Wernhardt and M. Rosenberg. *Rev. Sci. Instrum.* 58(2), 276 February (1987).
23. McDonald, D.G. *Appl. Phys. Lett.* 50(12), 775, 23 March (1987).

19

Computer Modeling of Solidification Processes in Metal Casting/Joining

O. PRABHAKAR, R. KRISHNAKUMAR, K. MADHUSUDANA AND P. NAGENDRA

Department of Metallurgical Engineering
Indian Institute of Technology, Madras, India

ABSTRACT

A software package CADCAST (Computer Aided Design in Casting) for the simulation of solidification processes based on the finite element method has been developed. This package has been used in three areas, casting, surface treatment processes and flash-butt welding.

Solidification of long freezing range Al alloy was simulated and compared with experimental results. Feeding efficiency factors were numerically computed and used to predict mechanical properties. Isotherm movements were graphically observed. It was found that the feeding efficiency as influenced by the pasty zone (FEP) correlates well with the casting soundness. The computed liquidus freeze waves agree well with the experimentally determined curves and the solidus freeze waves differs slightly, but the agreement is satisfactory.

Surface treatment process using TIG and Laser were simulated using the software package. The dimensions of the remelted tracks were obtained numerically and were found to agree reasonably with the experimental values.

The finite element package was also used to predict Heat Affected Zone (HAZ) in a flash butt welding process. Theoretical temperature distribution was used to predict HAZ width and compared with experimental values.

1. INTRODUCTION

Computer simulation of manufacturing processes has assumed greater significance in recent years. This is due to the advancement that has taken place in computer hardware and software as well as the numerical techniques used for the solution of differential equations. The growth of the computer technology is well-known. But the more important development that has kept pace with this growth is in the area of numerical mathematics.

Although numerical techniques like finite element method are being used in the area of design and analysis over the past two decades, its application in the area of manufacturing are of recent origin. Many problems are yet to be sorted out and are still in the researchers bench. But a few problems have been solved and can be used directly to predict certain important parameters which are of consequence to the quality of the product. One such area where significant development has taken place is computer modeling of casting/welding processes.

A suitable software to simulate and model the solidification process in casting and welding has been developed at the Department of Metallurgical Engineering, Indian Institute of Technology, Madras. The package is essentially based on 2D heat transfer analysis using finite element (FE) formulations. The package could be used in casting and in certain welding processes. This paper details the mathematical aspects of the package and three areas in which the package was successfully employed.

2. LITERATURE REVIEW

Major numerical schemes for process modeling and simulation comprise finite difference, finite element and boundary element methods [1].

Finite difference method is the oldest of these techniques. The first major use of this technique for solidification was reported by Henzel and Keverian [2]. They essentially extended the technique of Dusinberre [3]. Implicit, explicit and alternate direction method [4] have been used for solidification. Finite difference method has been used even a decade back to solve problems of practical interest like simulation of a pressure diecast connecting rod [5] and gravity die casting aluminum piston [6]. Morrone et al. [7] simulate the solidification of fundamental shapes like 'T' and 'L' cast in sand moulds. Jeyarajan and Pehlke [8–10] in a series of papers documented the potential of solidification simulation as a design tool. FDM has also been used for solidification simulation by Davies and Moe [11] for feeding range determination in gravity die casting, Niyama [12] for shrinkage prediction and by Ebisu [13] for morphological studies in centrifugal cast steel hubs. An example of a large-scale design can be found in the steel roll design by Stoer and Brody [14].

Finite element method is relatively of recent origin. It can handle curved boundaries very elegantly compared with finite difference method and hence can be used in general-purpose codes. It was first used by Sneider [15] for simulation of solidification and later by Mathew and Brody [16] and Morgan et al. [17]. Brimacombe [18] modeled the two dimensional heat flow and stress generation in a steel ingot using the finite element method. They assessed the accuracy, stability and cost of different numerical methods. They also compared the results of standard finite element method with the matrix method, alternating direction implicit method and analytical solutions. More recently Richter and Sahm [19] developed computer codes for examining directional solidification of gas turbine wheels.

3. FINITE ELEMENT METHOD

The Finite Element Method is an approximation procedure for solving differential equations of boundary and/or initial value type in engineering and mathematical physics. In FEM the region of interest is discretized into a number of domains or elements bounded by the nodes. An approximate solution is assumed based on certain conditions and the governing partial differential equation is replaced by a set of simultaneous equations with unknown parameters at nodes, as the variables to be solved.

The Fourier heat conduction equation has been universally applied to simulate solidification phenomena. In its general form it is given by the parabolic equation

$$\frac{\partial}{\partial x}\left(K_x\frac{\partial T}{\partial x}\right) + \frac{\partial}{\partial y}\left(K_Y\frac{\partial T}{\partial y}\right) + Q = C\frac{\partial T}{\partial t} \tag{1}$$

where k_x, k_y are thermal conductivities in x and y directions, Q is the internal heat generation with one or both of the following conditions specified on the boundary:

$$T = T(x,y) \text{ on } S_1$$

$$K_x\frac{\partial T}{\partial x}n_x + K_y\frac{\partial T}{\partial y}n_y + q + h\,(T - T_\infty) = 0 \text{ on } S_2 \tag{2}$$

where $T(x,y)$ is a specified boundary temperature distribution n_x, n_y are the direction cosines of the outward normal; q is the heat loss from the boundary due to conduction; and $h\,(T - T_\alpha)$ is heat loss due to convection from the boundary subject to the initial condition

$$T = T(x,y) \text{ at } t = 0 \tag{3}$$

Detailed derivation of element equations are given elsewhere [20]. Summarizing the results of the derivation, the above partial differential equation reduces to

$$[K]^e \{T\}^e + [C]^e \{\dot{T}\}^+ \{F\}^e = 0 \tag{4}$$

The matrix $[K]$ is the conductance matrix, $[C]$ is the capacitance matrix and the vector $\{F\}$ is the force vector. Their terms are given by

$$K_{ij} = \iint_{D^{(e)}} \left(k\frac{\partial N_i}{\partial x}\frac{\partial N_j}{\partial x} + k\frac{\partial N_i}{\partial y}\frac{\partial N_j}{\partial y}\right) dx\, dy + \int_{S_2^{(e)}} h\,N_i\,N_j\,ds$$

$$C_{ij} = \iint_{D^{(e)}} \rho c\,N_i\,N_j\,dx\,dy$$

$$F_i = - \iint\limits_{D^{(e)}} QN_i \, dx \, dy - \int\limits_{S_2^{(e)}} h \, T_\infty \, N_i \, ds + \int\limits_{S_2^{(e)}} q \, N_i \, ds$$

Linear triangular element was chosen for the present analysis. The shape functions in natural coordinates were substituted in eqn. (5) and the element matrices were evaluated. The Element Matrix equations were then assembled into the global matrix. This would result in a system of equations similar to eqn. (4) except that the matrices $[K]$ and $[C]$ would have $n \times n$ terms where n is the number of nodes in the solution domain and $\{F\}$ would be an array of n terms, which is written as:

$$[K] \{T\} + [C] \{\dot{T}\} + \{F\} = 0 \tag{6}$$

3.1 Time marching schemes

Hogge [21] has summarized the two-level and three-level time integration schemes, which, when applied to the system of first order differential equations reduce them to a set of algebraic equations. Since the matrices $[K]$, $[C]$ and $\{F\}$ are temperature dependent, the resulting algebraic equations will have to be solved iteratively. A simplification can, however, be made by using a linearization technique.

Crank-Nicolson Scheme: $\quad [K] [(T_2 + T_1)/2] + [C] [(T_2 - T_1)/\Delta t] = [F]$

Dupont II Scheme: $\quad [K] [(3T_3 + T_1)/4] + [C] [(T_3 - T_1)/2\Delta t] = [F] \tag{7}$

Since the Dupont II scheme is not a self-starting scheme, the Crank-Nicolson scheme was used for the first time step. The application of the time marching schemes on eqn. (6) yields a set of algebraic equations of the type

$$[A] \{T\} = \{B\} \tag{8}$$

The resultant set of simultaneous equations were solved by the Gaussian elimination technique.

3.2 Latent heat modeling

The latent heat of melting is modeled as force vectors equally distributed on all the nodes of an element. The latent heat term is given by $\rho L A /3 \, \Delta t$ for a three-noded triangular element, where ρ is density, L is latent heat, Δt is length of time step and A is the area of the element. The latent heat of solidification is also modeled as a step function increase in specific heat.

4. THE SOFTWARE PACKAGE

Two versions of the software package were developed: one to run on IBMPC compatiables written in Turbo Pascal/Fortran 77; another version with more capabilities has also been developed to run on Micro Vax II written in Fortran 77

Table 1. Main features of the computer program developed for the simulation of
solidification of casting and welding process

Feature	Implementation North star dimension	Micro Vax II
1. Language	Turbo Pascal	Fortran 77
2. Data input	Interactive computer graphics mesh generation data entry through key board against prompts)	Interactive mesh generation (data input is either through tablet or key board).
3. Modeling Library	Linear triangular element (LTE)	LTE, four and eight non-isoparamatric elements three noded axisymmetry element.
4. Output	Graphics display of solidification front at each time step; liquidus/ solidus freeze wave display at the end of solidification.	Graphics display of isotherms at each time step; Generation of output file for the plotter plots.
5. Geometry	2-D Cartesian and cylindrical	2-D Cartesian and cylindrical
6. Time marching scheme	Crank-Nicolson and Dupont II	Crank-Nicolson and Dupont II
7. Latent heat	Effective specific method	Effective specific method
8. Solution procedure	Gaussian elimination	Band matrix solver
9. Supporting features	Use of commercial drafting package (AUTOCAD) for defining the part geometry	Drafting package developed for defining part geometry

along with Tektronix Graphics terminal. Table 1 shows the main features of the two versions. The various parts of the package are also described in Fig. 1.

4.1 Preprocessor

Data preparation for FE analysis can be extremely time consuming and any simplification that can be designed to reduce the analyst's burden should prove highly beneficial. It has been recognized that the FEM preprocessor must reduce the time spent on data preparation to a bare minimum. This is achieved if more data preparation aspects are left to the computer itself. For any FE analysis, the data structure can be identified to consist of:

 i) the mesh data comprised of coordinates of nodes and nodal configuration
 ii) thermophysical properties of material
 iii) initial and boundary conditions
 iv) other data as required by the program.

Of all the data types listed above, the geometrical data lends itself to input simplification through interactive computer graphics. The visual feedback greatly

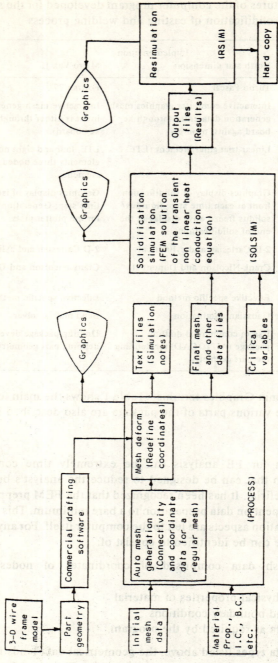

Fig. 1. The pre-processor, main module and post-processor of the software-CADCAST.

aids data preparation. Commercial drafting softwares are highly versatile and they help the analyst to draw various shapes and contours with ease. However, the actual analysis has to be carried outside the drafting environment. Further, commercial drafting softwares do not generate data which can be further used by the solution module. A preprocessor must therefore combine the advantages of a commercial drafting software with the flexibility of generating the data in the proper format which the solution module requires. In view of the above, a two-dimensional drafting software has been developed on TEK 4111 which overcomes the above drawbacks.

4.2 Interactive mesh deformation [22]

The preprocessor SOLSIM is based on the interactive distortion of a previously defined regular mesh. In any mesh comprised of linear triangular elements, the coordinates of any node can be redefined within limits without altering the nodal connectivity. Such deformation of an already existing regular mesh when carried out sequentially will result in faithfully reproducing any complex contour.

4.3 Data input

Other data like the material properties are input against prompts. All the parameters and their values are entered into a file whose name the user will have specified at the beginning. Text files are prepared and can be saved if desired, containing the values of all the parameters for any simulation run.

4.4 Main module

This part of the program does the finite element analysis of the problem. Data files previously prepared by using the preprocessor are the input files for this main module. A few key parameters like the time step value and pouring parameters are allowed to be input at this stage so that the same data can be used for different trial runs.

This module also creates output files so that the results of simulation can be examined later without resorting to detailed analysis each time. The computer asks for an output file name which is supplied at the beginning of simulation. The output is in graphic form with temperatures only at key locations being displayed. The position of the solidus isotherm is displayed at the end of each time step so that the solidification sequence can be seen by the user on CRT.

4.5 Postprocessor

The postprocessor reconstructs the results part of the analysis only, again with emphasis on graphics display. This provides for running the simulation at leisure for any re-examination. Also provided in the postprocessor is the facility for taking hard copy outputs of the display on a dot matrix printer/plotter. It has facilities for displaying isotherms and freeze waves separately.

412

5. SOLIDIFICATION MODELING AND ASSESSMENT OF THE SOUNDNESS OF CASTINGS

5.1 Solidification modeling

The CADCAST was used to simulate the solidification of Al-3% Cu-4.5% Si alloy cast in sand mould. The thermophysical properties of the mould material and the Al-3% Cu-4.5% Si alloy are given in Tables 2 and 3 respectively. Experimental results already published in the literature were used to validate the model. In addition the soundness of the casting was also predicted. For this the model for assessing the soundness proposed by Pathak [23] was considered. The following three factors were taken to govern the efficiency of feeding in a casting during solidification, ultimately controlling the mechanical properties.

1) *Directional solidification*: A clear idea of the extent of directional solidification in the casting can be obtained from the nature of the movement of the solidus isotherm.

2) *The extent of the pasty zone after the exit of the liquidus isotherm*: The contribution of this factor to feeding efficiency has been termed FEP (Feeding Efficiency as influenced by the Pasty Zone).

3) The time allowed for interdendritic feeding during the last stages

Table 2. Thermophysical properties of the mould material

Property	Value	
Thermal conductivity (cal/cm sec C)	0.0018	
Specific heat (cal/g C)	0.18 + 0.0001875 (T-30)	for T<400 C
	0.255	for T>400 C
Density (g/cm)	1.58	

Table 3. Thermophysical properties of Al-3% Cu-4.5% Si alloy

Alloy state	Property	Value
Solid	Thermal conductivity (cal/cm sec C)	0.463
	Specific heat (cal/g C)	0.278
	Density (g/cm)	2.690
Liquid	Thermal conductivity (cal/cm sec C)	0.220
	Specific heat (cal/g C)	0.300
	Density (g/cm)	2.560
Mushy	Thermal conductivity (cal/cm sec C)	0.340
	Specific heat (cal/g C)	1.219
	Density (g/cm)	2.625

of solidification (determined by the slope of the solidus isotherm): The contribution of this factor to feeding efficiency has been termed FET (Feeding efficiency as influenced by the time allowed during the last stages of solidification).

Soundness modeling

The FEP, FET parameters can be used to predict the percentage porosity and mechanical properties of a casting solidifying under the given heat transfer conditions. It is apparent that in order to make this an effective 'design' method, it should be possible to incorporate the same in numerical techniques. The FEP and FET for two dimensional freeze waves can be computed numerically and hence it is suggested that they could be used as relevant parameters for CAD of castings. They cover the entire spectrum of casting techniques from massive sand castings to gravity die castings including chill castings.

5.2 Two-dimensional freeze waves

Experimental techniques of thermal analysis restrict temperature measurements to only a few locations and, naturally, there is no information available on the temperature field existing throughout the body of the casting. Previous experimenters, examining the feeding behaviour of plate-cylinder combinations invariably measured the temperatures along the centre line of the casting and the solidification front was assumed flat i.e., the heat transfer was one-dimensional for them. A significantly different picture would result if a two-dimensional (2-D) heat transfer is considered. The liquidus and solidus fronts would show a definite gradient in the direction perpendicular to the centre line also. A better picture of the front shape and position could be obtained only with numerical simulation of solidification under 2-D heat transfer conditions.

Figure 2 shows a slice of plate casting taken along its axis at different time intervals during solidification. Both the liquidus and solidus fronts are shown curved with the 'channel' mouth opening towards the feeder. As time progresses, a trough-like surface travels through the casting and finally exits at the riser end, provided directional solidification is ensured. The surface generated by this trough-like solidification front in the x-y-t axis is termed as the two-dimensional freeze wave.

Feeding efficiency as influenced by the pasty zone (FEP)

It can be assumed that the difficulties in feeding are encountered only after the exit of the liquidus isotherm in the casting and only when mass feeding and interdendritic feeding are the only major feeding mechanisms. For efficient feeding, the pasty zone during the important stages of solidification should be small. The feeding efficiency as influenced by the pasty zone is defined as the inverse of the average 'residence time' of the mushy zone after the entire zone has become mushy. However, when these feeding efficiency factors are applied to a

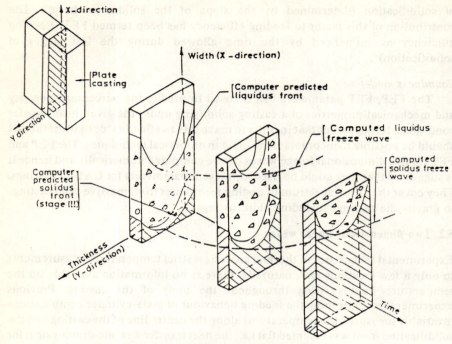

Fig. 2. 2-D liquidus/solidus freeze waves.

finite section of a casting, it would be desirable to take into account, the physical dimensions of the casting under study. The FEP is defined as [24]

$$FEP = \frac{W \, d_o}{\int_{t_1}^{t_f} a_p \, dt} \qquad (10)$$

Where a_p is the area of the mushy zone, t_1 is start of solidification time, t_f is end of solidification time, W is width of the casting and d_0 is the length of the casting under consideration.

With increase in FEP value it has been reported that mechanical properties are improved under different casting conditions. The FEP for given solidification time can be improved by quickening the onset of the progress of the solidus isotherm in the casting. Any adjustment that allows liquid feeding to continue for a longer duration and in turn delays the exit of the liquidus isotherm from the casting is bound to improve the feeding efficiency.

Feeding efficiency as influenced by the available time for interdendritic feeding (FET)

The final level of shrinkage porosity in the casting is determined by the conditions existing during the last stages of solidification. It has been reported that the time available for interdendritic feeding during the final stages of solidification is one of the factors that affects the feeding efficiency. The feeding efficiency as influenced by the interdendritic feeding time (FET) can be improved if, for a sweep of solidus isotherm $d(ds)$ across the casting, the time elapsed (dt) is increased. This condition should exist during the entire period of the movement of the solidus isotherm across the casting. Thus a low average value for $d(ds)/dt$ is preferable for efficient interdendritic feeding where ds/dt is the slope of the solidus isotherm. For two-dimensional case ds/dt is replaced by da_s/dt where a_s is the area under the solidus isotherm.

The FET for is defined by eqn. 11 [24].

$$\text{FET} = \frac{W \cdot d_o}{(da_s/dt)_{av}} \qquad (11)$$

5.3 Solidification times

The validity of the solution algorithm and the accuracy of the computations were checked against the experimental data on solidification time of the castings. The thermocouple locations in the experimental castings [24] were in the middle of the FEP zones along the centre of the casting. The time to reach solidus for all these thermocouple locations were extracted from the appropriate cooling curves. The theoretical solidification times for these thermocouple locations were estimated by interpolation since the simulation runs gave the temperatures at nodes in between two thermocouples.

The computed solidification times were plotted against the experimentally determined ones in Fig. 3. It is seen that the computed and experimental solidification times of the castings agree. An error analysis of the estimated values gave

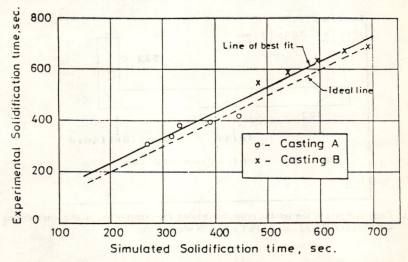

Fig. 3. Comparison of computed and experimentally determined solidification times.

416

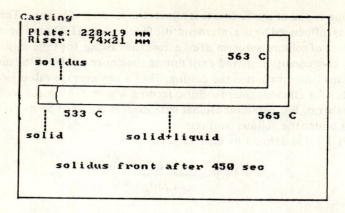

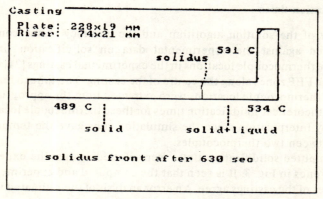

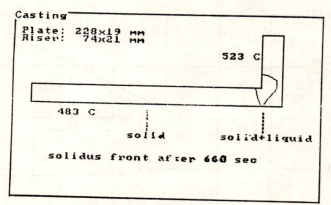

Fig. 4. Computer plot of the solidus isotherm showing the extent of the solid portion during the solidification of sand cast Al-3% Cu-4.5% Si alloy casting.

a standard error of estimate of 31.06 sec. The error is about 4–5% of the total solidification time.

Solidification profile

Hard copies of the CRT display during the solidification simulation of the Al-3% Cu-4.5% Si alloy plates cast in dry sand molds at three stages are given successively in Fig. 4. This sequence of displays clearly shows the shape and location of the hot spot. Any design modifications like providing chills, padding, hot tops, insulating sleeves etc., including changing the feeders can be easily accomplished interactively and the effect of these design changes on the solidification pattern assessed visually.

Freeze waves

The liquidus and solidus freeze waves along the length of the casting also were obtained as graphic display. The experimental cooling curve data were superimposed on these plots for comparison. Figure 5 is the hard copy of such a display for one of the castings. It is clearly seen that the general pattern of solidification through the casting centre is simulated well.

FEP and strength

In order to compute the mechanical properties cast by different authors, a factor UTS/UTS* was used. UTS is the ultimate tensile strength and UTS* is the minimum UTS value to declare the casting sound. The UTS/UTS* values for two castings were plotted against computed FEP values on a semi-log plot (Fig. 6).

6. SURFACE REMELTING PROCESS

Surface Treatment processes offer a solution to improve the wear resistance

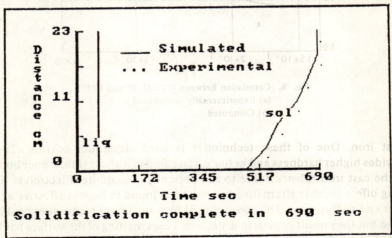

Fig. 5. Liquidus and solidus freeze waves in the sand cast Al-3% Cu-4.5% Si alloy casting.

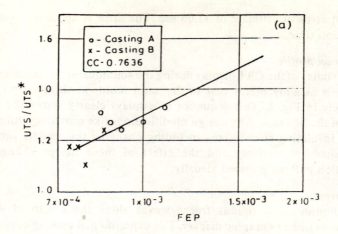

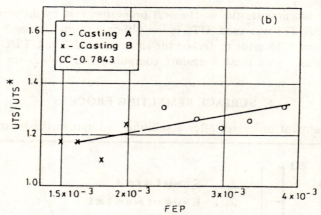

Fig. 6. Correlation between UTS/UTS* and FEP
(a) Experimentally determined
(b) Computed

of cast iron. One of these techniques is hard chromium coating. Though it provides higher hardness and better surface finish, it also exhibits poor bonding with the cast iron substrate due to the presence of graphite. Electroless nickel coating offers another alternative but has been found to be less effective against impact and sliding wear. The conventional heat treatment techniques could also be used but they result in pattern distortion. Laser melting of the surface has been reported to provide superior wear-resistant properties. TIG melting could also be used in a similar fashion to enhance the wear resistance of pattern grade cast iron. The authors have conducted experiments on TIG and Laser surface melting treatments earlier [25]. Subsequently they have tried to use the present package and predict the depth of the surface remelted track profiles and compared these with the results obtained in their earlier experiments.

A schematic diagram of the surface treatment is shown in Fig. 7. In order to calculate the temperature distribution in one particular cross-section, quasi-stationary conditions are assumed. In other words it is assumed that no heat transfer takes place in the X3 direction. Due to the symmetry in the problem only one half of the plane was considered for the analysis and the central plane was taken to be insulated. Boundary conditions are represented schematically in Fig. 7. The material properties used in this simulations are given [26].

Heat input to the workpiece

The basic heat input is the heat output from the laser or the arc. The net heat input to the work piece is given by

$$Q = \eta Q_L \quad \text{(Laser surface treatment)}$$
$$Q = \eta EI \quad \text{(TIG surface treatment)} \tag{12}$$

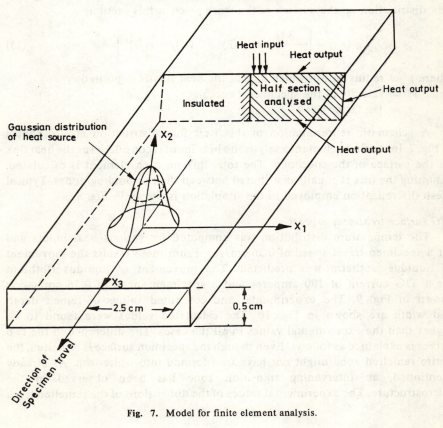

Fig. 7. Model for finite element analysis.

Fig. 8. Mesh discretization.

where Q is the heat input, η is the thermal coupling efficiency or efficiency of the process, and Q_L is the laser beam power, E is the input voltage and I is the input current. Heat input is assumed to be deposited on the surface as radially symmetrical normal distribution. This assumption results in the following heat flux distribution on the surface of the work piece at this section.

$$q\,(r,t) = \left[\frac{3\,Q}{\pi \bar{r}^2}\, e^{-3}(r/\bar{r})^2\, e^{-3}\,\{V(t-\tau)/\bar{r}\}^2 \right] \tag{13}$$

where r = radius within which 95% of the heat flux is deposited
  V = velocity of the specimen (workpiece)
  τ = lag factor $(\bar{r}/V)$

A schematic representation of this heat flux distribution is also shown in Fig. 7. In the finite element analysis the heat input is considered as the heat flux on the surface of the specimen. The total flux on each element is calculated, assuming the flux is equally distributed between the two binding nodes. Typical mesh discretization employed in the simulation is shown in Fig. 8.

TIG surface treatment process

The temperature distribution was computed for various heat inputs and for a specimen travel speed of 0.36 cm/sec. From these results the movement of liquidus isotherm was predicted. The movement of liquidus isotherm for a TIG current of 100 amperes and a specimen speed of 0.36 cm/sec is shown in Fig. 9. The experimental and computed values of remelt depth and width are shown in Fig. 10. The calculated values were found to be higher than the experimental values in all the cases. The difference in the two curves is explained as follows. Even though the specimen surface is remelted, the entire remelted zone might not have transformed into white iron. As already mentioned, an intervening transition zone has been observed in the microstructure. The experimental values of the dimensions of the remelted track

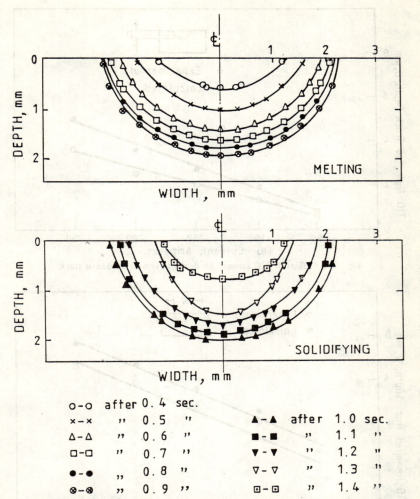

Fig. 9. Liquidus isotherm movements at different time intervals—TIG surface treatment.

given in Fig. 10 do not take into account this transition zone as it was found difficult to delineate the same clearly for measurements in the microstructure in many cases. This explains the discrepancy between the experimental and computed values of the remelt zone dimensions.

Laser surface treatment process

The mesh discretization used for FEM analysis is shown in Fig. 8. The programme developed estimated the temperature profiles in the specimen. From the above data, the movement of the liquidus isotherm depicting the melting and solidification is plotted as shown in Fig. 11. From this the

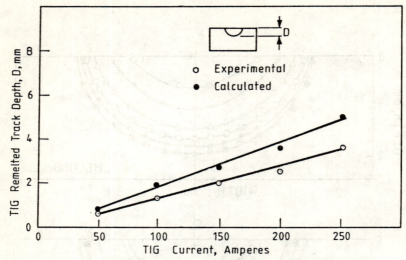

Fig. 10a. Effect of TIG current on depth of remelt hardened track.

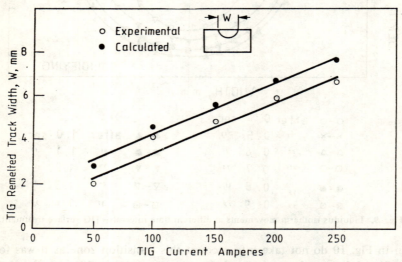

Fig. 10b. Effect of TIG current on width of remelt hardened track.

depth of remelted track can be estimated theoretically for a given heat input and interaction time. The temperature distribution was computed for different heat inputs ranging from 1.0 kW to 5.0 kW and for different interaction time from 0.07 seconds to 0.21 seconds. In the above calculations different values of heat input to the specimen, eqn. (12), were assumed and the corresponding remelt track depth was calculated with the help of the FEM programme developed. From the two values, namely heat input assumed in the calculations

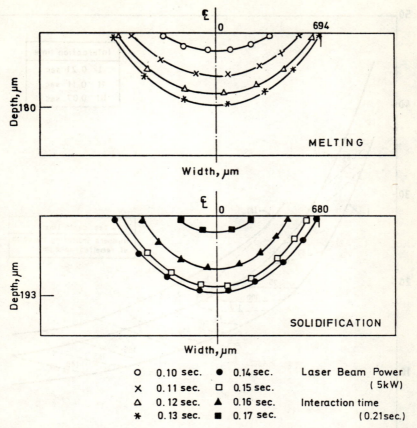

Fig. 11. Liquidus isotherm movements at different time intervals—laser surface treatment.

and the laser beam power to yield the same track depth, the ATCE Factor defined by eqn. (12) was calculated. It was observed that the ATCEF decreases with an increase in heat input or interaction time. The ATCEF versus heat input with the interaction time as the parameter is shown in Fig. 12. Foundrymen can utilize Fig. 12 to determine the laser treatment parameters once the desired remelt depth is fixed. The same figure also gives the ATCEF which can be utilized for two-dimensional FEM analysis to extrapolate or verify the theoretical values.

7. FLASH BUTT WELDING PROCESS

Flash Butt welding finds application in such diverse fields as automobile, railways, tool production and oil industries. Components manufactured include wheel rims, railway tracks, pipes, drill bits, etc. In spite of its importance as a manufacturing method, little work has been done to mathematically model the process and to optimize the process variables or predict the weld characteristics.

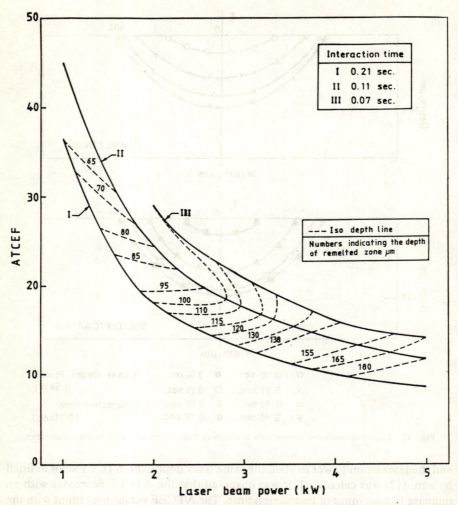

Fig. 12. Relationship between ATCEF and lasebeam power, remelted depth and interaction time.

While optimizing the process parameters, one of the factors that needs to be monitored for quality welds, is the heat-affected zone (HAZ). The HAZ in steel welds, is defined as the portion of the base metal which reaches a temperature above 800°C during the welding process [27]. In the case of Flash Butt welds, the HAZ is a zone of steep mechanical property variations and hence affects the weld performance. In contrast to fusion welds, a wider HAZ zone width is preferred to enhance the weld quality in flash butt welding. Hence it is important to estimate the width of the HAZ of FB welds. With this in view, the HAZ width of flash butt welds and were estimated and compared with the experimental results.

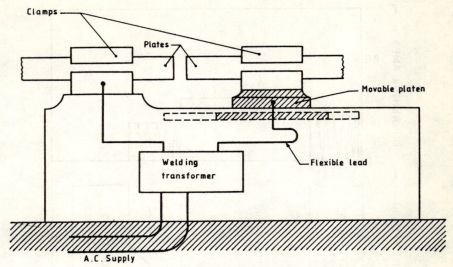

Fig. 13. Basic arrangement of flash butt welding.

In this paper, FBW of two plates was considered. The above mentioned CADCAST Two-dimensional FEM program was used to simulate the process. The variation of material properties, like thermal conductivity, specific heat, resistivity was considered in the simulation. The material properties used in this simulation study are given reference [27]. The process parameters are given in Table 4 for condition 1 and condition 2.

Figure 13 shows the basic arrangement of the Flash Butt welding process. A typical welding cycle of current and platen displacement with time is

Table 4. Process parameters used in the analysis—Flash Butt welding process

Parameter		Preflashing stage	Preheating stage	Flashing stage	Upsetting stage
Condition 1:					
Current	kA	35	50	30	0
Time sec	*	1.0	3.5	4.5	5.5
Length m	*	0.0040	0.0040	0.0090	0.0150
Condition 2:					
Current	kA	35	50	30	0
Time sec	*	1.0	5.0	6.5	7.5
Length m	*	0.0040	0.0040	0.0090	0.0150
* denotes cumulative values					

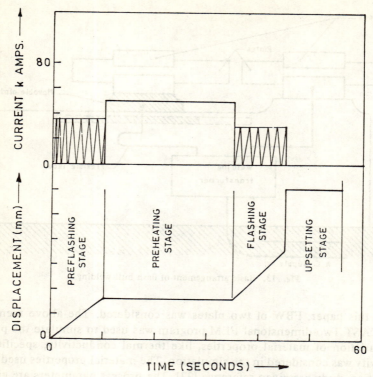

Fig. 14. Typical cycle showing variation of displacement and current with time.

as shown in Fig. 14. The boundary conditions for the present problem are shown in Fig. 15.

Mesh discretization

The mesh discretization used in this FE Analysis is shown in Fig. 16. As can be seen, a graded mesh was employed, the mesh in the plate side gradually increasing in size as it approaches the clamps. This kind of discretization is justified by the temperature distribution in the clamp side which did not increase above 150°C. The platen movement was calculated during every time step and the new geometry determined and mesh regenerated. Figure 16 represents the mesh discretization at the start of the computational process.

Temperature distribution

From the temperature outputs obtained at different time intervals and at different nodal points, Figs. 17a and 17b were drawn. Various stages of welding cycle times are indicated in the graphs. The temperature was found to increase under both conditions in preflashing, preheating and flashing stages. Subsequent to the end of the flashing stage, the plate was found to cool. It was observed that

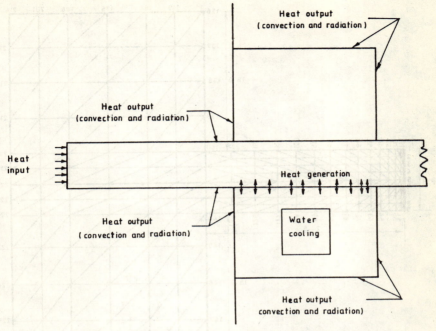

Fig. 15. Boundary conditions for finite element analysis.

the maximum temperature reached is 1150°C in condition 1 and 1580°C in condition 2. Conditions 1 and 2 represent two conditions where the specimen showed absence and presence of melting at the interface, respectively. In condition 2, melting was observed in the flashing stage. There was a sudden drop in temperature in a few elements and then an increase, when some of the elements reached liquidus temperature. This fluctuation might be due to a steep increase in electrical conductivity above liquidus. It should also be noted that elements which reached near liquidus temperature were removed from the plate in subsequent times. The same trend was not observed in the condition 1 where none of the nodes reached the liquidus temperature. It was also observed that the rate of cooling was higher for the elements which had crossed 1250°C due to the predominant radiative heat transfer.

Isotherms

The post-processor was used to analyse the computer output and to plot the isotherms. The isotherms for conditions 1 and 2 are shown in Figs. 18 a and b. Figure 18 clearly shows no oscillations in the solution. This conclusively proves the robustness of the Dupont scheme in spite of the fact that the problem is highly nonlinear. It is seen that the isotherms are parallel to the weld interface in the plate side. This may be due to uniform

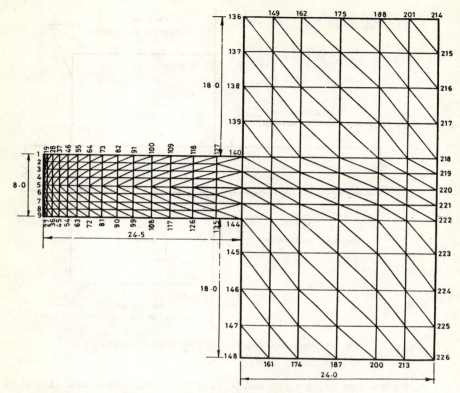

Fig. 16. Mesh discretization for FEM analysis.

internal heat generation and because the entire heat loss takes place within a short period and the impact of convective heat transfer due to the surrounding air is not realized. It appears that the water cooling was the main heat sink controlling the heat extraction rate.

Heat affected zone

The width of the HAZ depends upon its definition. The extremity of the Heat Affected Zone was generally associated with a minimum peak temperature that causes an observable microstructural change in the base metal [26]. The minimum peak temperature in the case of steels is A1 temperature. Due to non-equilibrium thermal cyclic nature, this temperature has been considered as 800°C. The same is shown in Fig. 17. The HAZ width was interpolated and computed, the values of which are given in Table 3 along with the experimental values. The computed HAZ widths were found to be higher than the experimentally observed values.

Figure 19 gives the hardness variations in the HAZ for the two conditions

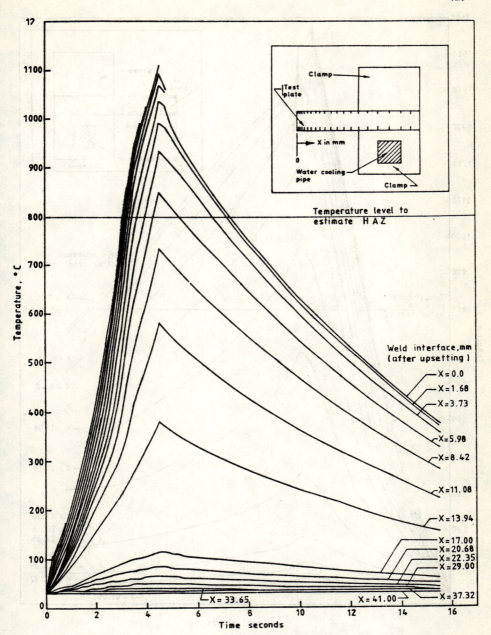

Fig. 17a. Variation of temperature with time at different locations—8.0 mm thick specimen, condition 1.

430

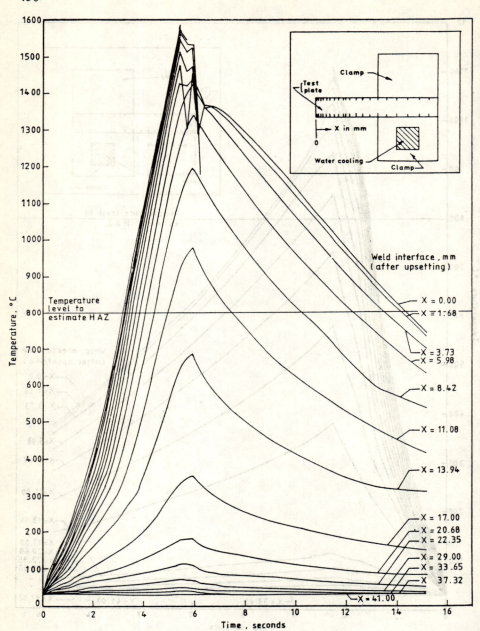

Fig. 17b. Variation of temperature with time at different locations—8.0 mm thick specimen, condition 2.

431

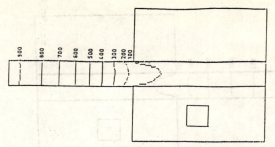

Isotherms at a time of 3.50 seconds from commencement of welding – Condition 1

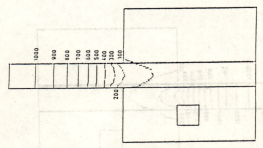

Isotherms at a time of 4.50 seconds from commencement of welding – Condition 1

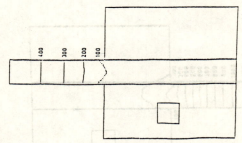

Isotherms at a time of 8.50 seconds from commencement of welding – Condition 1

Fig. 18a. Isotherms at different time intervals—condition 1.

432

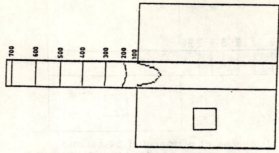

Isotherms at a time of 3.50 seconds
from commencement of welding – Condition 2

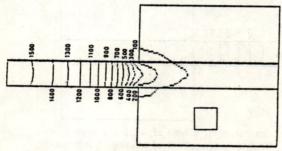

Isotherms at a time of 5.40 seconds
from commencement of welding – Condition 2

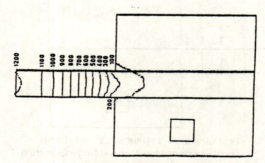

Isotherms at a time of 9.15 seconds
from commencement of welding – Condition 2

Fig. 18b. Isotherms at different time intervals—condition 2.

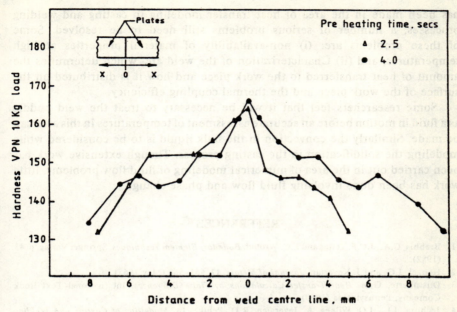

Fig. 19. Hardness variation across the welded zone.

Table 5. Comparison of HAZ widths obtained from different methods

Condition	HAZ from FEM mm.	HAZ from macro sample mm.	HAZ from hardness studies mm.
1.	7.043	6.25	6.5
2.	12.86	8.75	9.0

considered. The length over which the hardness is above the base metal hardness could also be taken as a criteria for assessing the HAZ width [26]. The values of the HAZ determined this way are also given in Table 3. It is observed that the macrostructural and the hardness based estimation of HAZ differ by a marginal amount. The computed values of the HAZ are determined by the criteria that a particular element exceeds a temperature of 800°C. This the authors feel explains the difference between the observed and the computed values.

8. SUMMARY

With the advent of digital computers with enormous calculational ability and graphics facility, it is attractive for the metal casting and welding engineers to incorporate analysis as a part of the design cycle. Although a great deal of progress

434

has been made in the area of heat transfer modeling in casting and welding processes, a number of serious problems still need to be resolved. Some of these problems are: (i) non-availability of material properties at high temperatures, and (ii) Characterization of the weld arc, which determines the amount of heat transferred to the work piece and how it is distributed on the surface of the work piece and the thermal coupling efficiency.

Some researchers feel that it will be necessary to treat the weld puddle as a fluid in motion before an accurate assessment of temperatures in this area can be made. Similarly the convection in the bulk liquid is to be considered while modeling the solidification in the casting process. Though extensive work has been carried out in the area of numerical modeling of fluid flow problems, little work has been done involving fluid flow and phase change.

REFERENCES

1. Brebbia, C.A., J.C.F. Telles and L.C. Wrobel. *Boundary Element Techniques*, Springer Verlag, p. 43 (1982).
2. Henzel, J.G. and J. Keverian. *Journal of Metals,* 17, 561–568 (May 1965).
3. Dusinberre, G.M. *Heat Transfer Calculations by Finite Diferences*, International Text Book Company, Pennsylvania (1960).
4. Jechura, J.L., J.O. Wilkes, A. Jeyerajan, R.D. Pehlke. In: *Modelling of Casting and Welding Processes,* Harold Brody and D. Apelian, Eds., TMS-AIME, pp. 73–82 (1981).
5. Weatherwax, R.B. and O.K. Riegger. *Trans. Soc. Automobile Engineers,* 80, sec 3, pp. 2112–2120 (1971).
6. Weatherwax R.B. and O.K. Riegger. *Trans. AFS,* 85, 317–322 (1977).
7. Morrone, R.E., J.O. Wilkes and R.D. Pehlke. *AFS Research Report,* (1972).
8. Jeyarajan, A. and R.D. Pehlke. *Trans. AFS.* 83, p. 405–412 (1975).
9. Jeyarajan, A. and R.D. Pehlke. *Trans. AFS.* 84, 647–752 (1976).
10. Jeyarajan, A. and R.D. Pehlke. *Trans. AFS.* 86, 457–464 (1978).
11. de L. Davies, V. and R. Moe. Proc. International Conference on Solidification, Sheffield, ASM, London, 1979.
12. Niyama, E. et al. Computer Simulation and Modeling Workshop, CIATF, Cairo, pp. 115–120 (Nov. 1983).
13. Ebisu, Y. *Trans. AFS.* 85, 643–654 (1977).
14. Stoehr, R.A. and H.D. Brody. *Solidification Contours of Selected Rolls,* The Roll Manufacturers Institute, Pittsburgh (1976).
15. Poirier, D.R. and C.A. Snieder. The Finite Element Technique Applied in Casting Solidification, Paper Presented at AIME Spring Meeting, Atlanta, Ga, USA (March 1977).
16. Mathew, J. and H.D. Brody. *Proc. Int. Conf. on Solidification,* The Metals Society, London, 244–249 (1979).
17. Morgan, K., R. Lewis and K.N. Seetharamu. *Simulation,* 55–63 (Feb 1981).
18. Thomas, B., I. Samarasekera and J.K. Brimacombe. *Modelling of Casting and Welding Processes II,* Op. cit., New Hampshire, pp. 3–12 (1983).
19. Richter, W. and P.R. Sahm. *Modeling of Casting and Welding Processes II,* Op. cit., pp. 251–257 (1984).
20. Huebner, K.H. *The Finite Element Methods for Engineers,* John Wiley and Sons Ltd., New York, USA (1975).
21. Hogge, M.A. In: *Numerical Methods in Heat Transfer,* R.W. Lewis, K. Morgan and O.C. Zienkiewicz, eds., John Wiley and Sons, pp. 75–90, 1981.

22. Prasannakumar, T.S. and O. Prabhakar. *AFS Transactions,* 715–718 (1986).
23. Pathak, S.D. and O. Prabhakar. *AFS Transactions,* 671–680 (1984).
24. Prasannakumar, T.S. Assessment of Soundness in Aluminium Alloy Castings: Through Process Modeling and Acoustic Emission Monitoring, Ph.D Thesis, IIT, Madras, 1986.
25. Chidambaram, S. et al. *AFS Transactions* (1987).
26. Venkataramani, R. Prediction of Heat-Affected Zone Widths in Flash-Butt Welds Using Finite Element Method. M. Tech Thesis, Department of Metallurgical Engineering, IIT, Madras, 1987.
27. Metals Handbook, Vol. 6, American Society of Metals.

20

Mathematical Modeling of Solidification Processes

JONATHAN A. DANTZIG

Department of Mechanical and Industrial Engineering
University of Illinois at Urbana-Champaign
1206 West Green Street
Urbana, IL 61801, U.S.A.

ABSTRACT

This article describes methods used for mathematical models to examine solidification processes, including formulations for the phase change problems. Several numerical methods are given, along with a discussion of their relative merits for different kinds of problems. Examples of mathematical models of foundry casting and continuous casting of steels are presented, with emphasis on the means for properly representing the physics of the problems in the boundary conditions. Finally, important problems remaining to be solved are outlined.

INTRODUCTION

The use of mathematical models to analyze, control and improve processes is becoming increasingly prevalent. The potential for these models to produce even small gains in efficiency in established processes is often seen as a motivating factor. There are other more important reasons, as well.

Mathematical models of complicated processes are often used to isolate and identify the effects of changes in individual variables on the products. Such changes can be very difficult to perform by themselves in an experiment, thus the model enables researchers to explore areas which might otherwise be unattainable. It is also possible with a good mathematical model to examine many more situations in a short time than could be considered in a comprehensive experimental program.

Finally, process models can serve as the link between what are often separate components in the engineering environment—the design and

438

manufacturing processes. Designers are afforded the opportunity to test the performance of their designs in the production environment, and process engineers can assess the effects of processing variables on the properties of the final product. This route promises to bring new products to market at greater speed and with higher overall quality than has previously been possible.

The promise stated above for mathematical models is predicated on the assumptions that they can accurately represent the physical processes involved, that calculations can be done in a timely and cost-effective manner, and that results can be interpreted in a reasonable way by the users of the models. The purpose of this paper is to examine the state of the art in mathematical modeling of solidification processes and to assess the degree to which these assumptions are satisfied in these models. This is not intended to be a comprehensive review of the area, concentrating instead on work in the author's laboratory on several different metallurgical processes. In particular, this paper will seek to identify the key features of mathematical models of casting processes, the means for assessing the validity of the models, and comment on needs for the future.

That the quality of a mathematical model can be judged by the quantitative agreement of its predictions with experimental observations is hardly a disputed point. The ease with which those results are obtained, measured by the difficulty involved in translating a physical situation into the model, and by the computational effort required to solve the resulting problem are also important features of the model. The art of the modeler, then, is to discard as many of the physical aspects of the process as possible for economy of solution, while retaining as much as necessary to adequately describe the process. The increasing power of computers along with their decreasing cost leads to continually changing views of the balance between these two competing goals.

Nevertheless, there are some features of mathematical models of solidification processes which are so basic that almost no simulation can fail to include them. Three primitives in this view would be: (1) accurate representation of geometry, (2) sensitivity to the thermophysical properties of the materials used in the process, and (3) adequate treatment of the evolution of latent heat of fusion during solidification. These factors are dominant in determining the progress and conditions of freezing in alloy systems, and thus they control the microstructure, properties and the appearance of defects.

The geometry of cast parts is critical to the progress of solidification. Owing to the nature of the diffusion equation, as discussed in the next section, the freeze front moves through the casting at a rate roughly inversely proportional to the square root of time. This was demonstrated experimentally under a variety of casting conditions by Chvorinov [1, 2] using both high and low thermal conductivity molding materials. Flemings [3] described the relation of these experiments to the governing heat flow equations, drawing upon analytical solutions for one dimensional heat flow dating back to Stefan [4, 6]. When heat flow occurs in more than one dimension, the freeze front moves away from all exterior surfaces, and the combined effect of several surfaces cannot be easily described analytically. Thus, the analytical solutions provide valuable insight into the behavior of solidifying metals, but they are extremely limited because no solutions exist for the realistic processes where geometry is complex, metals are cast with superheat, and latent heat is evolved over a range of temperatures in alloy systems [7].

For these reasons, most simulations of casting processes resort to numerical techniques. The two most common techniques are finite difference methods (FDM) and finite element methods (FEM) [8]. With a few notable exceptions, the FEM practitioners dominate the field because this technique is more readily adapted to problems in complicated geometries. Indeed, FEM simulation of casting processes has become increasingly widespread with the advent of powerful general purpose FEM codes running on engineering work-stations. This trend has been accelerated by the appearance of geometric modeling programs coupled to computer-aided design (CAD) packages, offering the opportunity for manufacturers to integrate design and manufacturing operations through the use of common databases.

The simulations proceed by specifying the geometry, material properties, boundary and initial conditions for the process. The corresponding heat flow equations are then integrated in time to determine the thermal history at various locations in the part. The solutions are no better than the input data, so that much of the work in this area in recent years has concentrated on finding appropriate values of the material properties, and proper specifications of the boundary and initial conditions.

Pehlke *et al.* [9] as part of a larger effort in applying CAD techniques to casting processes, have published a valuable compendium of thermal properties for many casting alloys. In some simulations, it is also of interest to determine thermal stresses resulting from solidification, which requires one to know the mechanical properties of materials over a broad range of temperatures, as well. These data are not nearly so well known for many important alloy systems, and this poses a serious limitation for the use of mathematical models. In the example given in the next section, the mechanical properties over the range of interest were available in the literature.

The specification of boundary and initial conditions remains an art, owing to the complexity of the physical processes which occur at interfaces. This is probably the most significant problem facing the engineer wanting to apply modeling techniques to manufacturing processes. Much of the work described in this paper focuses on efficient and appropriate representation of the existing boundary conditions.

MATHEMATICAL FORMULATION AND NUMERICAL METHODS

The mathematical representation of the heat transfer problem for solidification is given by the following equation [10]:

$$\rho \frac{d\varepsilon}{dt} = \nabla \cdot (k \nabla T) + Q \qquad (1)$$

where ρ is the density
  ε is the specific internal energy
  k is the thermal conductivity
  Q is the volumetric heating rate
  T is the temperature.

In an Eulerian representation, the time derivative is given by

$$\frac{d\varepsilon}{dt} = \frac{\partial \varepsilon}{\partial t} + \mathbf{u} \cdot \nabla \varepsilon \qquad (2)$$

where $\mathbf{u}$ is the velocity vector.

The specific internal energy is often decomposed into enthalpy plus the work of compressibility [11], viz:

$$H = \varepsilon + pV \qquad (3)$$

where H is the enthalpy
  p is the pressure
  V is the volume

for conventional casting processes, and for most metallic systems, the latter term is neglected. The enthalpy at a given temperature for an alloy which undergoes both congruent and extended single phase solidification is represented as

$$H(T) = H(T_o) + \int_{T_o}^{T} c_p \, dT + \int_{T_o}^{T} L_f \frac{df_s}{dT} \, dT + \int_{T_o}^{T} L_f \delta (T_m) f_m \, dT \qquad (4)$$

where f_s is the fraction solid at a given temperature
  f_m is the fraction of congruent melting phase
  c_p is the specific, or sensible, heat
  δ is the Dirac delta function
  L_f is the latent heat of fusion
  T_m is the congruent freezing temperature.

Equation (4) presupposes that the freezing temperature can be specified independently of the process dynamics, an assumption that will fail for processes which occur at high speed and materials with relatively sluggish kinetics.

To use eqn. (4) in eqn. (1) requires that a relation between the fraction solid and temperature be specified. Such relations can be derived from the redistribution of solute during freezing, and two extreme cases can be used profitably in many analyses. When the solidification process occurs slowly compared to the time scale for local diffusion in the solid (L^2/D), the familiar inverse lever rule accurately describes the fraction solid-temperature relationship (cf. 3). In that case, the mathematical form is

$$f_s = \frac{1}{1-k} \left\{ \frac{T-T_L}{T-T_o} \right\} \qquad (5)$$

where k is the segregation coefficient
T_L is the liquidus temperature
T_o is the freezing temperature for the pure phase.

In processes which occur at higher relative rates, limited diffusion occurs in the solid phase. It is typical of metal systems, however, that the interdiffusion coefficient is several orders of magnitude higher in the liquid phase compared to the solid. A limiting case can be identified, comprising infinite diffusion in the liquid and no diffusion in the solid, yielding the Scheil equation [cf. 3]

$$f_s = 1 - \left(\frac{T-T_o}{T_L-T_o} \right)^{1/(k-1)} \qquad (6)$$

Intermediate cases, where solute redistribution occurs to limited extent in both phases can be identified, and appropriate treatment for them can be found in the literature.

There are several numerical methods used for including the relations given above into the heat transfer equations. If the fraction of solidification which occurs as congruent freezing is small or zero, then the last term in eqn. (4) can be neglected, and the substitution of one of the relations for fraction solid versus temperature into the enthalpy, then differentiating with respect to temperature, yields a function which can be interpreted as a temperature dependent specific heat [12]. The range over which liquid and solid coexist, called the mushy zone, can then be identified at any time by its boundaries, the liquidus and solidus isotherms. On the other hand, if the congruent freezing process cannot be neglected, there is a singular surface associated with solidification which evolves latent heat as it moves according to the following eqn. (4).

$$k\nabla T . \mathbf{n}|_{\text{solid}} - k\nabla T . \mathbf{n}|_{\text{liquid}} = L_f v_n \qquad (7)$$

where $\mathbf{n}$ is the vector normal to the interface
v_n is the normal velocity of the interface.

Numerical techniques which follow the progress of this singular surface during the freezing process are known as front tracking methods.

Examples of both kinds of techniques abound in the literature [cf. 13], and the choice of method should depend on the particular problem at hand. Front tracking methods tend to require much more computation, and the programs can be very difficult to generalize to complex shapes, but they do afford greater accuracy in locating the phase boundary. Practitioners of the effective specific heat method can include congruent melting by "smearing" the transformation over a small range of temperatures to avoid the singularity in df/dT. This leads to an effective mushy zone in alloys where one should not exist, and care must be taken to avoid losing some of the heat of transformation in time integration or in spatial discretization. It is possible to avoid the singularity in the effective specific heat by instead using the enthalpy as an independent variable. Many examples of this method can be found, as well [13]. There are extra computations associated with the addition of a new independent variable, and in some algorithms, the enthalpy of freezing must be spread over a small temperature range to improve numerical stability.

An example of the convolution of the artificial freezing range introduced in the specific heat with the temporal and spatial discretizations can be seen by considering the one dimensional freezing of a eutectic iron alloy. The freezing of a eutectic iron slab 40 mm thick, poured into cold sand with a superheat of 55K, was simulated using varying numbers of elements through the half-thickness, and with several different time step sizes. The simulation was performed using three dimensional isoparametric linear elements in ANSYS [14], which employs a linear extrapolation procedure to evaluate temperature dependent properties at succeeding time steps, and does not iterate at each time step. The figure of merit used to evaluate the solution was the error in computed time to fully solidify the slab.

The results are summarized in Fig. 1, where each square represents a numerical experiment. It can be seen that higher mesh densities and smaller time steps improve the accuracy of the solution until four elements through the half-thickness and 200 time steps through the freezing range are reached. Further refinement beyond these values failed to improve the results significantly. However, coarsening either the spatial or temporal discretization above these limits led to rapidly deteriorating and unpredictable results.

The explanation for these results lies in the way in which the properties were evaluated at successive time steps in the solution. The elements were numerically integrated using an eight point rule, and the latent heat for iron was spread over 8K, centered on the actual eutectic temperature. If the temperature gradient is high, as it is early in the simulation, then the artificial freezing range may lie entirely between neighboring integration points. In that case, the simulation would use a much lower than appropriate specific heat, resulting in rapid progress of the mushy zone, without fully accounting for the latent heat. Similarly, the forward extrapolation in time can entirely miss the freezing range if the range is small and the time steps are large.

Various strategies to overcome this problem have been used. The most effective of these track the evolution of latent heat within the elements. The extra heat may be included as a heat source term [15], or it can be included in the specific heat. In the latter methods, an enthalpy-temperature curve is stored, and the specific heat is constructed from spatial gradients

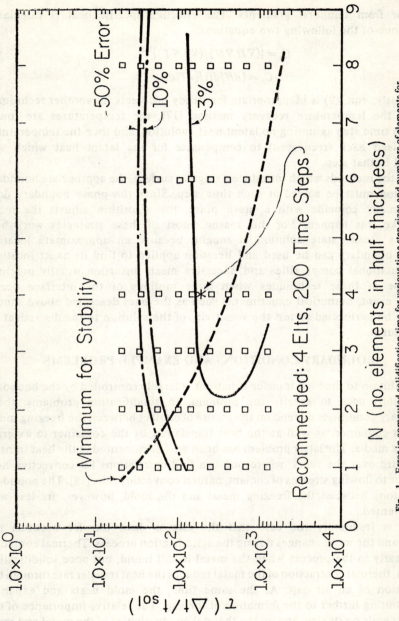

Fig. 1. Error in computed solidification time for various time step sizes and numbers of elements for an iron slab.

[16] or from temporal gradients. The effective specific heat is calculated using one of the following two equations:

$$C_p = \{(\nabla H.\nabla H)/(\nabla T.\nabla T)\}^{1/2} \tag{8}$$

$$C_p = (dH/dt)/(dT/dt) \tag{9}$$

Obviously, eqn. (9) is inappropriate for steady problems. In another technique, called the temperature recovery method [17], the temperatures are found at each time step assuming no latent heat evolution, and then the temperatures are raised back (recovered) to compensate for the latent heat which was missed in that step.

In the methods which introduce a singular surface, the appropriate boundary condition must be solved at each time step. Since the phase boundary does not usually coincide with a mesh point, the algorithm adjusts the mesh to make this happen. For this reason, most of these strategies work best when a steady state solution is sought, because an approximate location of the boundary can be used and iteration applied to find its exact location. Computational complexities and excessive mesh distortion usually preclude the use of these techniques when large motions of the interface occur. In any event, numerical experiments such as the ones described above should always be performed to test the sensitivity of the solution to the discretization procedures.

BOUNDARY CONDITIONS AND EXAMPLE PROBLEMS

The solution to the heat transfer equations is largely controlled by the boundary conditions used to specify the problem. In solidification problems, these boundary conditions depend on the nature of contact between the freezing metal and its container, as well as the heat transferred by the container to external cooling media. The latter problem has been well documented in the heat transfer literature over the years, where one can find correlations for convective heat transfer to flowing streams of coolant, natural convection, etc. [18]. The boundary conditions between the freezing metal and the mold, however, are less well documented.

It is frequently the case that the rate of heat transfer between the metal and the mold changes during the solidification process. Thermal contact is good early in the process when the metal is still liquid, but once solidification begins, thermal contraction of the metal reduces the heat transfer rate through the formation of an air gap. At the same time, the mold heats and expands, contributing further to the formation of the gap. The relative importance of the gap depends on its size, and on the thermal conductivities of the metal and mold materials. Two examples given below illustrate this point. In one case, foundry castings in sand molds, the gap is insignificant because the thermal conductivity of the sand is very poor, and the air gap corresponds to about one extra mm of sand

[21]. In the other, continuous casting of steel in water-cooled copper molds, the air gap almost exclusively controls the heat transfer.

In sand castings, the thermal conductivity of the mold material can be less than 1% of that of the metal [9]. In that case, the temperature gradients in the sand will be much steeper than in the metal, which requires use of a fine spatial discretization in the sand to resolve the gradient. This is a particular disadvantage because the mold is very extensive, and the solution there is usually of less interest. Thus the simulations can become prohibitively expensive due to the "excess baggage" in the problem.

A number of researchers have developed techniques for replacing the sand mold in these simulations in favor of a set of boundary conditions applied on the surface of the casting itself [7, 19, 20]. These methods all take advantage of the fact that the heat transfer can be represented locally by a function which depends on the local geometry. Probably the most versatile method for three-dimensional problems is called the boundary curvature method [7]. The surface of the casting is represented on an element by element basis by one of a set of trial functions which contain information about how heat flows into a sand mold from a curved surface. The local representation for curvature is selected by examining the local topography of the casting in the region affected by heat flow. The trial functions are cylinders or spheres, parameterized by the radius. These functions are precalculated in a manner described in the references and correlated to the simulation at hand by the surface temperature and the total heat flow through the surface. Graphical representation of a typical set of the trial functions are given in Fig. 2.

An example of such a calculation is shown in Fig. 3, a selected portion of the simulation of the temperature distribution in an engine block casting. The figure also shows the association of boundary curvatures with locations on the surface, found by an automated procedure described elsewhere. The requirements described earlier for the time and spatial discretization necessitated that the mesh have over 8108 nodes and 8665 elements, and that 500 time steps be used to cover the solidification and subsequent cooling of the part. This consumed approximately five cpu hours on a Cray X/MP 48, operating under CTSS. Earlier work demonstrated that the use of the boundary curvature method can reduce the total calculation time in three-dimensional problems by over 90% [21], indicating that this simulation would have been prohibitively expensive (> 500 cpu hours) without using the special boundary conditions.

In the continuous casting of steels, the formation of an air gap between the strand and the mold has long been recognized as the controlling factor in the heat transfer rate from the steel. It is quite difficult to measure this gap, however, because it it typically less than 0.2 mm and the steel surface temperature is nearly 1700 K. In recent work, this gap has been calculated for steel billets cast in round molds by computing both the temperature and stress fields in the billet and mold material, accounting for the thermal

446

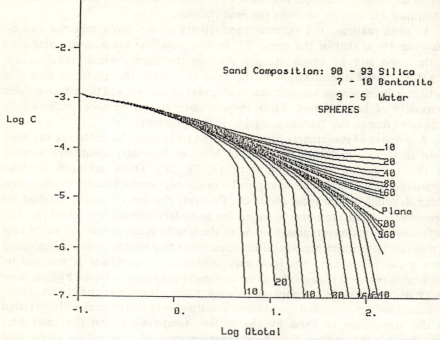

Sand Composition: 90 - 93 Silica
7 - 10 Bentonite
3 - 5 Water
SPHERES

Fig. 2. Computed heat transfer coefficients in W/mm²K vs. integrated heat flux in J/mm² for various spheres. Radii are given in mm, those above the plane being divergent, and those below being convergent.

distortions of both the billet and the mold [22]. The thermal and stress problems were assumed to be uncoupled except through the heat flow boundary condition, were

$$h_{\text{eff}} = \frac{k_g}{\Delta_g} + h_{\text{rad}} \qquad (10)$$

where h_{eff} is the effective heat transfer coefficient
k_g is the thermal conductivity of the gap medium
Δ_g is the size of the gap
h_{rad} is a contribution from radiation, usually negligible.

This decoupling is valid when the heat of deformation is negligible, which certainly applies to this case.

In the analysis, estimated values were used for the heat transfer coefficients between the strand and mold to arrive at a temperature distribution. The associated stress analysis, driven by the thermal distortions of the mold and strand was then solved. The local air gap comes from this analysis, which was then

447

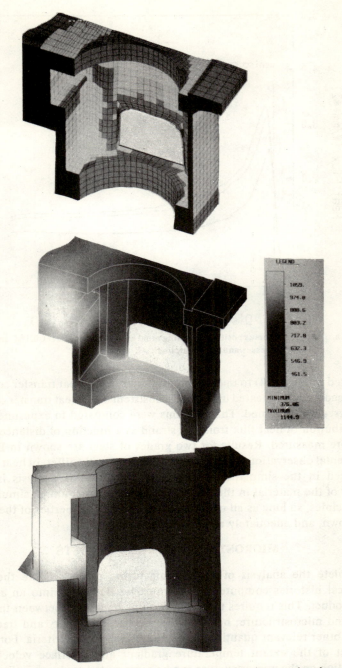

Fig. 3. Engine block simulation showing groups of elements with common assigned curvatures and computed temperature distribution one second after pouring.

448

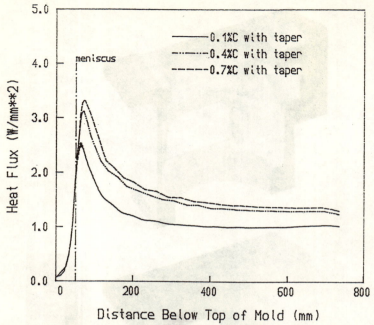

substituted into eqn. (8) to update the estimates of the heat transfer coefficients. This procedure was repeated until a self-consistent set of heat transfer coefficients and stresses was obtained. The solutions were compared to experiments where the distribution of heat flux from the strand as a function of distance down the mold were measured. Results for two grades of steel are shown in Fig. 4. The experimental observation that the two grades have vastly different heat fluxes was reproduced in the simulations. The importance of this analysis is that the behavior of the material in the manufacturing process could be simulated from first principles, so long as all of the relevant physical properties of the materials were known, and adequately represented in the model.

MICROSTRUCTURES AND DEFECTS

To complete the analysis of the manufacturing processes, the thermal and mechanical histories computed above must be translated into an assessment of the product. This requires that relationships be known between the thermal history and microstructure; between the mechanical history and fracture; and between other relevant quantities and appropriate failure criteria. For example, the effect of the extant temperature gradient and interface velocity during solidification on the microstructure can be represented as shown in Fig. 5 for the tin-lead system. Under conditions of slow growth in high temperature gradients,

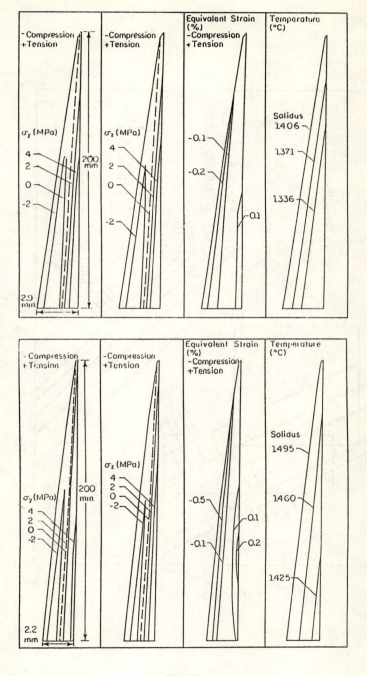

Fig. 4b.

450

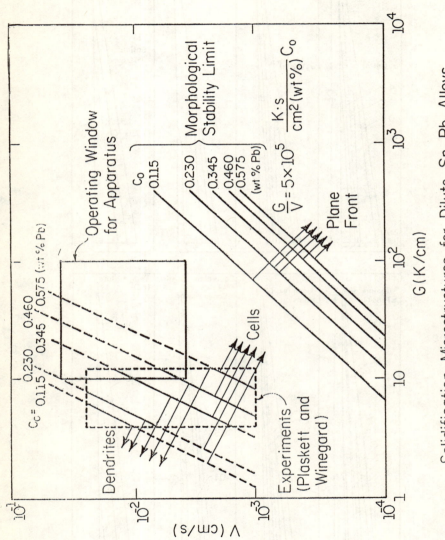

Fig. 5. Microstructure map for dilute alloys of Pb in Sn.

Solidification Microstructures for Dilute Sn-Pb Alloys

the solidifying interface is stable against local solute segregation and planar front growth occurs. As the interface speed increases and/or the temperature gradient decreases, the interface becomes unstable and cells or dendrites appear, depending on the degree of departure from stability. Superimposing the results of the thermal simulations described above on such maps provides the ability to predict the grain structure in the final microstructure.

Similar criteria can be advanced for the prediction of more complex defects in cast parts. Correlations between the temperature gradient during solidification and the porosity in steel castings have been given [23]. The analyses can also be examined to find the last place to freeze in complex-shaped castings to determine whether shrinkage cavities due to poor feeding will occur.

Mechanical failures can also be predicted if stress analyses have been performed. Nearly all alloys of interest become coherent after a substantial portion has solidified, say 60–80%. In the temperature range immediately below this, these alloys are susceptible to hot cracking. In the case of the continuously cast steel billets considered above, a cracking criterion was used which said that the billet would crack if the plastic strain exceeded 0.1% in the crack susceptible range [24]. Using this criterion, it was correctly predicted that 0.1% carbon steels would crack under the casting conditions described, but that 0.4 and 0.7% carbon steels would not (See Fig. 4).

DISCUSSION AND FUTURE DIRECTIONS

The role of experiments cannot be overestimated in any discussion of mathematical modeling of physical processes. As noted above, the true test of any mathematical model lies in its ability to properly predict the outcome of experiments. It is crucial to define and perform key experiments to provide these tests so that the simulations can be used with confidence to predict the results of other situations which were not covered in the experiments. The experiment also provides a reduction to practice of concepts which may only exist as model results up to that time.

The material properties which are used in the models also can only be reliably obtained in careful experiments. The models can be used to help identify the needed properties; the sensitivity of the analysis to that property being a guide to the precision required for the results. Similarly, the processing-property relations described above can usually be found experimentally much more easily than by theoretical prediction, because the physical processes are not completely understood. Finally, a good model of an experiment can be used to better define and characterize the experiment itself, improving the insight into the important mechanisms.

While the recent past has seen enormous growth in the area of numerical techniques and the sizes of problems which can be tackled, there remains much room for improvement. It has been said that mathematical simulations always take two to three hours on whatever computer is available, and that as computers become larger and more powerful, the engineer's appetite for problem complexity increases accordingly [25]. The author's experience cannot argue against this statement.

Some important problems in solidification processing simulation remain for which there is no satisfactory solution technique. One of the most important ones is the lack of robust and general techniques for solving mold-filling problems, where the filling, heat transfer, and sometimes chemical reactions occur simultaneously. Some good work has been done in this area, notably at the University of Pittsburgh [26], but to date most has been two dimensional. The numerical techniques used are likely to be prohibitively large and expensive in three dimensions, and new methods are needed.

There has been much recent activity in trying to predict the microstructure from the solidification analysis [27, 28]. One of the difficulties has been that the process dynamics control the temperature at which phase changes take place, and this makes the conventional methods of prespecifying the material properties as functions of temperature incorrect. While this could be viewed as just another nonlinear feature to be solved by iteration, there are fundamental problems here, as well. These dynamic problems are poorly understood, and many of the microstructure models are difficult to verify, and hence may not be worthy of the increased computation. This is a good example of an area where mathematical modeling in conjunction with a solid experimental approach would be valuable.

An important extension of mathematical modeling techniques will be to integrate them into the product engineering process. An adequate model of a manufacturing process can also be used to improve the control of the process, provided that real time sensing of important process variables can be made. The simulation can provide a common ground for the design and manufacturing engineers to collaborate on new products. Indeed, the use of common databases for these products could result in substantial savings for many manufacturers.

CONCLUSIONS

It is now possible to perform large-scale simulations of solidification processes, with realistic expectations that the results will correspond to observations in the real world. The simulations are no better than the material properties and boundary conditions which are entered into them, and these remain important areas for future research. Further developments in the area of prediction and control of microstructures await better experimental correlations between the fundamental variables in the process and the resulting structures.

The advent of more powerful computers makes possible the integration of design and manufacturing functions in a computer-integrated engineering environment. New techniques are needed to address important problems having to do with mold filling with simultaneous heat transfer.

ACKNOWLEDGEMENTS

The work described in this article was supported by the National Science Foundation, through the Division of Materials Research and the Production Engineering Program, Deere and Company, the University of Illinois Materials Processing Consortium, and the National Center for Supercomputing Applications. The work was performed by the author along with several graduate students, including Jeff Wiese and Jim Kelly. The author gratefully acknowledges their contributions.

REFERENCES

1. Chvorinov, N. *Giesserei,* **27** (10), 177 (1940).
2. Chvorinov, N. *Giesserei,* 27 (11) 201 (1940).
3. Flemings, M.C. *Solidification Processing,* McGraw-Hill, New York (1947).
4. Stefan, J. *Ann. Phys. and Chem.,* **42,** 269 (1891).
5. Adams, C.M. and H.F. Taylor. *Trans. AFS,* **65,** 170 (1957).
6. Neumann, F. See Riemann-Weber, *Die Partiellen Differentialgleichungen der Mathematischen Physik,* **2,** 121 (1912).
7. Dantzig, J.A. and S.C. Lu. *Met. Trans.,* **16B** (2) 195 (1985).
8. Dantzig, J.A. and J.T. Berry. eds., *Modeling of Casting and Welding Processes-II,* TMS-AIME, Warrendale, PA (1984).
9. Pehlke, R.D., A. Jeyarajan and H. Wada. *Summary of Thermal Properties for Casting Alloys and Mold Materials,* Univ. of Michigan, Ann Arbor, NTIS-PB83-211003 (1982).
10. Truesdell, C. *Rational Thermodynamics,* McGraw-Hill, New York, p. 28 (1969).
11. Progogine, I. and R. Defay. *Chemical Thermodynamics,* Longman, London pp. 66–67 (1962).
12. Morgan, K., R.W. Lewis and O.C. Zienkiewicz. *Int. J. Num. Meth in Engr.,* **12,** 1121 (1978).
13. Crank, J. *Free and Moving Boundary Problems,* Oxford University Press, Oxford (1984).
14. DeSalvo, G.S. and J.A. Swanson. *ANSYS Theoretical Manual,* Swanson Analysis Systems, Inc., Houston, PA (1983).
15. Rolph, W.D. and K.J. Bathe. *Int. J. Num. Meth. Eng.,* **18,** 119 (1982).
16. Lemmon, E. *Numerical Methods in Heat Transfer,* John Wiley and Sons, New York, 1981.
17. Hong, C.P., T. Umeda and Y. Kimura. *Met. Trans.,* **15B,** 101 (1984).
18. Rohsenow, W.M., J.P. Hartnett and E.N. Ganic. *Handbook of Heat Transfer,* 2nd Ed., McGraw-Hill, New York (1985).
19. Niyama, E. *J. Japan Foundrymen's Soc.,* **49,** 26 (1977).
20. Wei, C. and J.T. Berry. *Int. J. Heat and Mass Transfer,* **25,** 590 (1982).
21. Dantzig, J.A. and J.W. Wiese. *Met. Trans.,* **16B** (2), 203 (1985).
22. Kelly, J.E., K.P. Michalek, B.G. Thomas and J.A. Dantzing. *Met. Trans.,* **19A,** 2589 (1988).
23. Walther, M. In: *Modeling of Casting and Welding Processes,* op. cit.
24. Brimacombe, J.K. and K. Sorimachi. *Met. Trans.,* **8B,** 489 (1977).
25. Giamei, A.F. private communication.
26. Stoehr, R. and W.S. Hwang. In: *Modeling of Casting and Welding Processes,* op. cit.
27. Rappaz, M. and E. Blank. *J. Crystal Growth,* **74,** 67 (1986).
28. Rappaz, M. and Ph. Thevoz. *Acta Met.,* **35,** 1487 (1987).

21

Electromagnetics in Metallurgical Processing

K.V. RAMA RAO AND J.A. SEKHAR

Defence Metallurgical Research Laboratory, Hyderabad 500 258 India

ABSTRACT

This article is aimed at documenting the various innovations in metallurgical processing that have been brought about by electromagnetic devices. We have considered applications developed on account of the heating due to induced currents as well as on account of the forces developed on the body carrying the induced currents. A brief discussion on Maxwell's equations, forces and the resultant induced fluid flow has also been incorporated so that an engineering 'feel' for the magnitude of these may be possible. Finally we have reviewed the effect of electromagnetically-induced fluid flow on the metallurgical microstructure in large ingots and net-shape castings.

1. INTRODUCTION

There is a growing interest in the application of electromagnetics for metallurgical processing. Common applications are induction furnaces, induction stirrers, electromagnetic levitators and the use of electro-magnetic forces and resultant fluid flow for casting control, of these there are already a large number of electromagnetic furnaces available for commercial use. The main advantage of these devices is the absence of any physical contact between the material and the melting device, thus leading to clean heating and additionally, homogenization is achieved due to efficient stirring. Although these applications have been attracting the interest of a large number of metallurgists all over the world not much has been understood about the interaction of the various electromagnetic forces that are present and the flow patterns in the melt. In this report we have made an attempt to present a brief description of the basic equations and their solutions, a few mathematical models already in use, typical devices and the nature of metallurgical structure modifications obtained by electromagnetic stirring.

456

When an electrically conducting fluid flows in the presence of a magnetic field two kinds of effects are noticed: due to the motion of the electrically conducting fluid across the magnetic lines of force eddy currents are generated and the magnetic fields associated with these currents contribute to the changes in the original magnetic field, and secondly, the interaction of the currents and the magnetic field produces a body force called the Lorentz force.

To understand the nature and magnitude of these currents and the forces, one has to solve the electromagnetic equations (Maxwell's equations) and obtain the field and current distributions. Once these are known their effect on the fluid may be studied by solving the Navier Stokes equations in which the Lorentz force term is introduced. In the next section we present a brief discussion of these equations and the order of magnitude approximations for the various physical quantities like skin depth and fluid velocity.

2. CALCULATIONS OF FORCE FIELDS

To calculate the fields and the related forces, we consider an induction furnace as shown in Fig. 1. The equations governing the electromagnetic field are given by Szekely [1]

$$\nabla \times E = -\partial B / \partial t \tag{1}$$
$$\nabla \times H = J \tag{2}$$
$$\nabla \cdot H = 0 \tag{3}$$
$$B = \mu_e H \text{ and } J = \sigma (E + \mu_e V \times H) \tag{4}$$

When the melt velocity is constant and the flow is parallel to the field, the equation may be approximated as

$$J = \sigma E$$

Where E is the electric field V/m

B is the magnetic flux density Wb/m^2

H is the magnetic field intensity A/m

J is the current density A/m^2

μ_e is the permeability of the metal H/m

σ is the conductivity

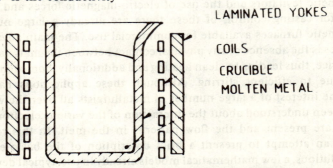

LAMINATED YOKES

COILS

CRUCIBLE

MOLTEN METAL

Fig. 1. Schematic sketch of a common induction furnace [1].

Eqns. (1-4) are the Maxwell equations. The Navrier-Stokes equations, governing the melt velocity V, are given by

$$\varrho \, (V \cdot \nabla) \, V = - \nabla P + \nabla . \tau + F_b \tag{5}$$

Where V is the velocity vector, P is pressure gradient τ is the stress tensor and F_b is the body force vector given by

$$F_b = J \times B \tag{6}$$

Eqns. (1-4) with appropriate boundary conditions have to be solved to obtain the current distribution. The eqns. (5 and 6) have to be solved to obtain the melt velocities and other fluid characteristics. The Lorentz force, may be written as

$$F_b = 1/\mu_e \, (\nabla \times B) \times B$$

$$= - \nabla \, (1/2\mu_e) \, |B|^2 + (1/\mu_e) B . \nabla B$$

Which shows that the Lorentz force is the sum of pressure $1/2\mu_e \; |B|^2$ transverse to the lines of force and a tension $\dfrac{|B|^2}{\mu_e}$ along the lines of force.

Now to understand the interaction of the magnetic field and the fluid flow, let us consider the flow and mixing in an induction furnace. Let us first find the order of magnitude of various quantities involved. Following Szekely and Chang [1], the equation of motion is written as

$$\varrho \, (V \cdot \nabla) \, V = - \nabla P + \nabla \cdot \tau + J \times B \tag{7}$$

If in eqn. (7) the inertial forces dominate, i.e., if the Reynolds number is very high, then one may neglect the pressure head and the viscous forces and, thus we have

$$2 \varrho \, (V \cdot \nabla) \, V = J \times B$$

Now $O \varrho \, (V \cdot \nabla) V = \varrho U_o^2 / L$ where U_o is some representative velocity and L is some representative length.

$$O \, (J \times B) = B_o^2 \, f \, L \, \sigma$$

where $f = w/2\pi$ is the frequency and on taking $B_o = J_o \, \mu_e \, L$ we have

$$\varrho \, U_o^2 / L = \sigma f \, L^3 J_o^2 \, \mu_e^2 \tag{8}$$
$$\text{or } U_o = J_o \, (\varrho^{-1} \, f \sigma L^2 \, \mu_e^2)^{1/2}$$

which indicates that the individual velocities are directly proportional to the coil current J_o and also to the square root of the frequency. This relationship between the coil current and the velocity has been verified by Dragunkina and Tir [6] The depth of penetration of the magnetic field into melt is given by

$$\delta = (2/ \mu_e \, \omega \sigma)^{1/2} \tag{9}$$

Table 1. Numerical values for velocity and skin depth as a function of frequency

Frequency	Velocity	Skin depth
25 Hz	0.055 m/sec	0.06 m
50 Hz	0.078 m/sec	0.048 m
60 Hz	0.085 m/sec	0.044 m
100 Hz	0.1 m/sec	0.034 m
1 KHz	0.3478 m/sec	0.011 m
3 KHz	0.5477 m/sec	6.28×10^{-3} m
10 KHz	1.0 m/sec	3.4437×10^{-3} m
450 KHz	6.70 m/sec	5.133×10^{-4} m

ϱ	density of the liquid	6.8×10^3 kg/m^3
σ	conductivity of the liquid	2.13×10^6 s/m
J	current	500 A
μ_e	magnetic permeability	1.26×10^{-6} H/m
L	length of the liquid column	1 m

In Table 1 we presented some typical values of the velocity and skin depth for various values of the frequency calculated using the above formula.

To solve the electromagnetic equations, hereafter called (e.m.e) a vector potential A, is introduced as

$$B = \nabla \times A \text{ and } E = -\frac{\partial A}{\partial t}$$

$$\text{with } \nabla \cdot A = 0$$

The e.m. eqns. (1) to (4) may now be recast as

$$\nabla^2 A = \mu_e \sigma \frac{\partial A}{\partial t} \tag{10}$$

In cylindrical coordinates assuming symmetry (assuming one component A_θ only) the equation may be written as

$$\frac{1}{r} \frac{\partial}{\partial r} \left(r \frac{\partial A_\theta}{\partial r} \right) + \frac{\partial^2 A_\theta}{\partial z^2} - \frac{A_\theta}{r^2} = \mu_e \sigma \frac{\partial A_\theta}{\partial t} \tag{11}$$

The magnetic field components are given by

$$B_z = \frac{1}{r} \frac{\partial}{\partial r} (r A_\theta) \tag{12}$$

$$B_r = -\partial A_\theta / \partial Z \tag{13}$$

$$J = -\sigma \partial A_\theta / \partial t \tag{14}$$

and the body force $J \times B$ is then given by

$$F_z = J B_r \tag{15}$$

$$F_r = J B_z \tag{16}$$

and
$$F_\theta = 0.$$

Eqns. (11) to (14) are to be solved with proper boundary conditions for obtaining the current density J and the magnetic field distribution B. In addition, the following relationships have to be solved separating two different media: (i) The normal component of the magnetic induction is continuous i.e., $\hat{\eta} \cdot (B_1 - B_2) = 0$, and any discontinuity in the tangential component is equal to the surface current J_s or $\hat{n} \times (B_1 - B_2) = \mu_o J_s$.

Once the force $J \times B$ is known, for obtaining the velocities one has to solve the turbulent Navier-stokes equations. For cylindrical symmetry they are given by:

$$\varrho \left(u_r \frac{\partial u_r}{\partial r} + u_z \frac{\partial u_r}{\partial z} \right) = -\frac{\partial P}{\partial r} + \frac{\partial}{\partial r} \left\{ \frac{u_e}{r} \frac{\partial}{\partial r} (r\, u_r) \right\} + \frac{\partial}{\partial z} \left(u_e \frac{\partial u_r}{\partial z} \right) + J_\theta B_z \tag{17}$$

$$\varrho \left(u_r \frac{\partial u_r}{\partial r} + u_z \frac{\partial u_z}{\partial z} \right) = -\frac{\partial P}{\partial z} + \frac{1}{r} \frac{\partial}{\partial r} \left(u_e r \frac{\partial u_z}{\partial r} \right) + \frac{\partial}{\partial z} \left(u_e \frac{\partial u_z}{\partial z} \right) + J_\theta B_r \tag{18}$$

Combining these two equations one obtains:

$$\frac{\partial}{\partial z} \left(\frac{\zeta}{r} \frac{\partial \psi}{\partial r} \right) - \frac{\partial}{\partial r} \left(\frac{\zeta}{r} \frac{\partial \psi}{\partial z} \right) + \frac{1}{\varrho} \left[\frac{\partial}{\partial r} \left\{ \frac{\mu_e}{r} \frac{\partial}{\partial r} (r\, \zeta) \right\} \right.$$

$$\left. + \frac{\partial}{\partial z} \left\{ \frac{\mu_e}{r} \frac{\partial}{\partial z} (r\, \zeta) \right\} \right] + \frac{1}{\varrho} \left\{ \frac{\partial}{\partial z} (J_\theta B_z) + \frac{\partial}{\partial r} (J_\theta B_r) \right\} = 0$$

Here ζ and ψ are the vorticity and the stream functions, given respectively by:

$$\zeta = \frac{\partial u_r}{\partial z} - \frac{\partial u_z}{\partial r}, \quad u_r = \frac{1}{r} \frac{\partial \psi}{\partial z}, \quad u_z = -\frac{1}{r} \frac{\partial \psi}{\partial r}$$

These equations and the effective viscocity $\mu_e = \mu + \mu_t$ have to be solved subject to the following constraints:

    i) Both velocity components are zero at the solid boundaries.

    ii) No shear stress is transmitted through the free surface or the velocity gradient is zero.

    iii) Symmetry is observed about the central line.

To solve these equations one first solves the Maxwell's equations to obtain the Lorentz force-field. Once the magnetic field is known, then the equation of motion (19) is solved numerically with equation for turbulent viscosity. The solution procedure is documented by Chang [2].

Here, following Szekely and Chang [1], we will consider two basic results of induction processing depending on the nature of the driving magnetic fields: (i) a travelling wave magnetic field, and (ii) a stationary magnetic field. The coil current for a travelling magnetic field is given by:

$$J_\theta = J_o e^{i(wt - kz)} \tag{20}$$

While for a stationary magnetic field it is given by

$$J_\theta = J_o{}^* e^{iwt} \sin kz$$

Where t is the time w is the angular frequency, k is the wave number and z is the axial coordinate. The flow pattern generated by the two types of fields are given in Fig 2. It is seen from these figures that for travelling magnetic fields two axisymmetrically located circulating loops are generated (Figs. 2 a and b) while for a stationary magnetic fields four such loops are generated (two in the upper half of the container and two in the lower half) (Fig. 2c). We also observe from these figures that the direction of circulation near the wall coincides with the direction of the travelling magnetic field. It is seen that the lower the frequency the deeper the penetration i.e. the longer the wavelength the deeper the penetration.

3. FURNACE TYPE APPLICATIONS

In this section we discuss the salient features of two types of furnaces, namely one in which the coil configurations are such the field generated

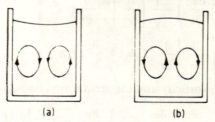

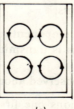

Fig. 2. Schematic of connective patterns during induction melting (a) and (b) are for travelling magnetic field in upward and downward direction respectively, and (c) for stationary magnetic field [1].

is a travelling magnetic field and one in which it is a stationary magnetic field.

Travelling magnetic fields in induction furnaces are obtained by connecting polyphase exciting coils. A detailed discussion of the generation of these fields are given in Sundberg [3]. We assume that the coil configuration is such that the current is given by

$$J_\theta = J_o \, e^{i(wt-kz)}. \tag{21}$$

using this it is shown that the magnetic vector potential is given by

$$A_\theta(r, t) = A_o(r) \, e^{j(wt-kz)} \tag{22}$$

where r is the radial distance from the central line of the furnace and it may be shown that

$$A_\theta^*(r^*) = I_1 \, (\beta \, r^*)/\beta I_o(\beta) \tag{23}$$

where $r^* = r/R$ and the current is given by

$$J_\theta^*(r^*) = -jw\mu_e \, \sigma \, R^2 I_1/\beta I_o \tag{24}$$

with $J_\theta = J_\theta^* \, H\zeta, \phi/P$ where R being the radius
Using these quantities the Lornet'z force is obtained and is

$$F_z^* = -J_o^* \, B_r \, {}^* \, e^{2i(wt-kz)} \tag{25}$$

$$F_r^* = J_o \, Bz^* \, e^{2i(wt-kz)} \tag{26}$$

Now using these quantities the Navier Stokes equations given by (7) are to be solved [2]. Experimental measurements on these furnaces have been conducted by Marr et al. [4, 5]. The stream line pattern is depicted in Fig. 2 (a and b) and it is seen that near the wall the velocity is in the direction of the magnetic field. It is also observed that the eddy diffusivity is very high at the centre of the vessel which is responsible for promoting the coalescence of the impurities and the rapid dispersal of the alloying elements.

For the case of a furnace driven by a stationary magnetic field, the current generated is given as:

$$J_\theta = J_o \sin \, kz \, e^{iwt} \tag{27}$$

The forces are given in this case as

$$F_z^* = - \, R_E \, \{(iw\mu_e \, \sigma \, k^2 \, R^2 \, I_1 \, \bar{I_1})/(4\beta\bar{\beta} \, I_o\bar{I_o})\} \sin \, 2kz^*$$

$$F_r^* = R_E \, \{(iw\mu_e \, R^2 \, I_1 \, I_o)/(2\beta \, I_o\bar{I_o})\} \sin^2 \, kz$$

Again using these two components, the Navier-Stokes equations are solved.

The streamline pattern are given in Fig. 2(c). It is seen here that there are four circulating loops two in the upper half of the cylinder and two in the lower half. Dragunikna and Tir [6] have conducted experiments and made a comparison of

462

the calculated parameters with the experimental values and obtained qualitative agreement. In these furnaces the role of the system geometry i.e., the placement of the coils and shields play an important role and they have been discussed by Lavers [7]. Lavers concludes that the electromagnetic pressure is very sensitive to the partitions of the coil and the screen. Small variations in the coil radius have brought about large changes in the e.m. pressure and it is preferable to have the coupling coil and the screen as tight as possible. However the pressure is less sensitive to the changes in frequency. In a recent paper Meyer et al. [32], have studied electromagnetic parameters in an inductively stirred melt both in the presence and absence of shields. A good agreement has achieved between the theoretical and experimental values. It has been shown that the use of shields is a useful way of modifying the flow patterns in these systems.

4. DEVICES APPLICATIONS

4.1 Stirring in bath

Manual stirring in a molten bath is inconsistent with cleanliness, is time-consuming and because it has to be carried out through open furnace doors, leads to high heat losses. Additionally there is always some iron pick-up in the melt from the dissolution of the stirring tools. On the other hand, the potential advantages of electromagnetic stirring are [5]: (1) improved internal structure, (2) improved internal cleanness, (3) improved uniformity of composition and mechanical properties, and (4) alleviation of operational constraints. The structural improvements result mainly because induced motion inhibits formation of columnar zones. There is also reduction in the amount of central porosity. A review of the literature on this subject has been presented by Marr [5]. In Fig. 3 a sketch of the stirrer taken from Bamji [8] is shown. This operates by what is known as the jumping force principle technique. The stirrer is made up of a coil loaded along a long laminated steel core. A length of the core is immersed in

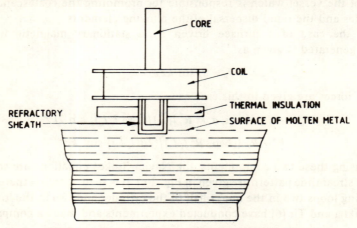

Fig. 3. Schematic of apparatus for stirring in bath [7].

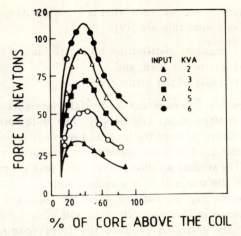

Fig. 4. Calculated profile of force distribution as function of core position [7].

the molten metal and is repelled away from the coil by electromagnetic fields. This sets in the circulation of molten metal continuously. Thus a fast and relatively cheap stirring process is achieved. Bamji has also determined the magnitudes of the stirring forces for various locations of the placement of the coil and also for various values of input power. The results are summarized in Fig. 4. It is seen from this figure that the stirring forces increase directly with increase in the input power; (KVA) and the positioning of the coil has an appreciable effect on the force produced. For maximum force the coil is placed such that one-third of the core is above the centre of the coil. Travelling wave electromagnetic stirring devices have been used previously by Sundberg [9]. Hurtuk and Tzavaras [10] have used linear induction type stirrers in continuous castings. Rotary type stirrers which use polyphase windings have been discussed by Alberny [11] and Hanbitzer [12] for continuous casting applications. Lavers [7] has examined one and two dimensional models to determine the magnitudes of forces produced on long cylinders as functions of frequency and cylinder dimensions. Gaule et al. [13] Burke and Dudley [14] and Geolini [15] deal with current distributions and the forces in these systems. The current pattern for different inductors are presented in Figs. 11–13. Figure 14 shows the decay of magnetic flux with thickness of the medium.

4.2. Electromagnetic levitation

Levitation is a technique by which a body is suspended in a state of stable equilibrium when it is subject to the influence of electromagnetic forces against gravity. The suspended body is to remain either free from material contact with its environment as in a vacuum—or is to be exposed to a material contact that is not essential to keep the body stably suspended. If the levitated body is melted and

464

still remains in equilibrium, we have what is known as levitation melting. The parameters that play a vital role are [19].

1) the magnetic pressure distribution on the surface of the load,
2) the bulk stability of the load, and
3) the surface stability of the load.

The magnetic pressure balances the hydrostatic pressure developed by the interaction of the molten metal and gravity. The bulk stability is concerned with the levitator's ability to keep the load, as a whole suspended in a position of equilibrium, i.e., the load will be restored to its position of equilibrium if disturbed slightly. The surface stability is concerned with keeping the shape of the surface intact when the material is in a molten state.

A levitator essentially consists of a set of coils fed by currents. The metallic load is placed in the magnetic field and eddy currents are generated which melt the body and also produce the force required to levitate the body.

A single frequency levitator is one in which the forces are developed essentially by a single frequency magnetic field. A rough schematic of this situation is shown in Fig. 5. A levitator of this type was first developed by Wronghton and Okress [16]. Extensive bibliographic references are given by Geary [17]. All these levitators can only lift loads of very small quantities since they suffer from non-uniform fields and also because the force component in the axial direction is small.

To overcome these drawbacks, multifrequency levitators have been proposed. In multifrequency levitation, the three requirements mentioned earlier are satisfied. A hydrostatic distribution of pressure is achieved by the superposition of magnetic pressures contributed by magnetic fields of different frequencies. Sagardia [18] has shown that this is possible as there is no net force reaction between fields and currents of different frequencies. Sagardia and

Fig. 5. Schematic of single frequency levitator. As pointed out in the text these levitators are limited to lifting only small masses.

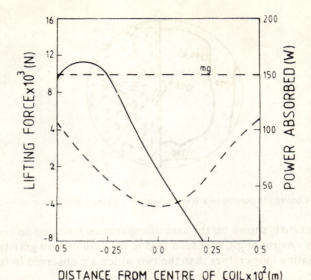

Fig. 6. Calculated profile of lifting forces on a levitated metal sphere as a function of position in coil [30].

Segsworth [19] have constructed levitators which can support kilogram quantity loads. This was achieved by using multifrequency (at least three) orthogonal currents to support the body as well as to prevent any surface pertubation from amplifying and destroying the equilibrium levitated shape of the liquid metal. El-Kaddah and Szekely [20] have presented a mathematical model to calculate the force field of liquid flow and the temperature profiles in a levitated metal sphere. The computed results are found to be in good agreement with the experimental values. In Fig. 6. [20] are presented the lift force and power absorbed as a function of the distance from the centre of the coil. For stable levitation the lift force should be greater than the weight of the specimen (mg). They concluded that in a position closer to the lower coil a stable configuration is possible. They have also presented the convection pattern and the temperature profiles within the sphere for a number of input conditions. In Figs. 7 and 8 taken from their investigation, the

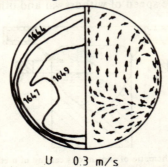

U   0.3 m/s

Fig. 7. Convection pattern in a levitated sphere [30].

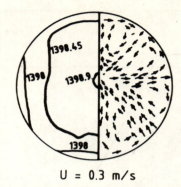

U = 0.3 m/s

Fig. 8. Convection pattern in a levitated sphere in zero gravity environment [30].

convection pattern is shown for the case of a specimen levitated on earth and also in zero-gravity environment. It is seen that in the case of zero-gravity case there are four circulating loops rather than the two which are observed in the presence of gravity.

4.3. Ingot forming in continuous casting

One of most interesting applications of electromagnetic forces on molten melts has been the development of the electromagnetic mould for the casting of metals. The basic configuration is shown in Fig. 9 [21] and consists of an induction coil with a cooling system around the metal column being cast. In principle, the operation of the mould is quite simple. The inductor operating at an appropriated frequency and current level, produces pinching forces which keep the melt standing free on the solidified portion of the billet. If the system parameters are properly choosen the billet can be withdrawn from the coil steadily with continuous addition of molten metal from the top. In this application screens may be conveniently positioned to shape the free-standing melt. We note also that certain net shape superalloy casters also utilize such free-standing melts for turbine blade superalloy casting. Meier [22] and Lavers [23] have discussed the estimation of various parameters like the frequency of the current, the billet size, the speed of withdrawal and other related quantities for

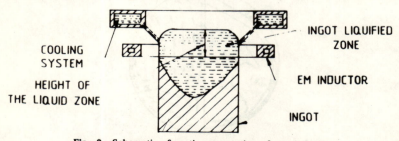

Fig. 9. Schematic of continuous casting of metals [19].

steady state continuous casting. In a recent paper Meyer et al. [33] have presented a theoretical formulation to study the process of electromagnetic casting of aluminium. A good agreement has been obtained between experimental and theoretical predictions.

4.4. Cross-sectional area reduction of conducting solids by electromagnetic techniques

Reduction of cross-sectional area of solids is achieved traditionally by mechanical means. This has many limitations and its utility is limited to brittle materials. There is the risk of contamination from die materials particularly when hot working activities are performed. However in the case of electromagnetic cross-sectional area reduction, these limitations have been overcome [24]. In this technique an electromagnetic field is employed to achieve localized melting and containment of small zone of input material to achieve a dieless reduction in a cross-sectional area. This is applicable to all electrically conducting materials. The principle is depicted in Fig. 10. Here a solid load A is fed at a controlled speed into an appropriately designed induction coil. The coil may be about a longitudinal arms also. The material is melted in the electromagnetic field produced by the coil and forms a solid liquid zone (E). This liquid has a cross-sectional area less than the input stock (A). The zone is primarily suspended by levitation forces. A starter number (B) having a cross-section geometry similar to.

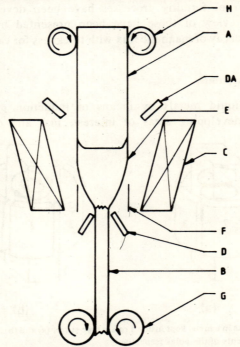

Fig. 10. Schematic of the electromagnetic wire drawing technique [22].

468

the desired product is inserted into the liquid zone (E) optionally held for a short period of time and then continuously withdrawn using a withdrawal mechanism. A more-or-less steady state process is achieved through the continuous replenishing of input using a feed mechanism. To improve the process control the final product optionally may be cooled.

4.5. Electromagnetic metal forming

When two current-carrying conductors are in close proximity, they are acted upon by mutual attraction or repulsion forces which may be sufficiently high to deform one or both of them. In the magnetic metal-forming process, one conductor is a high current coil, the other conductor is the work piece. A bank of high energy storage capacitors is discharged through the first coil producing a high frequency oscillatory current and this induces eddy currents in the work piece. This process is used mainly to work fairly light components formed from sheet metal or thin sheet metal or thin wallled tube. At present this process is used for a variety of assembly operations such as shrinking copper bands on to electrical insulators and swaging end fittings in tubes. The forming of parts with complicated geometries from flat sheet is achieved by driving the material with the induced electromagnetic currents on to a female die of the desired shape. It has been reported [25] that typical magnetic field strengths may be of the order of 300 kgf which give rise to surface pressures on the work piece of about 20–40 kg/mm². A number of forming processes have been developed using this techniques and a review of these have been presented by Benines et al. [29]. In Fig. 10, the decay of magnetic flux with thickness for various materials is shown.

4.6. Electromagnetic pumps

The transport of liquid metals by means of induction pumps is one of the more rapidly developing areas of interest. Flat and cylindrical linear

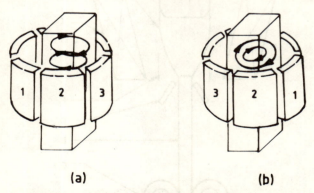

(a)　　　　　　(b)

Fig. 11. Induced currents in a metal kept inside a rotary type stirrer (a) and (b) refer to sequential and cross arrangements of the poles respectively.

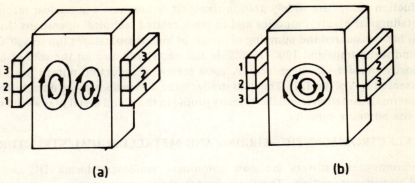

Fig. 12. Induced currents in a metal kept inside a linear type inductor (a) and (b) refer to sequential and cross arrangement of poles respectively.

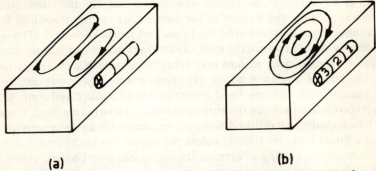

Fig. 13. Induced currents in a metal kept inside a tubular type conductor (a) and (b) refer to cross and sequential arrangements of the poles respectively.

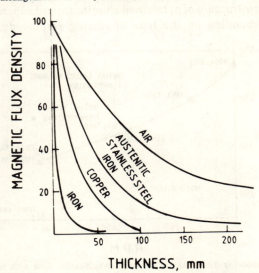

Fig. 14. Decay of magnetic flux density with thickness of the medium.

470

induction pumps are widely used to transport liquid metals in nuclear reactors, metallurgical plants, foundries and in other related technical operations. Kirko [26] has considered the pumping of metals at high temperatures (up to 600°C in cylindrical pumps and 1000–1500°C in flat ducts) and also for the case of high velocity transfer (25–30 m/sec). A good review of induction pumps has been presented by Vol'dek [27]. The field strength, the dimensions of the duct, the type of thermal insulation layer, the velocity profile in the liquid metal play a vital role on the pumping capacity.

5. ELECTROMAGNETIC STIRRING AND METALLURGICAL STRUCTURES

Electromagnetic stirrers are now commonly employed during DC casting and continuous casting. Their use in net shape castings are as yet limited but progressively increasing. In India however their use is uncommon and the effects of stirring on metallurgical structure are largely undocumented.

Stirrers kept along the strands of the continuous casting station of ferrous materials induce non-axisymmetric fluid flow and this is known to: (1) decrease central porosity as well as clean the melt, (2) decrease macrosegregation (however sometimes a wrong stirring pattern may actually increase macrosegregation, (3) increase the size of equiaxed zones, (4) refine grains and inclusions, and (5) disperse phases (such as ferrite) and other constituents more uniformly [28-30]. The direct correlation between the stirring patterns and structure modifications is as yet to be adequately established, however an acceptable design seems to be the placing of a linear inductor slightly below the region where the first skin forms.

Electromagnetic casting of aluminium ingots has also become commercial [28] and is associated with the complete elimination of expensive scalping. A recently completed detailed investigations on the effects of electromagnetic stirring on the microstructure of net shaped aluminium and superalloy casting [29] concluded that depending on the type of stirring employed: (i) temperature

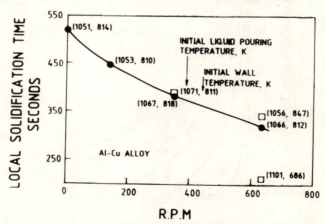

Fig. 15. Plot showing the change of the local solidification time with rotation [28].

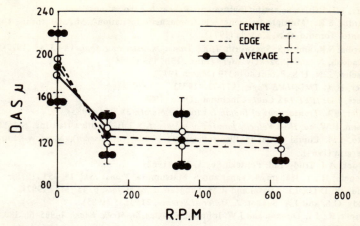

Fig. 16. Plot shown the change of the dendrite arm spacing in an Al-Cu with rotation [28].

gradients may be stabilized across the radius of the casting, (ii) grain as well as dendrite size may be refined, and (iii) grain size distribution may improve. Some interesting new observations were that stirring may actually lead to an improvement in the heat transfer coefficient at the mould metal interface. This is shown in Figs. 15 and 16 which show that the local time of solidification and dendrite arm spacing reduce with increased rotation.

ACKNOWLEDGEMENT

This review has been possible only because of the support and permission of Dr. P. Rama Rao Director, DMRL. The authors are grateful for the same.

REFERENCES

1. Szekely, J. and C.W. Chang. *Iron and Steel Making* 3, 190 (1977).
2. Chang, C.W. Ph.D. Thesis, State University of New York at Buffalo (1977).
3. Sundberg, Y. IEEE Spectrum, 79 (May 1967).
4. Marr, H.S. *Iron and Steel Int.*, **51**, 2 (1978).
5. Marr, H.S. *Iron and Steel Int.*, **52**, 29 (1979).
6. Dragunkina, N.I. and L.L. Tir. *Magnitnaya Gidrodinamika*, **2,** 137 (1966).
7. Lavers, J.D. *IEEE Trans. Industry and Applications* **IA-17**, 427 (1981).
8. Bamji, P.J.F. *Electric Power and Applications*, **1**, 4 (1978).
9. Sundberg, Y. paper 609, VII U.I.E. Congress, Warsaw, (Sept 1972).
10. Hurtuk, D.J. and A.A. Tzavarao. *Met. Trans.* **8B**, 243 (1977).
11. Alberny, R. et al. *61st Nat. O.H. & Bos Conf.*, Chicago, (April 1978).
12. Hanbitzer, W. *Electrowarme. Int.* **37** (B6), 297 (1979).
13. Gaule, G.K., J.C. Yarwood and D.E. Taylor. U.S. Patent 4236570 (Dec 1980).
14. Burke, P.E. and R.F. Dudley. *IEEE Trans. Industry Appl.* **IA**, 565 (1971).
15. Geolini, F. *Supt. IAS Conf. Record*, pp. 115/- 457, Toronto, Canada (1978).
16. Wronghton, D.M. and E.C. Okess. Westinghouse Electric Corporation U.S. Patent 2686864 (1951).

472

17. Geary, P.J. British Scientific Instruments Research Associations 1964.
18. Sagardia, S.R. "Multiple frequency electromagnetic levitation", M.A.Sc. thesis, University of Toronto, Toronto, O.N. Canada (1971).
19. Sagardia, S.R. and R.S. Segsworth. *IEEE Trans. Industry and Appl.* **IA-13,** 49 (1977).
20. El-Kaddah, N. and J. Szekely. *Met. Trans.,* **15B,** 183 (1984).
21. Getselev, Z.N. U.S. Patent 4014379 (March 1977).
22. Meier, H.A. *TNS-AIME Paper,* **A77-72,** (1973).
23. Lavers, J.D. *IEEE IAS Conf.* Cincinnati, Ohio (1980).
24. Sekhar, J.A. *Transactions of Indian Institute of Metals,* **38,** 194 (1985).
25. Wilson, F.W. ed. *High Velocity forming of metals,* A.S.T.E. Prentice Hall Inc. N.J. (1964).
26. Kirko, I.M. Liquid Metal in Electromagnetic Fields (in Russian), Moscow-Leningrad, Izd., Emergiya (1964).
27. Vol'dek, A.I. Trudy Tall, Politekh Ses **A, 3,** 1 (1962).
28. Takenchi, H., Y. Ikehara, T. Yanai and S. Matsumura. *Trans. ISIJ,* **18,** 353 (1978).
29. Goodrich, D.G., J.L. Dassel and R.M. Shogean. *J. of Metals,* p. 45 (May 1982).
30. Reddy, G.S. and J.A. Sekhar. *J. Material Science,* **21,** 3535 (1985).
31. Benines, K., J.L. Dunxan and J.W. Johnson, *Proc. Instn. Mech. Engrs.* 1–965–66, **180,** PT 1 No. 4, 93, (1965-66).
32. Mayer, J.L., N.El. Kaddah, J. Szekely, C.Vives and R. Ricou. *Met. Trans.* **18B,** 529 (1987).
33. Mayer, J.L., N.El-Kaddah, J. Szekely, C.Vives and R. Ricou. *Met. Trans.* **18B,** 539 (1987).